MW00760451

Introduction to
Process
Geomorphology

Introduction to
Process
Geomorphology

Vijay K. Sharma

CRC Press
Taylor & Francis Group
Boca Raton London New York

CRC Press is an imprint of the
Taylor & Francis Group, an **informa** business

CRC Press
Taylor & Francis Group
6000 Broken Sound Parkway NW, Suite 300
Boca Raton, FL 33487-2742

© 2010 by Taylor and Francis Group, LLC
CRC Press is an imprint of Taylor & Francis Group, an Informa business

No claim to original U.S. Government works

Printed in the United States of America on acid-free paper
10 9 8 7 6 5 4 3 2 1

International Standard Book Number: 978-1-4398-0337-0 (Hardback)

This book contains information obtained from authentic and highly regarded sources. Reasonable efforts have been made to publish reliable data and information, but the author and publisher cannot assume responsibility for the validity of all materials or the consequences of their use. The authors and publishers have attempted to trace the copyright holders of all material reproduced in this publication and apologize to copyright holders if permission to publish in this form has not been obtained. If any copyright material has not been acknowledged please write and let us know so we may rectify in any future reprint.

Except as permitted under U.S. Copyright Law, no part of this book may be reprinted, reproduced, transmitted, or utilized in any form by any electronic, mechanical, or other means, now known or hereafter invented, including photocopying, microfilming, and recording, or in any information storage or retrieval system, without written permission from the publishers.

For permission to photocopy or use material electronically from this work, please access www.copyright.com (http://www.copyright.com/) or contact the Copyright Clearance Center, Inc. (CCC), 222 Rosewood Drive, Danvers, MA 01923, 978-750-8400. CCC is a not-for-profit organization that provides licenses and registration for a variety of users. For organizations that have been granted a photocopy license by the CCC, a separate system of payment has been arranged.

Trademark Notice: Product or corporate names may be trademarks or registered trademarks, and are used only for identification and explanation without intent to infringe.

Library of Congress Cataloging-in-Publication Data

Sharma, Vijay Kumar, 1944-
    Introduction to process geomorphology / Vijay K. Sharma.
        p. cm.
    Includes bibliographical references and index.
    ISBN 978-1-4398-0337-0 (hardcover : alk. paper)
    1. Process geomorphology. I. Title.

GB402.S48 2010
551.41--dc22   2009047097

Visit the Taylor & Francis Web site at
http://www.taylorandfrancis.com

and the CRC Press Web site at
http://www.crcpress.com

Contents

Preface

Morphologic attributes of the contemporary geomorphic landscape and its component form elements evolve from the activity of environment-regulated process domains acting in the present, and local and regional-scale process rates governed by conditions external and internal to geomorphic systems. These process–response systems establish landscape characteristics in equilibrium with the movement, storage, and transfer of energy and matter between interrelated and interdependent components of the system. Characteristics of geomorphic landscapes also evolve through time, suggesting that conditions of equilibrium and disequilibrium between opposing forces persist in the process–form adjustment. In certain instances, therefore, landscapes and their constituent landforms inherit the imprint of past process activities, and still others are relics of processes no longer active in the present.

Geomorphic processes are essentially a manifestation of various forms of molecular and gravity stress on rocks and sediments. These stress forms produce a variety of strains known by the processes of weathering, mass movement, and erosion and deposition of the Earth's materials. Studies on process dynamics are rooted in the principles of mechanics and fluid dynamics. These principles provide the foundation for understanding the process dynamics in terms of semiempirical, empirical, mathematical, and theoretical postulates on the behavior of bodies under the action of the forces that produce changes in them, and explain the process–form relationships. Studies on process-form relationship have evolved through a great deal of fieldwork, and theoretical and experimental research, which suggest that landscape characteristics and component landform attributes evolve from the effects of several process combinations acting simultaneously on the Earth's materials of complexly related mechanical, chemical, and biochemical attributes. Further, a given morphology can evolve from several causes, and a given environmental change can affect the evolution of landscape in many different ways. Hence, hypotheses on process-form relationships are built around one or the other related aspect of the morphodynamic controls of landforms.

The introductory chapter of the text examines the philosophy of an open system approach for the explanation of contemporary process–form relationships, evolution of landscape through time, and application aspects of climate–process systems; thresholds; and the frequency concept of geomorphic processes for the explanation of landscape and its component form elements. The landscape and constituent depositional landforms evolving over a period of time also carry the signature of the environmental change, introducing a temporal change in the process domain or process intensity, or both. Interpretation of this signal in the environment-sensitive indices of the landscape provides a timeframe for the type, magnitude, and duration of the environmental change. The chapter, therefore, introduces the theory and procedure for establishing the chronology of events through which the landscape has passed to reach the present state of development.

Several aspects of the morphology and geological structure of the Earth owe to the internal forces of the Earth. The internal geologic processes of the Earth are due largely to plate tectonic activity. Hence, Chapter 2 is devoted to basic postulates of the theory of plate tectonics and associated aspects explaining the relief, major tectogenic features, and distribution of volcanic and seismic activity of the Earth. Later sections of the chapter discuss the evolution of rocks and their geomorphic significance, deformation dynamics of rocks and sediments, and general aspects of slope instability.

The dynamical system of weathering is far more complex than what we knew about it a decade ago. The chapter on weathering explains how and why earth materials alter in a particular manner to a residuum of original matter and decomposition products in equilibrium with the environmental stress of the biosphere. Chapter 3, on weathering, also highlights the evolution of soils as the end product of weathering, and describes duricrusts or indurated surfaces formed at or near the Earth's surface by the weathering of rocks and soils formation. Recent researches in chemical sciences and biotechnology suggest the role of heterogeneous chemical activities, photochemical processes, and bacterial activity in the transformation and mobilization of a variety of chemical elements and compounds in the Earth's environment. Therefore, the dynamics and consequences of the above processes to weathering are also introduced in this chapter.

Internal attributes of the earth materials and a variety of external controls of shear stress variously affect the mass movement activity called slope failure. Chapter 4 discusses these aspects, classifies the mass movement activity from the viewpoint of the dynamics of slope failure, and cites suitable field studies for the given type of failure.

The behavior of frictionally dominated open channel flow and morphologic activity of highly organized fluvial systems is better understood in theory and test conditions than in field situations. Hence, Chapter 5, on fluvial processes, provides a background to the hydraulic principles governing the nature and type of fluid flow, and highlights complex mechanisms of sediment transport in alluvial channels. The empirical relations governing the hydraulic geometry of streams and processes of channel patterns that are central to fluvial geomorphology are also discussed with suitable experimental and field data.

The sediment charge of drainage systems is deposited as flood plains, alluvial fans, and river deltas. Chapter 6 summarizes field and laboratory observations on processes of flood plain deposits, morphodynamics of flood plains, environmental controls of abandoned flood plains called fluvial terraces and their geomorphic significance, fan sediments, alluvial fan processes and morphologic evolution of alluvial fans, and morphodynamics of river deltas.

Chapters 7 through 11, on processes and landforms of glacial, periglacial, aeolian, karst, and coastal environments, respectively, are composed of two parts. The earlier part of each chapter provides perspectives on process dynamics, and the latter part evaluates comparative advantages and disadvantages of competing hypotheses on the morphogenesis of erosional and depositional landforms of the given environment.

Human activities alter the magnitude and direction of process rates manifold, affecting the symbiotic association between biotic and abiotic components of the landscape, stability of the landscape and its component landforms, and the quality of the environment. The manner of disturbance due to human actions and their effect on geomorphic systems is reviewed in Chapter 12 on applied geomorphology, to highlight the emerging concerns in man–environment relationships. The chapter includes varied topics on land resource planning and management, site selection, environmental impacts of resource use, and effects of land use and land use change on the quality of surface and subsurface water resources of the Earth.

Vijay K. Sharma
Kurukshetra, Haryana, India

Acknowledgments

I sincerely thank Dr. Milap Chand Sharma, CSRD/School of Social Sciences, Jawaharlal Nehru University, New Delhi; Dr. Surender Singh, Shivaji College, University of Delhi; and Dr. Anup Singh Parmar, D. N. College, Hisar, for useful comments on earlier drafts of the text. The completion of this work was made possible by the encouragement of my family members when I needed it most.

1 Process Geomorphology

Gravity and molecular forms of environmental stress produce a variety of strains in rocks and sediments that behave as elastic, plastic, and fluid substances (Strahler, 1952). The *gravity-controlled stress* utilizes solar energy to initiate the movement of wind, water, and ice on slopes. These three media are agents of the modification of the earth's surface forms, and considered fluids as their motion can be analyzed in terms of the *fluid mechanics*. The *molecular stress* depends on environmental factors and geologic properties of the earth materials. This form of stress, which can be expansive, ionic, and sorptive, weakens the bond strength of mineral grains and produces a variety of mechanical, chemical, biochemical, and other complex forms of surface and near-surface stress in rocks and their mineral constituents. The gravity-controlled stress produces strain in the earth's materials, which variously manifests in geomorphic processes of mass movement, erosion, and deposition of sediments. The gravity and molecular-controlled processes are broadly understood in theory and in outline, but the manner of their effect on complexly related aspects of the earth materials remains largely obscure in the earth's environment.

NATURE OF PROCESS GEOMORPHOLOGY

Process geomorphology identifies process domains and quantifies process rates for explaining the origin of landforms. *Process domain* refers to a specific process activity, such as deflation, solution, or frost shattering, which exceeds other concurrently functioning processes at the site by higher frequency of occurrence. Contrary to expectations, however, diverse processes as disintegration, decomposition, and biochemical weathering domains co-dominate in areas of both moisture scarcity and optimum moisture conditions (Pope et al., 1995). Thus, rocks in cold region environments are pulverized by frost disintegration, hydration shattering, and chemical alteration of the constituent minerals and mineral aggregates. Flood plains similarly evolve by co-dominant vertical and lateral accretion processes of sedimentation within, along, and beyond channel margins.

The evolution of geomorphic landscape depends on *process rates* that are governed by conditions external and internal to the *geomorphic systems*. Process rates provide meaning to the amount of work done by a specific process activity and, thereby, to the development of landscape attributes in diverse geomorphic environments.

Just as processes evolve landforms, so do the developing form affects process dynamics at a local scale. A change in any component of the landscape initiates a sequence of change in related form elements, affecting process rates in many different ways. Local steepening of channel gradient at constant width thus increases the flow velocity and, thereby, causes bed scour as a compensatory *feedback mechanism*. Feedback mechanisms come into existence when the established order of

1

relationship between forces of stress and resistance to change is disturbed, affecting process rates at local and regional scales. The internal adjustment between process rates and form attributes is viewed in terms of critical conditions of landform stability called *geomorphic thresholds*.

Scale of observation is basic to the nature of hypotheses on process-form relationships (Chorley, 1978). At the smallest arbitrary scale, continents and ocean basins emerge as *first-order landforms* of the earth explained by the internal geologic processes of the earth. At the scale of continents or oceans, however, the morphology of *second-order landforms* of the continents as mountains, plateaus, hills, and plains is a composite response of endogenous forces and exogenous processes. At the largest scale of observation, the morphology of *third-order landforms* of the continents as erosional and depositional forms of fluvial, glacial, periglacial, aeolian, karst, and coastal environments evolves by the activity of exogenous processes that may be epicyclic, episodic, or variable in time. Hence, the scale of inquiry distinctly affects the manner of landform explanation.

Process geomorphology is concerned with the understanding of process mechanics in terms of known laws in the earth and planetary sciences, and the manner of landform evolution (Thornes, 1980). Processes are instrumented in field, a few require calibration in test conditions, and others are understood in theory only, providing empirical and mathematical bases for the explanation of landforms. Laboratory models on rates of aeolian abrasion and recession of sea cliff, and theoretical models on frost crack propagation in rocks, subglacial erosion and deposition, and creep of sand beneath saltation layer provide *critical conditions* for the initiation and cessation of a particular geomorphic activity in the environment and the manner of landform evolution.

An inquiry into the process-form relationship implies that geomorphic landscape and its component form elements are in adjustment with contemporary process domains and process rates. Landforms, however, evolved over a period of time during which process domains and process rates had been affected by the *environmental change*. Hence, explanation of landforms also requires integration of time-dependent contemporary process data and historical process data reconstructed from signals of the past environment in the sedimentary record of landforms (Sugden, 1996).

SYSTEMS CONCEPT IN GEOMORPHOLOGY

The systems approach provides an ideal framework for the explanation of landforms, which evolve by the flow and transformation of energy and exchange of material across real or arbitrarily defined limits of natural environments. The manner of interaction between energy and material gives isolated, closed, and open types of systems. An *isolated system* does not permit the exchange of energy or material across its boundary and, therefore, is hard to find in the earth's environment. A *closed system* allows the exchange of energy but not of the material across its defined limit, such that its instant state depicts the utilization of initial input of energy through time. An *open system* is sustained by continuous flow and exchange of energy and material, evolving interrelated and interdependent objects and their form attributes in adjustment with the contemporary inflow and outflow of energy and material across a specified physical space. Hence, open systems are destroyed when energy or material or both are denied to them. Pond, forest, and drainage basins are all open systems of continuous through flow and exchange of energy and material across the given boundary. *Geomorphic systems* are open systems.

GEOMORPHIC SYSTEMS

Geomorphic systems function by the exchange of energy and material in an organized manner. They derive energy from *solar radiation* reaching the earth, of which much is lost through infinite states of transformation, and little less than 1% is actually available for performing the geomorphic work (Kirkby, 1990; Thorn and Welford, 1994). Inputs of energy and material, however, do not leave geomorphic systems in exactly the same form (Wilcock, 1983). In drainage systems, a part of precipitation escapes as evapotranspiration and the products of weathering pass as dissolved load, suspended load, and bedload of the system.

Geomorphic systems store energy and material at a certain level of optimum magnitude. The energy and material of *exogenous processes* are held as pools of detention and retention storage at or near the surface of the earth, and of *endogenous forces* remained locked in deep-seated rocks within the subcrust. These forms of energy and material are released in a manner that does not affect the normal functioning of geomorphic systems, implying that input and output rates of energy and material are balanced over longer than shorter periods of time. A pulsed released of energy and material at times as in landslides or seismic activity, however, brings about a catastrophic change in systems' form attributes. Geomorphic systems embody the principles of dynamic equilibrium, feedback mechanism, optimum magnitude, and relaxation period.

EQUILIBRIUM

The concept of equilibrium is primarily rooted in physics and chemistry. The equilibrium in physics denotes a *balance of forces* between interacting objects treated discretely, and in chemistry it is synonymous with the *thermodynamic equilibrium* between groups of molecules. Thus, the concept of equilibrium conveys different meanings across the broad spectrum of sciences (Thorn and Welford, 1994). The equilibrium concept is also central to geomorphology, where its *regulative principles*, rather than structural content, are of prime significance to the explanation of process-form adjustment and the interpretation of geomorphic landscape.

Equilibrium in Geomorphology

The concept of dynamic equilibrium was first applied to drainage basins and later to the interpretation of geomorphic landscape of humid environments. Hence, much of the literature is replete with application aspects of the concept to fluvial systems (Huggett, 1980; Howard, 1982, 1988; Knighton, 1984; Culling, 1987, 1988; Church and Slaymaker, 1989; Montgomery, 1989; Kirkby, 1990; Lamberti, 1992; Pizzuto, 1992; Thorn and Welford, 1994). Equilibrium in geomorphology implies some sort of balance between the geomorphic system and its environment, giving dynamic, steady-state, and quasi-equilibrium frames of reference for the system state. *Dynamic equilibrium* is understood as a state of adjustment between the process activity and components of landscape (Hack, 1960), as a state of balance between process rates and the environment of a geomorphic system (Ahnert, 1967), and as a state of instantaneous adjustment between the shear stress of contemporary processes and shear resistance of the earth materials (Abrahams, 1968). Hence, dynamic equilibrium implies a state of delicate balance between contemporary processes and their landforms. As explained later in this section, *disturbed systems* pass through a state or states of *punctuated equilibrium* between process rates and landforms. Hence, dynamic equilibrium is also viewed as a *function of time*.

Schumm and Lichty (1965) introduced the concept of cyclic, graded, and steady time for the state of equilibrium between the fluvial activity and components of landscape *evolving through time*. Schumm (1977) subsequently provided arbitrary time limits for each of the equilibrium states. *Cyclic time* is the geologic time of 10^5 to 10^7 years beginning with the *cycle of erosion* during which the morphology of drainage basins progressively losing relief through time, become adjusted to the average erosive energy of the system. *Graded time* refers to 10^2 to 10^3 years of time in which lower-order components of the system, like hillslopes, achieve optimum inclination with the erosive forces despite lowering of the basin relief. *Steady time* is a reference period in days or months in which the lowest-order components, such as channel gradient, strive for adjustment with the contemporary flow and exchange of energy and material across the system boundary. Hence, *steady-state equilibrium* is independent of the past conditions through which landscapes arrive at the present state of gradual development (Chorley, 1962). Interrelated *channel variables*, which mutually adjust to variations in discharge in channels developed in part by the earlier flows, also depict a state of quasi-equilibrium among the interacting components. *Quasi-equilibrium* refers to a uniform rate of energy expenditure among channel variables and least expenditure of energy in the performance of geomorphic work (Leopold and Langbein, 1962; Langbein and Leopold, 1964; Scheidegger and Langbein, 1966; Huggett, 1980). These energy principles are referred to again in Chapter 5.

PERTURBATION OF GEOMORPHIC SYSTEMS

Landscapes evolve about the average inflow and outflow of energy and material across the boundary of geomorphic systems. A disturbance in the functional energy by the effects of climate change, tectonism, and human activities, however, alters the established order of equilibrium between the flow and exchange of energy and material by which geomorphic systems strive for a new state of equilibrium by passing through phases of reaction and relaxation time (Graf, 1977; Bull, 1991; Rinaldo et al., 1995). *Reaction time* is the time taken by the landscape to absorb the effect of external stimulus, and *relaxation time* is the time taken by the landscape to establish characteristic form attributes in relation to the energy imposed on the system (Figure 1.1). Disturbed systems thereafter evolve at the average energy level, which persists throughout the period of steady-state equilibrium. The duration of reaction and relaxation time, however, is unique to each system. It varies with the magnitude of external stimulus and the sensitivity of landscape to change (Brunsden and

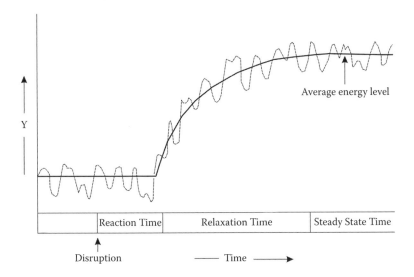

FIGURE 1.1 Graphical representation of the response of a geomorphic system subjected to disruption. Solid line indicates mean condition, dashed line represents actual values. A system parameter Y (some dimensional or spatial characteristic of the system such as length or width) is initially in a steady state. After the climatic or human disruption, new conditions are internalized by the system upon disruption of its functioning called the reaction time. During relaxation time, the system adjusts to the new conditions. The rate law provides a model for the system change during this period. A new steady state is eventually established, resulting in new dimensional characteristics as signified by a new mean value for Y. (Source: Figure 1 in Graf, W. L., "The Rate Law in Fluvial Geomorphology," *Am. J. Sci.* 277 [1977]: 178–91. With permission from Yale University.)

Thornes, 1979; Brunsden, 1988; Thomas and Allison, 1993), and with the inherent instability of certain geomorphic systems (Church and Slaymaker, 1989). *Sensitivity* refers to the extent of change effected in components of the landscape of disturbed systems. In general, systems of higher than lower capacity for storage of energy are relatively less sensitive to change. Some geomorphic systems, however, remain continuously unstable over long periods of geologic time. The Alpine system, in particular, had been inherently unstable on account of tectonic activity and other causes since the Holocene. As evidence, the glacier-fed large drainage basins of the Canadian Cordillera reportedly do not release the large pool of glaciogenic sediments in proportion to discharge as efficiently as their subbasins do (Church and Slaymaker, 1989).

Prediction of Relaxation Time

Relaxation time for the disturbed geomorphic systems follows the rate law governing the decay rate of *radioactive elements*. The *rate law* predicts that 50% mass of the radioactive nuclides exponentially decays into next radioactive nuclei in a specified time called the *half-life period*. The decay rate is given as

$$A_t/A_o = (1/2)^{t/T}$$

in which A_o is the amount of radioactive matter originally present, A_t is the amount remaining after a decay time (t), T is the half-life period, and 1/2 is the rate of change in the chemical composition of radioactive nuclide. *Dendrochronological data* from trees within a gully of recent accelerated rate of erosion in the Denver area of Colorado similarly suggest headward extension of the channel at an *exponential rate*, developing one-half, three-fourths, and seven-tenths of the length, respectively, in 17, 34, and 51 years of gully initiation (Graf, 1977). The above observations suggest that disturbed systems initially approach *steady state* rapidly, the rate of adjustment progressively diminishes during the relaxation period, and morphologic changes occur slowly nearer to the condition of steady-state equilibrium.

FEEDBACK MECHANISMS

Components of geomorphic systems are interdependent, such that a change in any form element due to internal or external causes or both initiates a change in other interacting components. This process of self-regulation in the component form elements is governed by positive and negative feedback mechanisms in the system. Feedback mechanisms are positive and negative in nature. *Positive feedback mechanisms*, which proceed in the same direction of change for all components of the landscape, account for temporal changes in the form regularity of geomorphic systems. In the *normal cycle of erosion*, the fluvial landscape gradually downwears in the direction of *base level of erosion* by the positive feedback in process-form adjustment. *Negative feedback mechanisms*, however, force changes that do not move in any one direction of change. The irregular thalweg of *valley glaciers*, explained in Chapter 7, develops from local deformation of the ice by compressing and extending flow mechanisms at the glacier-bedrock interface. The rotating ice deepens the trough but loses capacity to deepen the floor further when the flow geometry achieves equilibrium with the subglacial form. Once developed, the irregular thalweg remains stable despite short- and long-term variations in the thickness of glacier ice. Hence, negative feedback mechanisms maintain the stability, regularity, and complex structure of the geomorphic systems.

OPTIMUM MAGNITUDE

The principle of optimum magnitude states that the interrelated components of geomorphic landscapes strive toward optimum magnitude of development. In fluvial landscapes, basin relief through *geologic time*, hillslope gradient through *graded time*, and channel gradient through *steady-state time* attain optimum magnitude of development in relation to the average energy of geomorphic systems.

EQUIFINALITY

The concept of equifinality states that the attributes of spatially arranged component form elements of natural systems persist through time (Culling, 1987). Equifinality also implies that a particular system condition can be reached in many different ways. Thus, a given effect can have several causes and a given environmental change can have many different effects (Wilcock, 1983). The power-law *hydraulic geometry*, discussed in Chapter 5, suggests that stream velocity can increase by the increase in channel depth, increase in channel slope, or decrease in channel roughness elements, and by any combination of these effects. Hence, interdependent relationships in geomorphic systems make the prediction of *cause and effect* difficult indeed.

THRESHOLDS

Thresholds denote a threshold of change in the process activity or of the stability of geomorphic form. Reynolds number and Froude (pronounced "frood") number are best-known thresholds in *hydraulics*. The *Reynolds number* differentiates laminar and turbulent flows by viscous and inertial forces, and the *Froude number* distinguishes tranquil and rapid flows by the gravity effect on the fluid flow. A few other thresholds commonly referred to in geomorphology may also be mentioned. *Static* and *dynamic thresholds* respectively quantify the limiting condition for entrainment and continuous transport of sediments by the wind, the *angle of internal friction* just exceeding the slope inclination provides a limiting value for the stability of loose sediments on slopes, and *thermal limits* govern the cold region processes and provide limiting conditions for the distribution of Holocene coral worldwide.

Thresholds are extrinsic and intrinsic to the geomorphic systems (Schumm, 1981). *Extrinsic thresholds* refer to the external effect of climate change or tectonism or both on the functional energy and balance of forces in geomorphic systems. Most landform attributes, however, do not differentiate the climatic signature from the tectonic signal (Derbyshire, 1999).

Intrinsic thresholds (geomorphic) are internal to the process dynamics. They describe the limiting condition of shift between system states. Several geomorphic thresholds, like bankful discharge, channel slope, sediment load, and resistance to flow, have been suggested at which *channel patterns* change in the plan view. At constant discharge, laboratory channels change from straight to meandering and from meandering to braided at two thresholds of *sediment load* in the system (Figure 1.2). Channel patterns also change with a change in the *resistance to flow* on account of long-term variations in the sediment transport rate and, consequently, the evolution of bedform morphology. Coleman and Melville (1994) observe that the channel pattern in experimental alluvial streams changes at a critical height of bedform features. In certain conditions, the *basin area* upstream of gully heads becomes a threshold for the trenching of discontinuous gullies in valley fills (Figure 1.3), and so does the *biomass density*, reaching a low of 5 kg m^{-2} of the channel floor in sand-bedded streams (Graf, 1982). Geomorphic thresholds also exist at the microscopic level of landscape organization (Culling, 1988). The *creep deformation* of soil particles, discussed in Chapter 4, occurs at a threshold of activation energy in the system.

CLIMATE AND PROCESSES

Process domains broadly derive identity from the climatic regime. Hence, a certain set of climatic conditions support particular process domains and landscape characteristics that are different from the geomorphic landscapes of other climatic

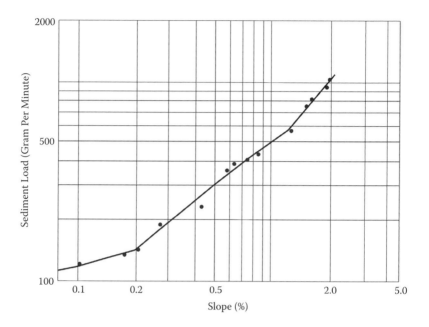

FIGURE 1.2   Relation between rate at which sand was fed into experimental channels and slope of surface on which channels were formed. Breaks in slope of line indicate threshold values of sediment load or slope on which channel pattern changed. (Source: Figure 2 in Schumm, S. A., and Khan, H. R., "Experimental Studies of Channel Patterns," *Nature* 233 [1971]: 407–9. With permission from Macmillan Magazines Ltd.)

conditions. On this analogy, Peltier (1950) proposed nine idealized climate-process systems called *morphogenetic regions* by the amplitude of mean annual temperature and mean annual rainfall amount (Figure 1.4). In essence, morphogenetic regions define limiting conditions for the existence and intensity domains of the earth surface processes. The concept of morphogenetic regions thus implies that landscape characteristics are in dynamic equilibrium with the contemporary processes and process rates (Table 1.1).

Climate, however, is only one among the variables that distinguishes regional landscape characteristics. Several related studies have shown that tectonic, structural, and lithologic aspects of terrain are variously more significant to the evolution of landscape characteristics of certain regions than are the climate-driven processes and process rates (Carson and Kirkby, 1972; Bradshaw, 1982). Hence, the relationship between climate and landscape evolution is imperfect at local and regional scales. The *relict landforms* similarly suggest out-of-phase association of the elements of landscape with the climate of areas in which they exist.

GEOMORPHIC RESPONSE TO CLIMATE

Although certain aspects of landscape characteristics are broadly related to the present-day climate, the manner of geomorphic response to climate and climate change is far from understood (Graf, 1977; Eybergen and Imeson, 1989; Rinaldo et al., 1995;

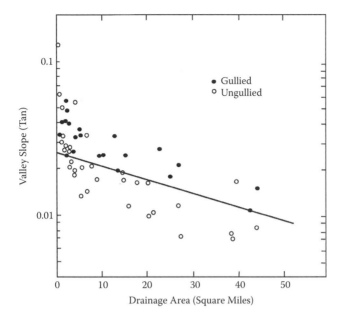

FIGURE 1.3 Relation between valley slope and drainage area, Piceance Creek basin, Colorado. Line defines threshold slopes for this area. (Source: Figure 1 in Schumm, S. A., "Geomorphic Thresholds and Complex Response of Drainage Systems," in *Fluvial Geomorphology*, ed. M. Morisawa [London: George Allen & Unwin, 1981], 299–310. Originally from Patton [unpublished MS thesis, 1973]. With permission from George Allen & Unwin.)

Whipple et al., 1999). Drainage density is one such geomorphic property that varies more with the climatic regime or *erosive potential* of climate than with other factors of terrain (Peltier, 1962). In general, the *drainage density* is highest in semiarid climates, decreases in humid temperate zones, and increases somewhat in humid tropical regions in close agreement with the pattern of mean annual effective rainfall (see Figure 6.10). In statistical terms, 93% variation in the drainage density is due to climate alone (Gregory, 1976), provided climate and terrain remain stable for a sufficiently long period of time (Rinaldo et al., 1995). Over shorter timescales, however, the dependence of form elements on erosive potential of climate becomes obscure and even reversed in semiarid and arid environments (Derbyshire, 1999).

Hillslope geometry is also governed by the erosive potential of regional climate (Toy, 1977). Thus, hillslopes that are shorter and steeper with smaller radii of curvature of the convex segment in arid climates become progressively longer and flatter toward the humid climatic regime (Figure 1.5). Studies subsequently referred to in the following section, however, suggest that the erosive potential of climate itself becomes a negative factor in the development of relief in high-energy alpine environments.

In certain environments, the relationship between *climate change* and evolution of geomorphic systems is indirect or even suspect. Studies from the recently deglaciated landscape of Alaska suggest that progressive dilution of nutrients in the

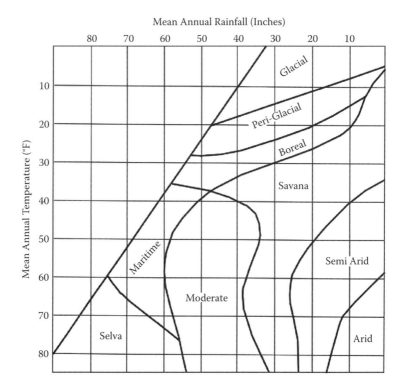

FIGURE 1.4   Climatic controls of morphogenetic regions. (Source: Figure 7 in Peltier, L. C., "The Geographic Cycle in Periglacial Regions as it is Related to Climatic Geomorphology," *Ann. Assoc. Am. Geogr.* 40 [1950]: 214–36. With permission from Routledge Publishers.)

water column of 10,000- to 12,000-year-old freshwater boreal lakes is related more to *successional changes* in vegetation and soil characteristics than to the warming of Holocene climate (Engstrom et al., 2000). Hence, the evolution of *aquatic systems* depends more on the physical and biological manifestation of climate in terrestrial environments (King, 2000). In contrast, the *pollen record* from eastern Amazon suggests that the Holocene warming of climate has had little effect on the *succession rate* of vegetation communities (Bush et al., 2004). Hence, natural systems do not always appear to embody identical signatures of a gradual environmental change.

In recent years, attempts have been made to theoretically evaluate the effect of climate-driven *process rates* on the evolution of landscape characteristics. Molnar and England (1990) predict that a glacial climate is more erosive than the humid climate. Hence, *isostatic compensation* resulting from *offloading* accentuates the mountain relief that furthers the erosive potential of climate. Other theoretical models to the contrary, however, suggest that a humid climate is more erosive than the glacial climate (Summerfield and Kirkbride, 1992), and that glacial erosion does not accentuate relief (Whipple et al., 1999) or add to the cooling of climate (Small, 1999). There is unanimity, however, that the greater erosive potential of climate irreversibly dissects hillslopes to optimum-sized smaller segments, restricting the development of relief in high-energy alpine environments (Schmidt and Montgomery, 1995;

TABLE 1.1
Climatic Controls and Morphologic Attributes of Morphogenetic Regions

Morphogenetic Region	Range of Mean Annual		Process Activity
	Temperature (°F)	Rainfall (in.)	
Glacial	0–20	0–45	Glacial erosion, nivation, wind action
Periglacial	5–30	5–55	Strong mass movement, moderate to strong wind action; weak effect of running water
Boreal	15–38	10–60	Moderate frost action, moderate to slight wind action, moderate effect of running water
Maritime	35–70	50–75	Strong mass movement, moderate to strong action of running water
Selva	60–85	55–90	Strong mass movement, slight effect of slope wash, no wind action
Moderate	35–85	35–60	Maximum effect of running water, moderate mass movement, slight frost action in colder parts, no significant wind action except on coasts
Savanna	10–85	25–50	Strong to weak action of running water, moderate wind action
Semiarid	35–85	10–25	Strong wind action, moderate to strong action of running water
Arid	55–85	0–15	Strong wind action, slight action of running water and mass movement

Source: Peltier, L. C., "The Geographic Cycle in Periglacial Regions as It Is Related to Climatic Geomorphology," *Ann. Assoc. Am. Geogr.* 40 [1950]: 214–36. With permission from Routledge Publishers.

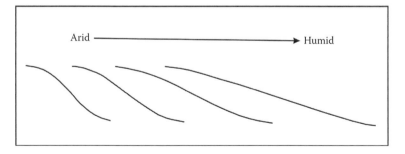

FIGURE 1.5 Conceptualized hillslope profiles. (Source: Figure 4 in Toy, T. J., "Hillslope Form and Climate," *Bull. Geol. Soc. Am.* 88 [1977]: 16–22. With permission from Geological Society of America.)

Whipple et al., 1999). Hence, geomorphic controls override the effect of climate on the development of relief.

FREQUENCY CONCEPT OF GEOMORPHIC PROCESSES

Earth surface processes are characteristic of a certain frequency and magnitude of occurrence, such that high-magnitude events are infrequent and low-magnitude events reoccur frequently in space and time. General observations suggest that high-magnitude, low-frequency events as stream floods seemingly perform a larger amount of work than frequent channel discharges below the bankful stage. In fluvial systems, the work done by geomorphic processes is conveniently expressed by the amount of solute and clastic load transported in the fluvial flow.

The geomorphic work of events is tied down to the magnitude of opposing driving and resisting forces in the system. Sediments are entrained in the fluid flow when the magnitude of fluid force or shear stress just exceeds the resisting force due to the bed and bank materials. This quantity is called *effective stress*. Leopold et al. (1964) predict that the quantity of sediment moved by the fluid flow varies with the lognormal effective stress as

$$Q = s(\tau - \tau_c)^n$$

in which Q is the sediment transport rate, s is the shear strength of materials, τ is the shear (or fluid) stress on the mobile bed, τ_c is the threshold stress for initiation of grain movement, and n is the exponent. The term $\tau - \tau_c$ then is the effective stress. The above relationship suggests that the sediment transport rate increases manifold with a small increase in the effective stress. Hence, high-magnitude stream flow events transport a large amount of sediment per unit of time. By comparison, the low-magnitude events are so frequent that their *cumulative work* exceeds the work of high-magnitude events of low recurrence interval. Leopold et al. (1964) present data on the temporal distribution of discharge and sediment load in selected streams of the United States, which similarly suggest that the cumulative volume of sediment moved is higher for frequent low and intermediate channel discharges than for infrequent overbank flows.

The statistical recurrence interval of extreme events of specified magnitude can be predicted, provided observations are spread over a long period of time. The *recurrence interval* is given as

$$T = (N + 1)/M$$

in which T is the average recurrence interval, N is the period of observation, and M is the rank of extreme event in a series of observations arranged in descending order of magnitude. The recurrence probability of events is given as 1/T, which suggests that high-magnitude events are infrequent and low-magnitude events are frequent in occurrence (Table 1.2).

TABLE 1.2

Probability of Instantaneous Peak Floods at Masani Barrage on the Inland Sahibi Basin in Southern Haryana, India (1965–1984)

Discharge Magnitude (cumec)	Rank Interval (T)	Recurrence (1/T)	Probability
3028.00	1	21.0	0.048
827.00	2	10.5	0.095
657.13	3	7.5	0.143
451.52	4	5.2	0.190
434.12	5	4.2	0.238
433.09	6	3.5	0.286
268.86	7	3.0	0.333
258.05	8	2.6	0.381
254.70	9	2.3	0.429
231.03	10	2.1	0.476
216.32	11	1.9	0.524
189.02	12	1.7	0.571
142.01	13	1.6	0.619
140.79	14	1.5	0.667
139.57	15	1.40	0.714
119.54	16	1.31	0.762
84.11	17	1.23	0.810
60.35	18	1.16	0.857
40.39	19	1.10	0.905
35.23	20	1.05	0.952

Source: Parmar, A. S., "Flow Patterns in Sahibi Basin, Haryana" (MPhil dissertation, Kurukshetra University, 1986).

The frequency concept provides a framework for comparing the work of a known process domain in diverse geomorphic environments. *Water hardness* data from the karst terrain of different climatic environments (Chapter 10) suggest that the rate of limestone/dolomite dissolution is highest in the tropics than in other environments. For equal magnitude of discharge, however, the dissolution rates do not differ much between climatic provinces (Smith and Atkinson, 1976). Rapid evolution of the tropical landscape was similarly once attributed to the high rate of *desilication* (Chapter 3) of terrain. By the equal magnitude of stream flow, however, the rate of silica loss is nearly the same in the tropics and other environments (Douglas, 1969). Hence, differences in the state of landscape development across lithologies and climatic regimes require an alternative explanation. The frequency concept also finds applications in the evaluation of *process thresholds* (Ahnert, 1987; Derbyshire, 1999), and in river training and site planning projects.

ENVIRONMENTAL CHANGE

Components of landscape and landscape characteristics evolve over a long period of time. Hence, landscape attributes inherit the imprints of past process domains and process rates. Biotic components of the landscape that evolve through *natural selection* and *succession* also affect the biochemical processes and, thus, the evolution of landscape through time. Interpretation of the landscape, therefore, requires appreciation of the environmental change. Environmental change, in general, refers to a broad spectrum of temporal changes in climate, sea surface level, continents and seafloor, and flora and fauna (Goudie, 1977). These changes can be reconstructed with confidence to the beginning of the *Quaternary period.*

Interpretation of the condition of the geologic past seeks to establish the cause and magnitude of change, and the geologic date for the initiation and cessation of the given environmental change. Signatures of the environmental change are preserved in *environment-sensitive* indices of certain biotic and abiotic components of the geomorphic systems. The theory, application aspects, and regional significance of some such *proxy indicators* for the interpretation of the environmental change are outlined in this section. Reliable *documentary evidence* for a specified period of written history also provides data of interpretative significance to the study of proximal earth surface events (Nash, 1996; Derbyshire and Goudie, 1997).

DENDROCHRONOLOGY

Absolute chronology of proximal climatic events extendible back in time to the age of trees can be reconstructed from the *tree-ring calendar.* Tree rings evolve by the addition of woody material to the xylem each year. The woody material develops large-sized cells at the beginning of the growth season, which progressively become smaller through time and eventually terminate as thick-walled minute cells at the end of the growth season. This sudden change in the annual appearance of wood tissues constitutes a tree ring. The ring counts in tree stems thus equal the age of trees. The ring widths, however, are sensitive to moisture surplus and deficiency, and thermal amplitude of the environment during the growth cycle of trees. In a general term, larger ring widths denote more optimum moisture and temperature conditions for the growth of trees than do the rings of smaller width.

Dendrochronology has been applied to date various kinds of archaeological remains, such as the pottery made by Pueblo Indians in the Rio Grande Valley of southwestern United States (Leopold et al., 1964). The technique, though, finds best applications in the interpretation of *landform change* and *variability of climate* for the last 3000 to 4000 years at the most. The ring width data from trees along the bed of a gully of accelerated rate of erosion in the Denver area of Colorado suggests that landform change follows the *rate law* (Graf, 1977), discussed in the section on perturbation of geomorphic systems. The ring width and wood density data on *Pinus sylvestris L.* (Scot pine) from northern Sweden suggest that summers were cooler by 0.5°C, and −1°C below the average thirteen times during the last millennium (Briffa et al., 1990). The air temperature calibrated from the ring widths of *Pinus wallichiana* (Blue pine) in the sub-Himalayan environment of northwestern India suggests a positive *ice budget* in 1970–1976, 1981–1984, and 1989 AD for otherwise retreating glaciers of the region (Bhattacharyya and Yadav, 1996). The *oxygen isotope* record in the woody material of trees over ecologically distinct sites in high mountains of northern Pakistan suggests that the last 1000 years of the twentieth century had been the wettest phase of climate on account of human interference with the environment (Treydte et al., 2005).

POLLEN

Pollen grains are microspores of seed plants. They are practically indestructible and remain indefinitely preserved in conventional and extreme environments as signatures of the past climate. Pollen-related reconstruction of the environment is based on the assumption that airborne pollen suites are derived from local vegetal associations in the vicinity of ideal accumulation sites in lakes, swamps, and bogs. The frequency distribution of pollen sequence through the depth of the deposit yields data on the paleoenvironment and *habitat* of vegetation communities of the area.

Palynology is ideal for the reconstruction of the *Quaternary climate*. The generic profile of the pollen sequence in the lakebed sediments of southern Thar Desert unveils a continuous record of the variability of Holocene climate for the northern plains of India (Singh, 1971). The pollen data suggest that the climate was intensely arid about 10,000 BC, moderately arid between 7500 and 3000 BC, and severely arid between 800 and 500 BC. Further, short humid phases of climate between 8000 and 7500 BC and in 1800 BC, recording respectively 25 and 50 cm more rainfall than the present-day amount, punctuated the long sequence of Holocene aridity in the northern plains of India.

The pollen preserved in aquatic systems holds immense potential for the reconstruction of *Late Pleistocene climate*. A few pollen-based studies from the Himalayan realm suggest a deglaciation date of 9500 BC at an elevation of 3120 m above the msl (mean sea level) at Toshmaidan in Kashmir Valley (Singh and Agrawal, 1976). The climatic signal of pollen suites from Tso Kar (Kar Lake) at 4535 m above the msl in central Himalayas of northwestern India suggests a warm and moist climate at 30 to 28 thousand years (ka), 22 to 18 ka, 16 ka, and 10 ka for the region (Bhattacharyya, 1989). The moist phase of climate at 10 ka for the northwestern Himalayas, however, coincides with the intense dry phase of the climate in the northern plains of India. The asynchronous moist and dry phases of the climate to the north and south of the Himalayas probably suggest fluctuation in the moisture-bearing monsoon advective system, rather than a change of climate per se (Benn and Owen, 1998; Owen et al., 2001).

The pollen profile from Kashiru Swamps, Burundi, unveils a paleotemperature record for the tropics (Bonnefille et al., 1990). The climatic signal of pollen sequence suggests a *late glacial advance* 21,500 years ago in east Africa, and a temperature $4 \pm 2°C$ below the freezing point for the event, as opposed to the accepted average temperature change of −5 to −7°C required for the onset of glaciation.

RADIOMETRIC METHODS

Radiometric methods utilize the decay property of unstable natural and induced radioactive elements, and ratios of stable chemical isotopes of the same element for

TABLE 1.3
Half-Life of Primary
Radioactive Elements

Element	Half-Life (year)
Natural Radioactive	
Uranium-238	4.5×10^9
Uranium-235	7.1×10^8
Thorium-232	1.39×10^{10}
Rubidium-87	5×10^{10}
Potassium-40	1.3×10^9
Induced Radioactive	
Carbon-14	$5,730 \pm 40$

the reconstruction of a paleoenvironment. *Natural radioactive elements* emit radiation of their own, and *induced radioactive elements* acquire radiation from the transmutation of nonradioactive elements (Table 1.3). *Chemical isotopes* fractionate by the physiochemical processes of enrichment that involve a vapor-liquid-solid state of water transformation in the environment.

Natural Radioactive Elements

Natural radioactive elements possess *unstable nuclei*. They spontaneously decay by emitting α and β particles. The decay decreases the number of atoms of radioactive elements, evolving disintegration products that behave chemically in a manner different from preceding and succeeding radioactive products. The disintegration eventually reaches the end product of stable nonradioactive elements. The time for half of the atoms of a given radioactive element to decay into the next disintegration product is called *half-life period*, which, depending upon the radioactive species, varies from a few days to millions of years. The radiometric dating of rocks and sediments is based on the molar ratio of the daughter to the parent nuclide in the specimen.

Several radioactive series are used for the age determination of rocks and the reconstruction of ancient environments. Among these, the uranium-thorium series has been widely used for calibration of the age of the earth. The method has also been applied to sediments of diverse geomorphic environments for the interpretation of paleoenvironment. Among others, the desert varnish on lag deposits of Colorado Plateau is 300,000 years old, suggesting a drier and windier environment for the Late Pleistocene (Knauss and Ku, 1980). The uranium-thorium dating of older marine sediments suggests the midpoint of $13,500 \pm 250$ years BP for the penultimate deglaciation in the northern hemisphere, a period of time consistent with the tilt and wobble of the earth's orbit around the sun and coral-based reconstruction of the sea level estimate (Henderson and Slowey, 2000).

Radiocarbon Method

Radiocarbon (^{14}C) is continuously produced in the *upper atmosphere* by the bombardment of nitrogen (^{14}N) with high-energy cosmic neutrons (n) as

$$^{14}N + n = {}^{14}C + H^1$$

The radioactive carbon incorporates into the atmospheric carbon dioxide and becomes a part of a wider *food chain system*. The intake of ^{14}C by living plants and animals, however, ceases when they die or become isolated from the atmosphere, such as by burial. The stored ^{14}C in organic remains decays thereafter with the emission of β radiation, evolving a stable nonradioactive nucleus of nitrogen as

$$^{14}C \rightarrow {}^{14}N + \beta$$

The radiocarbon decay has a *half-life period* of 5730 ± 40 years. The age of organic materials, such as wood, charcoal, bone, and shell, is determined from the amount of radioactivity they retain. The radiocarbon method is suitable for the reconstruction of events generally not more than 30 ka old (Deevey, 1952).

The carbon-14 method of environmental reconstruction is ideal for dating Late Quaternary climatic events. The radiocarbon *deglaciation date* of 9160 ± 70 years BP from the base of a peat deposit in the northwestern Himalayas (Owen, 1998) is nearly consistent with the 9600 years BP degalciation date established by other methods on marine sediments in the Gulf of Mexico (Emiliani et al., 1975). The organic matter in *morainic deposits* is 23,000 and 16,000 years old in Tibet (Derbyshire et al., 1991) and 18,100 to 15,700 years old in the northern Karakoram Mountains to the south of Tibet (Li and Shi, 1992), suggesting a phase of late glacial advance in this part of the world. Organic remains in valley fills (Chapter 6) likewise provide radiocarbon dates for the environment favoring channel aggradation.

CHEMICAL ISOTOPES

Physiochemical processes of enrichment and depletion of geochemical isotopes, and their application for dating climatic events, are discussed in several publications (Hamilton and Farquhar, 1968; Ferronsky and Polyakov, 1982; Bowen, 1991). In general, temperature-related processes of evaporation and condensation in the seawater, and sublimation and solidification of the ice, determine the chemical isotopic composition of oxygen and hydrogen in the seawater and the glacier ice. Thus, the isotopic ratios of two elements are sensitive to the *ambient temperature* of the environment. They are also an appropriate measure of the thermal amplitude of the environment. Oxygen exists as isotopes of ^{16}O, ^{17}O, and ^{18}O, and hydrogen as isotopes of hydrogen (^{1}H), deuterium (^{2}H), and tritium (^{3}H). Records of ^{18}O/^{16}O and ^{2}H/^{1}H in the seawater and the glacier ice are proxy for *paleotemperatures* of the earth's environment. The isotope method is applicable for dating climatic events generally not more than 30,000 years old.

Global temperature fluctuations during the *Great Ice Age* affected isotopic ratios of oxygen and hydrogen in the seawater and the glacier ice. Experimental studies suggest that the concentration of ^{18}O relative to ^{16}O decreases with decreasing temperature in ice sheets but increases in carbonate-secreting marine organisms called *foraminifera*. The $^{18}O/^{16}O$ data from several sources suggest that the Pleistocene ice grew to massive thickness by slow accretion over thousands of years but thawed quite abruptly within a decade (Bowen, 1991) about 9600 BC (Emiliani et al., 1975), oceans held 4 to 5% more volume of water in the interglacial than glacial phases of the Ice Age climate (Bowen, 1991), the sea level fluctuated by about 100 m during the glacial-interglacial phases of the Pleistocene climate (Shackleton and Opdyke, 1973), and continental ice of the last phase of Pleistocene glaciation 500,000 years ago readvanced five times with a periodicity of about 10,000 years (Schrag, 2000). The ocean-based oxygen isotope paleotemperature record suggests that the extent of Pleistocene aridity and thickness variation of high-latitude loess varies directly with the expanse of continental ice (Shroder et al., 1989; Kemp et al., 1995; Derbyshire, 1999). The record of $^{2}H/^{1}H$ from ice cores in Greenland and Antarctica suggests that the climate of the last glacial maxima warmed 1000 to 2500 years earlier in Antarctica than in Greenland (Blunier et al., 1998). Hence, the asynchronous pattern of Pleistocene climatic change for the two hemispheres requires reassessment of the dynamic behavior of the global climatic system (Steig et al., 1998; Henderson and Slowey, 2000; Scharg, 2000).

Thermoluminescence and Related Dating Methods

Thermoluminescence (TL) is a powerful research tool for the reconstruction of the environment. It is based on interpretation of the record of natural and induced radioactivity in the earth materials. McDougall (1968) and Hurford et al. (1986) discuss the source of TL in the earth materials, laboratory procedures for calibration of the age of rocks and Quaternary sediments, and application aspects of the technique for interpretation of the environment.

Igneous rocks are naturally thermoluminescent. They had trapped uranium and thorium in the structure of their imperfect mineral lattices at the time of formation. This *metastable* locked radiation escapes the mineral structure, and emits glow, when rock specimens are laboratory heated to a very high temperature. The glow intensity varies with the activation energy and duration of heating, producing thermoluminescent *glow curves* that are diagnostic of the age and thermal history of rocks.

Most recent thermal, crystallization, and sun-bleached events also lend to the TL dating. *Contact heating* of the Quaternary sediments by lava flows and firing events of archaeological pottery de-trap the geologic TL. The sediment and pottery, if subsequently buried, acquire radioactivity from the environment of the burial site. *Crystallization events*, like precipitation of tufa and gypsum, begin with *ab initio* zero radioactivity. These chemical sediments, till excavation, also acquire radioactivity from the surrounding environment. *Sun-bleached* weathered and transported sediments rapidly de-irradiate. The sediments, if subsequently buried, likewise acquire radiation from the surrounding environment.

TL-related optically simulated luminescence (OSL), infrared simulated luminescence (IRSL), and electron spin resonance (ESR) methods of environmental
reconstruction apply to a variety of sediments and geologic environments. The OSL
employs argon laser, IRSL uses an infrared source, and ESR utilizes a microwave
signal in a high magnetic field to stimulate luminescence from clastic sediments. In
comparative terms, the OSL technique is particularly useful for dating sun-bleached
events, the IRSL method is ideal for dating sediments of diverse geologic environments, and the ESR technique is suitable for dating chemical sediments and organic
deposits older than Quaternary.

The OSL technique is routinely used for calibrating the age and paleoclimatic
environment of aeolian sand, coastal sand, loess, and lacustral deposits. The dated
stabilized dunes of the western desert of India are Late Pleistocene in age, and suggest evolution by a slow rate of sedimentation in the Holocene (Singhvi et al., 1982).
The coastal sand overlying the emergent continental shelf of eastern China is Early
Pleistocene (2.4 million years [Ma]) to most recent in age (2500 years BP). This
sand correlates well with the periods of Quaternary sea level oscillation, sequences
of marine transgression, and the age of high-latitude loess of eastern China (Mulin,
1986). The TL-dated reworked peri-desert loess of the northwestern Himalayas is 25
to 18 ka old (Owen et al., 1992), and correlative with the Late Pleistocene multiple
glaciation of the region (Richards et al., 2000). The OSL-dated glaciogenic sediments in the Himalayas suggest three major glacial advances, of which the 60 to 30
ka Late Pleistocene glacial advance had been the most extensive in the region (Owen
et al., 1992; Sharma and Owen, 1996; Owen, 1998). The glaciolacustrine deposits on
either side of the Indus Valley in the Gilgit region of the northwestern Himalayas are
38.1 ± 2.6 ka old (Shroder et al., 1989), suggesting ice damming of the channel at
1000 m above the msl (Richards et al., 2000).

The IRSL method has been used for dating sand, loess, and alluvial deposits.
The IRSL dates on loess-paleosol sequences of Tadijikistan, and the aeolian sand of
Indian Thar Desert, are consistent with the OSL dates on similar sequences of two
areas (Mavidanam, 1996). The IRSL-dated Indo-Gangetic alluvium is 20 to 1 ka old
across its lateral extent (Mavidanam, 1996).

SUMMARY

Process geomorphology studies the motion and behavior of geomorphic processes at
and near the surface of the earth, and explains the evolution of landscape and component landforms in terms of the process dynamics. Geomorphic processes comprise of the activity of wind, water, and ice. These agents of change are considered
fluids, as their motion can be analyzed in terms of the fluid mechanics. Processes
are quantified in the field, few activities that present difficulty of instrumentation are
calibrated in test conditions, and the dynamics of still others is understood in theory
only. The work of geomorphic processes varies directly with the process intensity
and recurrence frequency. The scale of observation essentially affects the identity of
landforms and hypotheses on the evolution of landforms.

An open systems approach provides an ideal framework for the explanation
of time-dependent evolution of landscape and time-independent development of

spatially contiguous form elements of geomorphic landscapes. The landscape and its form elements evolve in equilibrium with the through flow and continuous exchange of energy and material in geomorphic systems. The state of equilibrium is cyclic, graded, and steady with reference to time, and quasi-equilibrium with respect to the rate of energy expenditure among the interdependent variables of drainage systems. The attributes of geomorphic landscape and its components form elements that are governed by feedback mechanisms, optimum magnitude, and equifinality in process-form relationships.

External effects of climate change, tectonics, and human activities disturb the established order of exchange of energy and material in geomorphic systems. Disturbed geomorphic systems, therefore, pass through reaction and relaxation thresholds, called extrinsic thresholds, before reaching the steady-state equilibrium. The rate of process-form adjustment in systems disturbed by the external stimulus of energy is exponential in nature. Geomorphic thresholds are intrinsic to the process activity. The intrinsic thresholds are due to internal causes that affect the shear stress and, thereby, process rates in geomorphic systems.

Regional climate broadly regulates the earth surface process domains. Hence, geomorphic landscape is visualized to evolve in equilibrium with the contemporary process domains of respective climatic regimes. The hypothetical concept of mor-phogenetic regions discusses a broad-based climate-process-form relationship for nine hypothetical regions. However, the manner of geomorphic response to climate and climate change is far from simple. Most often, physical and biochemical controls of landscape override the effect of climate on process rates. Theoretical studies on process-form relationships also suggest that climate itself becomes a limiting factor in relief development of the high-energy alpine systems.

Landscape and component landforms evolve over a long period of time, during which process domains may have changed altogether or process rates may have varied considerably due to a change in the intensity of process activity. Hence, an explana-tion of landforms also requires evaluation of the environmental change. Several envi-ronment-sensitive proxy indicators are used for interpreting the type, magnitude, and duration of environmental change. Depending upon the nature of indicator source, dendrochronologic, palynologic, radiometric, chemical isotopic, and themolumines-cence dating techniques find applications in the reconstruction of the Quaternary environment. Dendrochronology interprets the ring width data for deducing the hydrologic and thermal amplitude of the proximal environment. A depth profile of pollen sequences in lakebed sediments and peat deposits unveils a continuous record of the magnitude of moisture and temperature change in the environment. A record of the radiocarbon in organic remains enables interpretation of the environment and environmental change to within the safe limit of 30 ka before present. The chemical isotope ratio of ^{18}O and ^{16}O in the ice and skeletal remains of deep sea organisms, called foraminifera, is a proxy for the Quaternary ambient temperature, sea level change, and chronology of climatic events. The isotope ratio of ^{2}H and ^{1}H in the seawater and the glacier ice is a measure of the amplitude of paleotemperature of the earth's past environment. The property of luminescence due to geologic and induced radioactivity in rocks and sediments finds applications in calibrating the age and paleoclimatic environment of the earth materials of diverse origins.

REFERENCES

Abrahams, A. D. 1968. Distinguishing between the concepts of steady state and dynamic equilibrium in geomorphology. *Earth Sci. Rev.* 2:160–66.

Ahnert, F. 1967. The role of the equilibrium concept in the interpretation of landforms of fluvial erosion and deposition. In *Evolution des Versants*, ed. P. Macar, 23–41. Liege: L'Université de Liege.

Ahnert, F. 1987. An approach to the identification of morphoclimates. In *International geomorphology*, ed. V. Gardiner, 159–88. Vol. 2. Chichester: John Wiley & Sons.

Benn, D. I., and Owen, L. A. 1998. The role of Indian summer monsoon and mid-latitude westerlies in Himalayan glaciation: Review and speculative discussion. *J. Geol. Soc. Lond.* 155:353–63.

Bhattacharyya, A. 1989. Vegetation and climate during the last 30,000 years in Ladakh. *Palaegeogr. Palaeobot. Palaeoecol.* 73:25–38.

Bhattacharyya, A., and Yadav, R. R. 1996. Dendrochronological reconnaissance of *Pinus wallichiana* to study glacial behaviour in western Himalaya. *Current Sci.* 70:739–43.

Blunier, T., Chappellaz, J., Schwander, J., Dällenbach, A., Stauffer, B., Stocker, T. F., Raynand, D., Jouzel, J., Clausen, H. B., Hammer, C. U., and Johnsen, S. J. 1998. Asynchrony of Antarctic and Greenland climate change during the last glacial period. *Nature* 394:739–43.

Bonnefille, R., Roeland, J. C., and Guiot, J. 1990. Temperature and rainfall estimates for the past 40,000 years in equatorial Africa. *Nature* 346:347–49.

Bowen, R. 1991. *Isotopes and climate*. London: Elsevier Applied Science.

Bradshaw, M. 1982. Process, time and the physical landscape: Geomorphology today. *Geography* 67:15–28.

Briffa, K. R., Bartholin, T. S., Eckstein, D., Jones, P. D., Karten, W., Schweingruber, F. C., and Zetterberg, P. 1990. A 1,400-year tree-ring record of summer temperatures in Fennoscandia. *Nature* 346:434–39.

Brunsden, D. 1988. Slope instability, planning and geomorphology in the United Kingdom. In *Geomorphology in environmental planning*, ed. J. M. Hooke, 105–19. Chichester: John Wiley & Sons.

Brunsden, D., and Thornes, J. B. 1979. Landscape sensitivity and change. *Trans. Inst. Br. Geogr.* 4:463–84.

Bull, W. B. 1991. *Geomorphic response to climatic change*. New York: Oxford University Press.

Bush, M. B., Silman, M. R., and Urrego, D. H. 2004. 48,000 years of climate and forest change in a biodiversity hot spot. *Science* 303:827–29.

Carson, M. A., and Kirkby, M. J. 1972. *Hillslope form and processes*. Cambridge: Cambridge University Press.

Chorley, R. J. 1962. *Geomorphology and general systems theory*. U.S. Geological Survey Professional Paper 500-B. U.S. Geological Survey.

Chorley, R. J. 1978. Bases for theory in geomorphology. In *Geomorphology: Present problems and future prospects*, ed. C. Embleton, D. Brunsden, and D. K. C. Jones, 1–13. Oxford: Oxford University Press.

Church, M., and Slaymaker, O. 1989. Disequilibrium of Holocene sediment yield in glaciated British Columbia. *Nature* 337:452–54.

Coleman, S. E., and Melville, B. W. 1994. Bed-form development. *J. Hydraul. Eng.* 120:544–60.

Culling, W. E. H. 1987. Equifinality: Modern approaches to dynamical systems and their potential for geographical thought. *Trans. Inst. Br. Geogr.* 12:57–72.

Culling, W. E. H. 1988. A new view of landscape. *Trans. Inst. Br. Geogr.* 13:345–65.

Deevey, E. S., Jr. 1952. Radiocarbon dating. *Sci. Am.* 206:24–28.

Derbyshire, E. 1999. Landforms, geomorphic processes and climatic change in drylands. In *Palaeoenvironmental reconstruction in arid lands*, ed. A. K. Singhvi and E. Derbyshire, 1–26. New Delhi: Oxford & IBH.

Derbyshire, E., and Goudie, A. S. 1997. The drylands of Asia. In *Arid zone geomorphology of desert environments*, ed. S. G. Thomas, 487–506. 2nd ed. Chichester: John Wiley & Sons.

Derbyshire, E., Shi, Y., Li, J., Zheng, B., Li, S., and Wang, J. 1991. Quaternary glaciation of Tibet: The geological evidence. *Quat. Sci. Rev.* 10:485–510.

Douglas, I. 1969. The efficiency of humid tropical denudation systems. *Trans. Inst. Br. Geogr.* 46:1–16.

Emiliani, C., Gartner, S., Eldridge, K., Elvey, D. K., Huang, T. C., Stipp, J., and Swanson, M. F. 1975. Palaeoclimatological analysis of Late Quaternary cores from the northeastern Gulf of Mexico. *Science* 189:1083–88.

Engstrom, D. R., Fritz, S. C., Almendinger, J. E., and Juggins, S. 2000. Chemical and biological trends during lake evolution in recently glaciated terrain. *Nature* 408:161–66.

Eybergen, F. A., and Imeson, A. C. 1989. Geomorphological processes and climatic change. *Catena* 16:307–19.

Ferronsky, V. I., and Polyakov, V. A. 1982. *Environmental isotopes in the hydrosphere*, English trans. S. V. Ferronsky. Chichester: John Wiley & Sons.

Goudie, A. S. 1977. *Environmental change*. Oxford: Clarendon Press.

Graf, W. L. 1977. The rate law in fluvial geomorphology. *Am. J. Sci.* 277:178–91.

Graf, W. L. 1982. Spatial variation of fluvial processes in semi-arid lands. In *Space and time in geomorphology*, ed. C. E. Thorn, 193–217. London: George Allen & Unwin.

Gregory, K. J. 1976. Drainage networks and climate. In *Geomorphology and climate*, ed. E. Derbyshire, 289–315. London: John Wiley & Sons.

Hack, J. T. 1960. Interpretation of erosional topography in humid temperate regions. *Am. J. Sci.* 258-A:80–97.

Hamilton, E. I., and Farquhar, R. M., eds. 1968. *Radiometric dating for geologists*. London: Interscience Publishers.

Henderson, G. M., and Slowey, N. C., 2000. Evidence from U-Th dating against Northern Hemisphere forcing of penultimate deglaciation. *Nature*, 404, 61-66.

Howard, A. D. 1982. Equilibrium and timescales in geomorphology: Application to sand-bed alluvial streams. *Earth Surface Processes and Landforms* 7:303–25.

Howard, A. D. 1988. Equilibrium models in geomorphology. In *Modelling geomorphological systems*, ed. M. G. Anderson, 49–72. Chichester: John Wiley & Sons.

Huggett, R. J. 1980. *Systems analysis in geography*. Oxford: Clarendon Press.

Hurford, A. J., Jäger, E., and Ten Cate, J. A. M., eds. 1986. *Dating young sediments*. CCOP Technical Publication 16. Bangkok: CCOP.

Kemp, R. A., Derbyshire, E., Meng, X. M., Chen, F. H., and Pan, B. T. 1995. Pedosedimentary reconstruction of a thick loess-palaeosol sequence near Lanzhou in north-central China. *Quat. Res.* 43:30–45.

King, G. W. 2000. A lake's life is not its own. *Nature* 408:149–50.

Kirkby, M. J. 1990. Landscape viewed through models. *Z. Geomorph.* 79:63–81.

Knauss, K. G., and Ku, T. 1980. Desert varnish: Potential for age dating via uranium series isotopes. *J. Geol.* 88:95–100.

Knighton, D. 1984. *Fluvial forms and processes*. London: Edward Arnold.

Lamberti, A. 1992. Dynamic and variational approaches to the river regime relation. In *Entropy and energy dissipation in water resources*, ed. V. P. Singh and M. Fiorentino, 505–25. Dordrecht: Kluwer Academic Publishers.

Langbein, W. B., and Leopold, L. B. 1964. Quasi-equilibrium states in channel morphology. *Am. J. Sci.* 262:782–94.

Leopold, L. B., and Langbein, W. B. 1962. *The concept of entropy in landscape evolution*. U.S. Geological Survey Professional Paper 500-A. U.S. Geological Survey.

Leopold, L. B., Wolman, M. G., and Miller, J. P. 1964. *Fluvial processes in geomorphology*. San Francisco: W. H. Freeman and Co.

Li, S., and Shi, Y. 1992. Glacial and lake fluctuations in the area of the West Kunlun Mountains during the last 45,0000 years. *Ann. Glacio.* 16:79–84.

Mavidanam, S. R. L. 1996. Studies on physical basis of luminescence geochronology and its applications. PhD thesis, Nagpur University.

McDougall, D. J. 1968. *Thermoluminescence of geological materials.* London: Academic Press.

Molnar, P., and England, P. 1990. Late Cenozoic uplift of mountain ranges and global climatic change: Chicken or egg? *Nature* 346:29–34.

Montgomery, K. 1989. Concepts of equilibrium and evolution in geomorphology: The model of branch systems. *Prog. Phys. Geogr.* 13:47–66.

Mulin, Z. 1986. On China's Quaternary stratigraphy. In *Dating young sediments*, ed. A. J. Hurford, E. Jäger, and J. A. M. Ten Cate, 5–26. CCOP Technical Publication 16. Bangkok: CCOP.

Nash, D. J. 1996. On the dry valleys of the Kalahari: Documentary evidence of environmental change in central southern Africa. *Geogr. J.* 162:154–68.

Owen, L. A. 1998. Timing and style of glaciation in the Himalaya. *Himalayan Geol.* 19:39–47.

Owen, L. A., Gualtieri, L., Finkel, R. C., Caffee, M. W., Ben, D. I., and Sharma, M. C. 2001. Cosmogenic radionuclide dating of glacial landforms in the Lahul Himalaya, northern India: Defining the timing of Late Quaternary glaciation. *J. Quat. Sci.* 16:555–63.

Owen, L. A., White, B. J., Rendell, H., and Derbyshire, E. 1992. Loessic silt deposits in the western Himalayas: Their sedimentology, genesis and age. *Catena* 19:493–509.

Parmar, A. S. 1986. Flow patterns in Sahibi Basin, Haryana. MPhil dissertation, Kurukshetra University.

Peltier, L. C. 1950. The geographic cycle in periglacial regions as it is related to climatic geomorphology. *Ann. Assoc. Am. Geogr.* 40:214–36.

Peltier, L. C. 1962. Area sampling for terrain analysis. *Professional Geogr.* 14:24–28.

Pizzuto, J. E. 1992. The morphology of graded gravel rivers: A network perspective. In *Geomorphic systems*, ed. J. D. Philips and W. H. Renwick, 457–474. Amsterdam: Elsevier.

Pope, G. A., Dorn, R. I., and Dixon, J. C. 1995. A new conceptual model for understanding geographical variations in weathering. *Ann. Assoc. Am. Geogr.* 85:38–64.

Richards, B. W., Owen, L. A., and Rhodes, E. J. 2000. Timing of Late Quaternary glaciations in the Himalayas of northern Pakistan. *J. Quat. Sci.* 15:283–93.

Rinaldo, A., Dietrich, W. E., Rigon, R., Vogel, G. K., and Rodriguez-Iturbe, I. 1995. Geomorphological signature of varying climate. *Nature* 374:632–35.

Scheidegger, A. E., and Langbein, W. B. 1966. *Probability concepts in geomorphology.* U.S. Geological Survey Professional Paper 500-C. U.S. Geological Survey.

Schmidt, K. M., and Montgomery, D. R. 1995. Limits to relief. *Science* 270:617–20.

Schrag, D. P. 2000. Of ice and elephants. *Nature* 404:23–24.

Schumm, S. A. 1977. *The fluvial system.* New York: John Wiley & Sons.

Schumm, S. A. 1981. Geomorphic thresholds and complex response of drainage systems. In *Fluvial geomorphology*, ed. M. Morisawa, 299–310. London: George Allen & Unwin.

Schumm, S. A., and Khan, H. R. 1971. Experimental studies of channel patterns. *Nature* 233:407–9.

Schumm, S. A., and Lichty, R. W. 1965. Time space and causality in geomorphology. *Am. J. Sci.* 263:110–19.

Shackleton, N. J., and Opdyke, N. D. 1973. Oxygen isotope and palaeomagnetic stratigraphy of equatorial Pacific core V.28-238: Oxygen isotope temperature and its volumes of a 10^5 year and 10^6 year scale. *Quat. Res.* 3:39–55.

Sharma, M. C., and Owen, L. A. 1996. Quaternary glacial history of NW Garhwal, Central Himalayas. *Quat. Sci. Rev.* 15:335–65.

Shroder, J. F., Khan, M. S., Lawrence, R. D., Madin, I. P., and Higgins, S. M. 1989. *Quaternary glacial chronology and neotectonics in the Himalayas of northern Pakistan*, 275–94. Special Paper 232. Geological Society of America.

Singh, G. 1971. The Indus Valley culture seen in the light of post-glacial and ecological studies in north-west India. *Archaeol. Anthropol. Oceania* 6:177–89.

Singh, G., and Agrawal, D. P. 1976. Radiocarbon evidence for deglaciation in north-western Himalaya, India. *Nature* 260:232.

Singhvi, A. K., Sharma, Y. P., and Agrawal, D. P. 1982. Thermoluminescence dating of sand dunes in Rajasthan, India. *Nature* 295:313–15.

Small, E. 1999. Does global cooling reduce relief? *Nature* 401:31–33.

Smith, D. I., and Atkinson, T. C. 1976. Process, landform and climate in limestone regions. In *Geomorphology and climate*, ed. E. Derbyshire, 367–409. London: John Wiley & Sons.

Steig, E. J., Brook, E. J., White, J. W. C., Sucher, C. M., Bender, M. L., Lehman, S. J., Morse, D. L., Waddington, E. D., and Clow, G. D. 1998. Synchronous climate changes in Antarctica and the north Atlantic. *Science* 282:92–95.

Strahler, A. N. 1952. Dynamic basis of geomorphology. *Bull. Geol. Soc. Am.* 63:923–38.

Sugden, D. E. 1996. The east Antarctic ice sheet: Unstable ice or unstable ideas? *Trans. Inst. Br. Geogr.* 21:443–54.

Summerfield, M. A., and Kirkbride, M. P. 1992. Climate and landscape response. *Nature* 355:306.

Thomas, D. S. G., and Allison, R. J. 1993. *Landscape sensitivity*. Chichester: John Wiley & Sons.

Thorn, C. E., and Welford, M. R. 1994. The equilibrium concept in geomorphology. *Ann. Assoc. Am. Geogr.* 84:666–96.

Thornes, J. B. 1980. Processes and interrelationships, rates and changes. In *Process in geomorphology*, ed. C. Embleton and J. Thornes, 378–87. New Delhi: Arnold-Heinemann.

Toy, T. J. 1977. Hillslope form and climate. *Bull. Geol. Soc. Am.* 88:16–22.

Treydte, K. S., Schleser, G. H., Helle, G., Frank, D. C., Winiger, M., Hang, G. H., and Esper, J. 2005. The twentieth century was the wettest period in northern Pakistan over the past millennium. *Nature* 440:1179–82.

Whipple, K. K., Kirby, E., and Brocklehurst, S. H. 1999. Geomorphic limits to climate-induced increases in topographic relief. *Nature* 401:39–43.

Wilcock, D. 1983. *Physical geography*. Glasgow: Blackie.

2 Geologic Processes and Properties of the Earth Materials

Geothermal heat is the principal source of energy for the internal geologic processes of the earth. These mega-scale or global tectonic processes explain major geologic features of the earth, and are also basic to the formation of rocks and rock-forming minerals and aggregates. The strength and deformation behavior of rocks and sediments are two basic attributes of significance to the stability of slopes.

ORIGIN OF THE EARTH

The solar system came into existence some 4.5 billion years ago from the cold cosmic dust comprising partly compressed gas and dust particles of homogeneous chemical composition surrounding the then youthful sun. The dust particles coalesced and condensed by the gravitational attraction, evolving members of the solar system that later individually developed a distinct identity by selectively losing chemical elements from the condensed mass. The earth similarly evolved from the collision of cold cosmic dust particles and their condensation by the inward force of gravitational attraction. The impact energy of colliding particles converted into thermal energy, heating the condensed mass. The heating of proto-earth caused gravitational fractionation, producing its density-differentiated *internal structure* of core, mantle, and crust. The process also concentrated the lower-density, heat-producing radioactive nuclides within the lithospheric crust. Thus, the earth continued to heat internally from the decay of *radioactive elements* even when the impact frequency of colliding cosmic particles had progressively declined during the condensation of cosmogenic particles. Internal heating of the earth also released volatiles trapped at low temperatures to escape through volcanoes and hot springs as water vapor, carbon dioxide, methane, and possibly ammonia, evolving the earth's atmosphere and hydrosphere.

FUNDAMENTAL DIVISIONS OF THE EARTH

Continents and ocean basins are two first-order relief features of the earth. Although they are individually distinct in geographic and geologic domains, they are related to each other by common internal processes of the earth. *Hypsographic data* for the earth suggest that continents lie at the mean height of 875 m above the msl (mean sea level) and ocean basins rest at the mean depth of 3800 m from the sea surface level (Strahler, 1969). Further, the continents are 35% and oceans 65% of the earth's area.

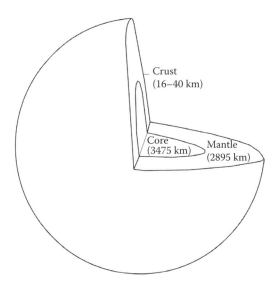

FIGURE 2.1   Layered structure of the interior of the earth, showing core, mantle, and crust.

Continents and ocean basins also differ in key geologic aspects. The *continental crust* is dominantly silica-aluminum rich, some 30 to 70 km thick, brittle in nature, and 2.7 in density composition. It is largely comprised of 2- to 3-billion-year-old crystalline rocks of stable *Precambrian shields*. The shield areas make the core of continents, or *cratons*, around which the younger sediment groups in parts of Australia, India, Africa, Canada, and Russia have deformed into mountain ranges. The continental crust carries a thick overburden of sediments, and is still evolving by the addition and deformation of younger sediments around the Precambrian shields (Stacey, 1992). The *oceanic crust* is 5 to 10 km thick. It is silica-magnesium rich, 3.0 g cm^{-3} in density composition, and built of *extrusive basalt* overlying equally thick *dikes* of intrusive origin (Miller and Ryan, 1977). The oceanic crust is 100 to 200 million years old, and overlain by fairly thick sediments.

INTERNAL STRUCTURE OF THE EARTH

The internal structure of the earth cannot be directly observed, except for a few kilometers of depth penetration in mining and drilling operations. It is, though, conveniently inferred from the propagation behavior of *earthquake waves* through the earth, suggesting that the interior of the earth is composed of density-differentiated concentric layers of core, mantle, and crust materials (Figure 2.1).

CORE, MANTLE, AND CRUST

The earth's core is 3473 km in radius, and composed of an iron-nickel alloy of 17 g cm^{-3} density composition. The core consists of an inner and outer shell of distinct physical properties. The inner core of 1370 km radius is solid, and the outer core of

2103 km radius behaves as a liquid. These properties are governed by the relationship between temperature and pressure conditions within the core.

The mantle is 2895 km in radius about the core. It is composed of olivine, pyroxene, and basalt minerals, giving a density of about 5 g cm^{-3} to the mantle material. The mantle temperature is about 5000°C at the core, 1350°C beneath the continental crust, and 1300°C at the bottom of the oceanic crust. The temperature and pressure conditions at the mantle-lithospheric crust interface, however, are such that the *upper mantle* undergoes discontinuous melting and plastic deformation much like that in butter or warm tar. This some 80 km thick and relatively mobile zone of weak strength is called the *asthenosphere.*

The crust is the outer shell of the earth to the depth of *Mohorovičić (Moho) discontinuity.* It has evolved from chemical differentiation in the upper mantle, and is volcanic in origin (Stacey, 1992). The *lithospheric crust* is continental and oceanic by chemical composition and evolutionary process. The properties of the two crusts were referred to above.

GEOTHERMAL HEAT

Geothermal heat is released from the decay of radioactive elements, igneous activity of the youngest continental crust, and hydrothermal cooling of the ocean floor. Of the three, the radioactive decay is the principal source of heat outflow from the interior of the earth. The radioactive elements are common in the continental crust and occur throughout the upper oceanic crust, where their decay releases the endogenic heat uniformly from the continental and oceanic crust.

The earth loses its internal heat at the average rate of 2.4 × 10^{20} cal year^{-1} (Stacey, 1992). The heat flux, however, is uneven from major geologic features of the earth. For continents, it is lowest from the Precambrian shields and highest from areas of Cainozoic volcanism and recent mountain building activity. For oceans, the heat loss is an order of magnitude higher than the world average from mid-oceanic ridges and exceptionally low from deep sea trenches at the edge of continents. Observations suggest that heat loss varies inversely with the age of continental rocks or the erosional loss of sediments and their radioactive elements from the earth materials (Table 2.1).

PLATE TECTONICS

Tectonics is the study of relative movement and deformation of the earth's crust, producing a variety of geologic structures, such as foliations, faults, and folds. Plate tectonics is that part of the tectonics that deals with the global process of lateral movement of lithospheric plates over the asthenosphere (Hobbs et al., 1976). The plate tectonic activity is ultimately tied down to the uneven heat flux from within parts of the earth's interior.

THE THEORY

The theory of plate tectonics is based on the *sea floor spreading hypothesis,* which envisages that the oceanic crust is created more or less continuously by the outflow of viscous magma from hot spots beneath the rift structure in mid-oceanic ridges,

TABLE 2.1

Variations of Continental Mean Heat Flow with Crustal Age

Crustal Age (10^6 y)	Mean Heat Flow (m Wm^{-2})
0–250	76 ± 53
250–800	63 ± 21
800–1700	50 ± 10
>1700	46 ± 16

Source: Stacey, F. D., *Physics of the Earth*, 3rd ed. (Kenmore, Queensland, Australia: Brookfield Press, 1992), 292. With permission from Brookfield Press.

Note: m Wm^{-2} = million watts per square meter; 1 W = 0.28846 cal cm^{-2}.

moved laterally outward over the asthenosphere by thermal convection through the depth of the upper mantle, and consumed by a like amount in the zone of subduction (Figure 2.2). The earth's lithosphere of continental crust, oceanic crust, or both is broken into seven major and several minor segments called plates, which ride over the partly molten weak asthenosphere, and move independent of each other in rate and sense of direction (Figure 2.3). The motion of one plate influences the motion of other plates, such that plates change shape and size through time. Dietz (1961, 1972), Hurley (1968), Dietz and Holden (1970), Dewey (1972), Toksöz (1975), McKenzie and Richter (1976), and Richter (1977) discuss fundamentals of the theory of plate tectonics. The theory and its geologic implications are also discussed in several texts on the subject (Press and Siever, 1986; Anderson, 1986; Stacey, 1992; Tarbuck and Lutgens, 1999; Garrison, 2001).

The sea floor spreading hypothesis visualizes that a hypothetical supercontinent called *Pangea* split at the beginning of Cretaceous period 165 Ma. The split opened a deep fracture pattern through to the depth of asthenosphere in the upper mantle (Figure 2.4(a)). These *tear zones* subsequently widened and created ocean basins between them. The rifting of continents and sea floor spreading thus are related activities, such that sea floor spreading begins when continental rifting ceases. This postulate, though, is not satisfied in New Guinea, where continental rifting and sea floor spreading are understood to be simultaneous activities (Taylor et al., 1995).

The rifting of continents lowered the melting point of the mantle material, and mobilized abundant magma possibly at the core-mantle boundary. The magma forced through the rift structure has evolved the *mid-oceanic ridge system* (Figure 2.4(b)) and the *new oceanic crust* of the earth (Figure 2.4(c)). Theoretical studies suggest that the magma either buoyantly rises from hot spots directly beneath the ridge structure in ridges and helps drive the oceanic plates apart (Spiegelman and Reynolds, 1999), or travels far and wide before converging at narrow ridge zones in response to the plates being pulled apart (Perfit, 1999).

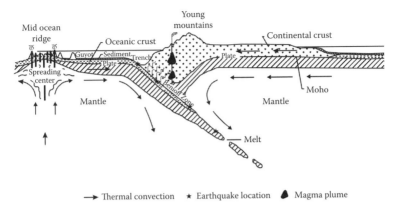

Thermal convection   ★ Earthquake location   ◢ Magma plume

FIGURE 2.2 Illustration of the concept of plate tectonics. The shaded zone below the Mohorovičić (Moho) discontinuity is the upper mantle portion, called the asthenosphere, which moves with the plates. The continental crust is some 35 km and the oceanic crust about 10 km thick at the Moho discontinuity in the earth. The plates are moved by thermal convection (shown by arrows) through to the depth of upper mantle rock. The oceanic plate plunges beneath the lighter-density continental plate in the zone of subduction. Deep sea trenches, high mountain chains, volcanoes, and earthquakes are associated with such regions of plate subduction. Other geologic features, magma plumes, and earthquake locations are also shown. (Source: Based on Figure 2 in Inman, D. L., and Nordstrom, C. E., "On the Tectonic and Morphologic Classification of Coasts." *J. Geol.* 79 [1971]: 1–21. With permission from University of Chicago Press.)

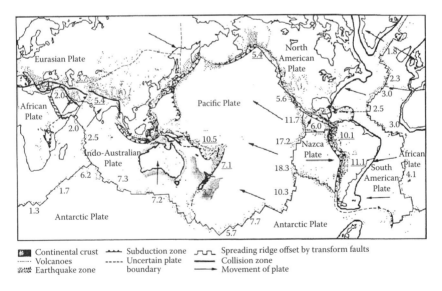

FIGURE 2.3 Configuration of world's major plates, indicating their name and boundaries, zones of collision, zones of subduction, and certain major geomorphological phenomena. The rates of sea floor spreading and plate convergence (in cm year^{-1}) are given. Note the very high rate of spreading in the Pacific. (Source: Figure 1.8 in Goudie, A. S., *The Nature of the Environment*, 3rd ed. [Oxford: Blackwell, 1993]. With permission from Blackwell Publishers.)

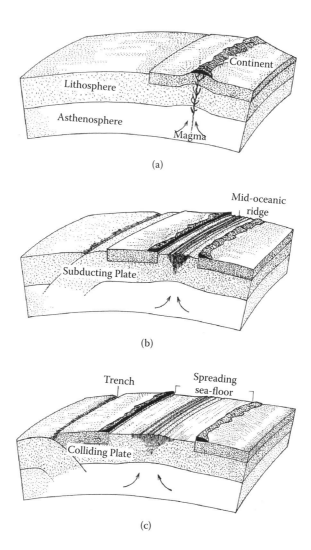

FIGURE 2.4 Theory of plate tectonics provides a mechanism for continental rifting. This process begins (a) when a spreading rift develops under a continent that is resting on a single crustal plate. Molten basalt from the asthenosphere spills out. (b) The rift continues to widen, separating the two parts of the continent further and creating an ocean between the two. As the new continent carried by the plate is rafted, a new ocean is created between the two landmasses, and a subducting plate comes into existence. (c) The diverging plates at the mid-oceanic ridge cause sea floor spreading. The continents impinge against the subducting plate, forming sea trenches. The mid-oceanic ridge with its active rift system always stays in the center of moving plates. (Source: Figure on p. 32 in Dietz, R. S., and Holden, J. C., "The Breakup of Pangaea," *Sci. Am.* 223 [1970]: 30–41. With permission from Scientific American © 1970.)

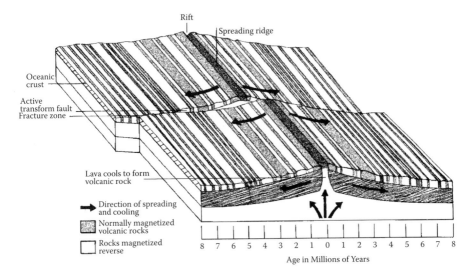

FIGURE 2.5 Illustration of the evolution of transform faults and fracture zones at the mid-oceanic ridge. The arrows show the direction of spreading and cooling of the oceanic crust. The line pattern shows magnetic strips on the ocean floor. The magnetic strips on the ocean floor are evidence of sea floor spreading. (Source: Figure on pp. 16–17 in Bisacre, M., Carlisle, R., Robertson, D., and Ruck, J., eds., *The Illustrated Encyclopedia of the Earth's Resources* [London: Marshall Cavendish Books Ltd., 1975–1984]. With permission from Marshall Cavendish Books Ltd.)

Mid-Oceanic Ridge

The mid-oceanic ridge is a 64,375 km long submarine network of topographic high on the *oceanic crust*. It circumnavigates the middle third of the Pacific, Indian, Atlantic, and Arctic oceans, maintaining an almost equal distance from the continents on either side. The ridge is 4500 m high above the sea floor, 150 to 1500 km wide at the base, 20 to 25 km wide at the crest, and carries an active central rift system (Ewing and Ewing, 1977). Also, numerous *transform faults* and their relics, called *fracture zones*, offset the ridge crest (Figure 2.5). These faults are caused by a change in the rate or direction of magma flow emanating from localized hot spots beneath the ridge structure. A few eminences of the ridge also steeply rise to above the sea surface level as *islands*. Iceland is one such hot spot that roughly bisects the rift structure beneath it.

Seafloor Spreading

Lateral movement of the magma at mid-oceanic ridges displaces the older oceanic crust outward over the asthenosphere and, thereby, causes the sea floor spreading. The evidence for sea floor spreading comes from a systematic pattern of linear *magnetic strips* of normal and reversed polarity in the oceanic crust (see Figure 2.5). These strips of roughly equal width and age on either side of mid-oceanic ridges become progressively older outward of the ridge system, which suggests a continuous release and cooling of the viscous magma in the then magnetic field of the earth that is known to have reversed polarity once in 5×10^5 years (Hurley, 1968; Stacey, 1992). The magma extruded in the present-day magnetic field of the earth provides normal and that released in times of reversed polarity of the earth negative magnetic anomaly to that part of the oceanic crust.

The *paleomagnetic record* in the oceanic crust suggests that it has been spreading at the rate of 10 to 150 mm year^{-1} for at least 100 million years (Ewing and Ewing, 1977). Other evidence from the remains of protozoa in deep sea sediments likewise suggests spreading of the North Atlantic and Pacific sea floors at the average rate of 20 and 100 mm year^{-1}, respectively (Press and Sievers, 1986). Submarine volcanic cones, called *guyots* and *seamounts*, that have moved away from associated mid-oceanic ridges, however, offer visible evidence for the sea floor spreading.

TECTONIC CYCLE

The tectonic cycle relates the morphodynamics of major geologic features of the earth to the creation, mobilization, and destruction of the oceanic crust. The oceanic crust, which is continually produced by the outflow of viscous magma from beneath the rift structure in mid-oceanic ridges, grows in size and solidifies in the lower-temperature zone outward of the ridge system. A difference of temperature between the top and bottom layer of released magma and that between the proximal and distal parts of the magmatic extrusion layer, however, sets a large-scale *thermal convection* in motion through the depth of asthenosphere on either side of the ridge crest. Thermal convection invites coherent motion of the material in the direction of thermal gradient (Niemela et al., 2000). Hence, the released oceanic crust is carried like a conveyor belt outward of the mid-oceanic ridge system. This crust is necessarily consumed by a like amount in the zone of *subduction*. Otherwise, the earth would have long deformed, and gone out of shape and size (Press and Siever, 1986).

Oceanic plates diverge at the mid-oceanic ridge system, evolving a *divergent plate boundary* at ridge crests. The leading edge of these higher-density divergent plates sinks beneath the lower-density continental plates. This process evolves a subduction zone of plate interaction and *subduction boundary*. Equal-density continental plates, however, collide and form a collision zone and *collision boundary* of plate interaction. The lithospheric plates converging at grazing angle only slide past each other. They develop a *transform fault boundary* at the surface.

Subduction Zone Processes

Creep movement of oceanic plates outward of the mid-oceanic ridge system, and *sliding* of plates in the subduction zone, changes the plate structure and affects the mantle equilibrium. In most cases, the sliding plates plunge into the subduction zone at a variable rate of plate descent. The plates of higher rate of descent, though, penetrate deeper into the mantle material, pass the 660 km *phase boundary*, intensely deform structurally at the leading edge, heat up efficiently, and melt to become a part of the mantle material (Lay, 1995; Taylor et al., 1995; Van der Hilst, 1995). The oceanic plates of higher subduction rate, such as in Fiji, Papua New Guinea,

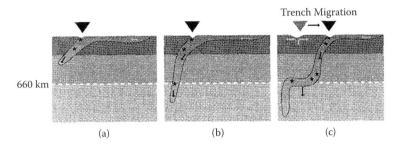

FIGURE 2.6 A sinking ocean plate in various stages of evolution of a subduction zone. Earthquake locations in the slab are shown by stars. Initially, the slab penetrates with relatively little deformation as the leading edge sinks through the upper mantle (a). When the leading edge encounters the 660 km phase boundary, which resists penetration to a certain extent, the earlier rapid slab motion is arrested and begins to buckle (b). In some subduction zones, the opening of a back-arc basin and seaward trench migration occur, which could result in the slab deflecting horizontally on the 660 km boundary (c). (Source: Figure on p. 115 in Lay, T. "Slab Burial Grounds," *Nature* 374 [1995]: 115–16. With permission from Macmillan Magazines Ltd.)

and New Zealand, buckle, overfold and even deflect horizontally while penetrating 800 km into the rigid mantle material (Figure 2.6). The sinking plates, in general, generate intermediate and deep-focus *seismic activity* of the earth, and evolve *sea trenches* adjacent to the continental margins and along island arcs. Some oceanic plates that deflect horizontally at the phase boundary, as in Figure 2.6, open back-arc basins and seaward-migrating trenches. A subduction zone of very high rate of plate penetration, called the *Benioff zone* (see Figure 2.2), is known for deep-focus seismic activity of the earth. The high incidence of earthquake activity at the Tonga Trench thus is the result of an exceptional 240 mm year^{-1} rate of Pacific Plate penetration beneath the continental Indo-Australian Plate (Bevis et al., 1995). The seismic activity of the earth results from the *frictional drag* of sliding oceanic plates. This activity melts and upwells the partly molten mantle material, and locally disturbs the mantle equilibrium (Spiegelman and Reynolds, 1999).

Sea trenches, in general, are thousands of kilometers long, hundreds of kilometers wide, and 3 to 4 km deep in the Pacific and Atlantic oceans. The Mariana Trench in the Pacific Ocean is 11 km deep, and the Puerto Rico Trench in the Atlantic Ocean penetrates 8.4 km to the depth of the sea floor.

Subduction zone processes raise mountains, evolve volcanoes, and recycle rock-forming minerals. The exceptional high degree of compression in subduction zones evolves mountains, such as the Andes system, along the continental margins of plate interaction. The compression of this order of magnitude lowers the melting temperature of mantle material, and drives water and other volatiles from subducting plates. The melt generates magma plumes (see Figure 2.2), evolving volcanoes on land and chains of volcanic islands offshore. The land-facing island arcs, like the Aleutians, Kuriles, Japan, and Lu-Chu, are roughly parallel to deep sea trenches. The subduction zones are also known for extreme temperature conditions in the subcrust. Therefore, juxtaposed continental rocks are *metamorphosed* by the effect of elevated pressure and temperature conditions in the subcrust, and sediments of the littoral origin arriving into the sea are recycled.

COLLISION ZONE PROCESSES

Collision zones are areas of extreme compressive stress in the lithosphere, and home to the young-fold mountain systems. Nearly 3000 km long and 250 to 350 km wide, the Himalayan system between Myanmar (Burma) in the east and Afghanistan in the west has similarly evolved from the collision between Indian and Eurasian plates. The Himalayas, however, are more complex in geology and evolution than the contemporary alpine systems elsewhere in the world (Gansser, 1964).

Himalayan Orogeny

Earth scientists hold that sediments comprising the Himalayan system had accumulated in geochemically distinct basins within the *Tethis Sea* that lay between Indian Plate in the south and the Eurasian Plate in the north. The geologic and geophysical evidence (Tapponnier and Molnar, 1976) suggests that the Indian Plate, initially moving northeast at the rate of 100 to 180 mm year^{-1}, had *subducted* beneath the Eurasian Plate 71 Ma before the collision. The collision slowed the subduction rate to 50 mm year^{-1}, generating further stress in the lithospheric crust. This stress is presently being relieved, causing the Himalayas to rise. This rise is 4 to 5 mm year^{-1} at the Pamir Knot to the west of the system (Arnaud et al., 1993). Geophysical data from the region also suggest that the compressive stress due to subduction caused the partly molten Tibetan subcrust to flow toward the eastern edge of the Himalayas, evolving the *Burma Syntaxis* (Coleman and Hodges, 1995; Searle, 1995). The *batholiths* and *extrusive volcanic rocks* of the Himalayan terrain attest to subduction, and evolution of 2500 km long *Indus Tsangpo suture* (ITS) zone and the *main boundary fault* (MBT) in the Central Himalayas of highest relief stand testimony to a subsequent collision between the Indian and Eurasian plates. The MBT is the relic of a deep fault into the upper mantle material (Gansser, 1964).

The tectonic model of Molnar and coworkers for the Himalayan orogeny predicts initial deep-wedge penetration of the Indian Plate in the northwest, and a slab-like penetration of the Indian Plate to the east of the system (Molnar and Tapponnier, 1975, 1978; Tapponnier and Molnar, 1976; Tapponnier et al., 1992). This form of penetration explains the general configuration of the Himalayan arc, and the tectogenic framework of Asia over much of which the stress of collision stands relieved (Figure 2.7). The wedge penetration has evolved linear faults of Heart in Pakistan and Altyn Tagh in Tibet, and the slab penetration has produced the curved faults of Kunlun and Kangtin in southeastern Tibet. The effect of compressive stress, though, had been most expressive in Tibet, where the continental crust shortened by 300 to 400 km, and the plateau gained height rapidly before collapsing and settling at about 5000 m above the msl 14 Ma ago (Coleman and Hodges, 1995; Searle, 1995; Sankaran, 1997). This stress was also relieved over China and Mongolia, evolving the Shanshi Graben and the Baikal rift zone as expressions of the *tensile stress* in the earth's crust. This stress has reactivated major precollision *strike-slip faults*, which are the known sites of deep-focus earthquakes in this part of the world.

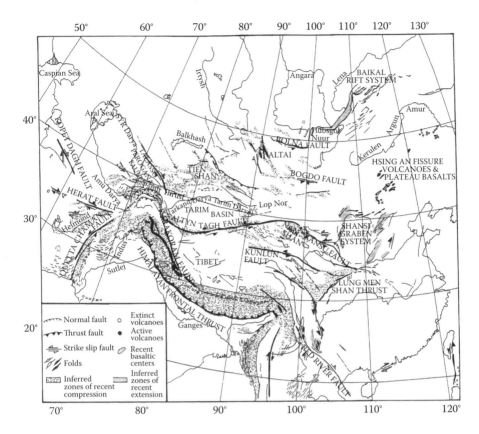

FIGURE 2.7 Simplified map of recent tectonics of Asia showing major faults, sense of motion, and regions of crustal thickening and thinning. (Source: Figure 1 in Tapponnier, P., and Molnar, P., "Slip-Line Field Theory and Large-Scale Continental Tectonics," *Nature* 264 [1976]: 319–24. With permission from Macmillan Magazines Ltd.)

SLIDING ZONE PROCESSES

Plates sliding past each other neither create nor destroy the lithosphere. The San Andreas Fault in central California is one such zone of plate interaction, where the oceanic Pacific Plate slides past the continental North American Plate at the rate of 34 mm year^{-1} (Press and Siever, 1986). This interaction zone is characteristic of a *transform fault boundary* at the surface. It is also an expression of the continuity of lithosphere between the rising and sinking cells of the thermal convection system.

PLATE BOUNDARY AND SEISMICITY

The geographic distribution of earthquake epicenters (Figure 2.8) coincides with plate boundaries earlier identified as divergent, subduction, collision, and transform. Earthquakes are shallow (<70 km), intermediate (70 to 300 km), and deep (>300 km) with respect to *focal depth*, such that earthquakes of deeper focal depth are higher

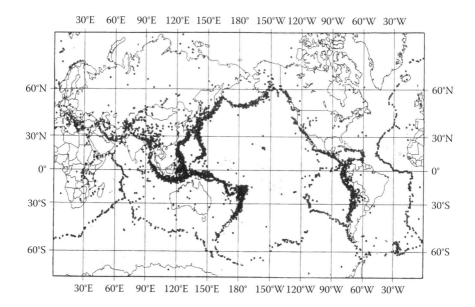

FIGURE 2.8   Distribution map of epicenters of twenty-nine-thousand earthquakes reported within the depths of 700 km between 1961 and 1967. (Source: Figure on p. 156 in Stacey, F. D., *Physics of the Earth*, 3rd ed. [Kenmore, Queensland, Australia: Brookfield Press, 1992]. With permission from Brookfield Press.)

in the magnitude of seismic activity. Focal depth is calculated from the behavior pattern of earthquake wave propagation through the earth, and the magnitude of the earthquake event is recorded on the Richter scale.

The global distribution of seismic activity (Stacey, 1992) suggests that less than 2% of all *shallow earthquakes* originate at divergent plate boundaries, 75% at subduction boundaries, and the remaining at collision and transform fault boundaries. The *intermediate-depth earthquakes* generally occur along deep sea trenches, at volcanic islands, and within collision boundaries of compressive stress release. The deep-seated earthquakes originate in *subduction zones* that are inclined at about 45°. They are also called *Benioff zones*. The deepest earthquake is known to have occurred at a depth of 720 km, suggesting thereby that deeper parts of the earth are aseismic (Stacey, 1992).

ROCKS

Rocks are an aggregate of a *mineral* or minerals. They are igneous, sedimentary, and metamorphic by the process of evolution. *Igneous rocks* are formed of magma, the denudation of which yields the solid and solute load for the evolution of sedimentary rocks. The igneous and sedimentary rocks subject to extreme temperature and pressure conditions within the subcrust alter in mineral composition, texture, and internal structure. These newer entities are called *metamorphic rocks*. The denudation of metamorphic rocks similarly provides the clastic and chemical fractions for the evolution of *sedimentary rocks*. This formation and transformation of rocks is

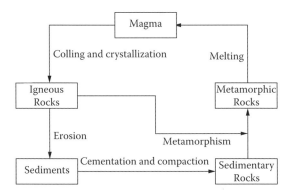

FIGURE 2.9   The rock cycle, illustrating the relationship between three main rock types: igneous, metamorphic, and sedimentary.

known as a *rock cycle* (Figure 2.9). Several texts on the subject discuss classification of rocks, and highlight the significance of rocks to the development of relief (Tyrrell, 1958; Krumbein and Sloss, 1963; Ernst, 1969; Sparks, 1975; Billings, 1977; Ehlers and Blatt, 1977).

IGNEOUS ROCKS

Igneous rocks solidify from a molten or partly molten *magma* that originates from heating down in the earth's crust or in the upper mantle, and ascends toward the surface by internal pressure in the subcrust. The magma rising through fissures and vents in the earth's crust either solidifies at the surface as *extrusive lava* or crystallizes within the subcrust as *intrusive lava*. The extrusive variety comprises basalt and andesite lava, and the intrusive type is basaltic and granitic in composition.

Extrusive Igneous Rocks

Extrusive igneous rocks are composed of the lava that reaches the earth's surface by either escaping quietly through deep fissures in the subcrust or rising violently through volcanic vents in the earth's crust. *Fissure eruptions* of the geologic past had repeatedly poured large volumes of thick viscous lava on continents and in ocean basins. The lava arriving at the continental crust crystallized into its constituent minerals in reverse order of mineral's melting point, and solidified into nearly horizontal massive layers of fine-textured basalt called *trap*. The trap makes plateaus of regional dimension, such as the Deccan of India, Karoo of South Africa, and Columbia of the United States. The quieter volcanic eruptions in oceans, though, cool efficiently in contact with water, and form a globular form of *basalt* deposits.

Volcanic eruptions, which are essentially a mixture of volcanic gases and liquid lava, eject pyroclastic fragments and volcanic ash. The fluid flows yield a large volume of subsilicic lava for building *shield volcanoes*, such as in Hawaii and other Pacific and Indian Ocean islands. The *viscous flows* escape explosively, and build conical hills by successive accretion of volcanic ash and lava of medium silica content. These hills of inter-layered lava and volcanic ash are called *stratovolcanoes*.

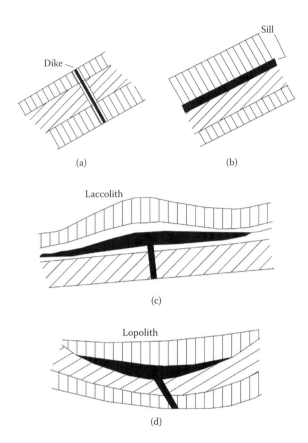

FIGURE 2.10 Schematic cross sections of various tabular and lens-like intrusives: (a) dike, (b) sill, (c) laccolith, and (d) lopolith.

Mt. Fuji in Japan, Vesuvius in Italy, and Mt. Rainier in the United States are examples of stratovolcanoes.

Intrusive Igneous Rocks

Intrusive igneous rocks are also known as plutonic rocks. They represent the magma that had invaded preexisting rocks, and crystallized slowly within the subcrust as dikes, sills, laccoliths, lopoliths, and batholiths (Figure 2.10). *Dikes* penetrate geologic strata and sills intrude bedding separations. Sills of a smaller size are *laccoliths* and of a larger size *lopoliths*. Batholiths and stocks extend to great depths in the subcrust as massive bodies of granite, of which batholiths are 100 km^2 or more in area and stocks smaller in size.

Geomorphic Significance

Thick *extrusive lava flows* make extensive plateaus. The erosion of plateau basalt leaves isolated tabular forms called mesas and butes, of which mesas are larger in size. The basalt cooling at the surface shrinks to some extent, and develops hexagonal joints that appear as *columnar joints* in lava cliffs. Lava flows rapidly cooling in the seawater similarly evolve spectacular six-sided columns of basalt, such as the famous Giants' Causeway in North Ireland. Massive extrusive lava flows that bury drainage systems cause *inversion of relief*. Areas of *intrusive lava flows* display local erosional eminences of volcanic necks, vertical and ledge-like erosional forms above sills and dikes, and regional domal topography above batholiths.

SEDIMENTARY ROCKS

Sedimentary rocks evolve from lithification and diagenesis of detrital and solute load of preexisting rocks, sediments, and remains of organic products that had once settled in *basins of sedimentation* on land and adjacent large-sized bodies of water. Lithification is the process of sediment compaction by own weight, and cementation of intergranular voids by the precipitates of dissolved matter in sedimentary basins. Diagenesis is a postdepositional phenomenon of sediment consolidation by compaction, cementation, and recrystallization processes. Sedimentary rocks are characteristic of a typical layered structure called bedding or stratification. These rocks are clastic and chemical in type.

Clastic Sedimentary Rocks

Clastic sedimentary rocks comprise a variety of sediment sizes present in sedimentary basins. They are accordingly rudaceous, arenaceous, and argillaceous by the constituent sediment size. The rudaceous variety comprises conglomerate and breccia. The arenaceous type contains sand or is built of sand. It represents dark-colored sandstone known as greywacke and a sorted variety of sandstone called arenite. The clastic sediments of rudaceous and arenaceous rocks are held together by the cementing effect of siliceous, calcic, or furruginous minerals present in the environment of sedimentary basins. The argillaceous group contains clay or has a notable proportion of clay in the rock.

Chemical Sedimentary Rocks

Chemical sedimentary rocks like limestone, chalk, and gypsum, and a variety of chemical sediments precipitate in sedimentary basins of specific chemical, biochemical, evaporite, and geochemical environment. Limestone *precipitates* from the solution of limestone and shell debris already present in sedimentary basins. Coral and chalk are products of a tightly controlled *biochemical environment* of the basins of sedimentation. Coral rocks are the skeletal remains of sedentary marine organisms called coral, and chalk is produced in the tests of sea-dwelling organisms called foraminifera. Gypsum, anhydrite, calcite, and halite are products of an *evaporite environment* typical of restricted water bodies of sizable extent. The siliceous, ferruginous, carbonaceous, and phosphatic sediments represent a certain *geochemical environment* of sedimentary basins. Siliceous sediments precipitate in oxygen-limited geosynclinal basins, ferruginous sediments segregate by bacterial leaching and biochemical activity in basins of anaerobic conditions, carbonaceous sediments crystallize in the alkali environment of sedimentary basins, and phosphatic sediments precipitate only after hydrogen sulfide consumes other ionic species in the environment of sedimentary basins.

Geomorphic Significance

Sedimentary rocks occupy two-thirds of the earth's surface (Ehlers and Blatt, 1997). These rocks are inherently weak in strength. In general, clastic sedimentary rocks are prone to *mass movement* activity, and a group of chemical sedimentary rocks that contain mineral calcite evolve *karst*.

METAMORPHIC ROCKS

Preexisting igneous and sedimentary rocks that were altered by new chemical substances in the subcrust or were affected by pronounced temperature or pressure changes within the earth's crust evolved new chemical suites. These rocks of thoroughly altered mineral composition, texture, and internal structure are called metamorphic rocks.

The above conditions of rock alteration are obtained in localized processes of contact and dislocation metamorphism, and regional processes of orogenic deformation of the earth's crust. *Contact metamorphism* (or metasomatism) is the process of mineral replacement along a narrow zone of magmatic intrusion, called aureole, into the body of rock where certain rock-forming minerals alter chemical composition by reacting with the solution of an external source. *Dislocation metamorphism* is the process of mineral alteration by the effect of pressure and frictional heat in the subcrust. The pressure strains mineral lattices and frictional heat alters the chemical composition of mineral suites. *Regional metamorphism* is synonymous with the orogenic deformation of the continental crust in subduction and collision zones of plate interaction, where extremely high pressure and temperature conditions alter the mineral suites of preexisting rocks into new chemical entities. Such rocks are classified by the degree of mineral alteration into low-, medium-, and high-grade metamorphics. *Low-grade metamorphic rocks* of the like of slate, phyllite, and low-grade schist evolve in the environment of high pressure and moderate temperature in the subcrust. These rocks comprise of complex hydrous calcium and sodium aluminosilicate minerals. *Medium-grade metamorphic rocks* evolve in a relatively high pressure and temperature regime of the subcrust. Schist, a prominent rock of this degree of metamorphism, is composed of amphibole and plagioclase minerals. *High-grade metamorphic rocks* evolve in uniformly high pressure and temperature conditions obtained in deeper parts of the earth's crust. Gneiss, a major rock of this grade of metamorphism, comprises interlocked pyroxene minerals.

Geomorphic Significance

Metamorphic rocks are some 15% of continental rocks. They comprise minerals and mineral aggregates that are resistant to *weathering* and *erosion* processes. Quartz, the dominant mineral of quartzite rock, is most resistant to alteration in the earth's environment. Hence, the rock appears as a residual within geomorphic landscapes. Gneiss is similar to plutonic rock in many respects, and evolves *inselberg landscape* in the manner of granite. Marble is an aggregate of the mineral calcite, which ordinarily is *soluble* in waters of the earth.

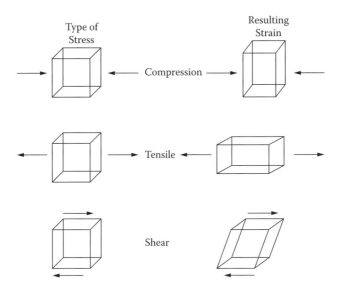

FIGURE 2.11 Common types of stress and resulting strain. (Source: Figure 3.5 in Keller, E. A., *Environmental Geology*, 2nd ed. [Columbus, Ohio: Bell & Howell, 1979].)

DEFORMATION OF ROCKS AND SEDIMENTS

Deformation behavior of the earth materials depends on the interplay between their strength attributes and the nature and magnitude of forces acting on them. The *strength* of a rock body varies with its mineral composition, fissure frequency, placement relative to the earth's surface, and climatic environment, and that of a sediment aggregate depends on interrelated components of solid, air, and water phases in the system. Rocks and sediments deform when the magnitude of internal and external forces, called *stress*, exceeds the strength of two materials. Stress is force per unit area, and expressed in Newton per square meter (NM^{-2}).

DEFORMATION OF ROCKS

Surface and near-surface rocks deform by the force of their own weight and the force of applied load. The applied forces of compression, tension, and shear above the threshold of rock strength change the shape, size, and volume of rocks. *Compressive stress* squeezes, *tensile stress* stretches, and *shear stress* produces planar deformation in rocks (Figure 2.11). For each type of applied load, the *rock strength* is experimentally determined as compressive, tensile, and shear, at which a stressed rock sample breaks. The deformation behavior of materials is in the ambit of *rheology*.

RESOLUTION OF STRESS

Rocks deform at certain threshold stress. Ramsay (1967), Sparks (1975), Hobbs et al. (1976), Embleton and Whalley (1980), and Billings (1997) discuss the stress

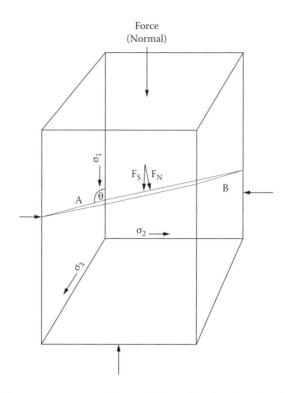

FIGURE 2.12   Stress components acting on the face of a cube-shaped rock.

deformation of rocks from general and theoretical viewpoints. The stress can easily be resolved into stress fields for cube-shaped rocks with mutually perpendicular principal axes along which the stress at any stressed point is zero (Figure 2.12). The stress *normal* to a cube face is σ_1, and two corresponding shear forces at right angle to the normal, and to each other, are σ_2 and σ_3. These stress axes are respectively major, intermediate, and minor by the stress magnitude in stress directions σ_1, σ_3, and σ_2. Rocks deform for unequal difference between the major (σ_1) and minor stress (σ_2) magnitudes. The quantity $\sigma_1 - \sigma_2$ is called *deviator stress*.

The angle of stress application is a major control of the magnitude of normal (σ) and shear stress (τ) in rocks. The stress at any given point along an imaginary plane, AB, inclined at an angle θ to the horizontal and oriented in such a way as to coincide with two normal stresses σ_1 and σ_2 of the cube, may be resolved into components of normal (F_N) and shear stress (F_S). Being a force per unit area, the stress normal to the inclined plane of an area (A_P) is $A_P\, \sigma \cos 2\theta$, and along the inclined plane $A_P\, \sigma \sin \theta \cos \theta$. These relations provide the magnitude of normal and shear stress at the inclined plane as

$$\sigma = \tfrac{1}{2}\,(\sigma_1 - \sigma_2)\cos 2\theta$$

$$\tau = \tfrac{1}{2}\,(\sigma_2 - \sigma_1)\sin 2\theta$$

Rocks deform when the magnitude of shear stress exceeds that of the normal stress. Applications of these relations for mass movement activity are addressed in Chapter 4.

STRAIN IN ROCKS

Forces of tension, compression, and shear strain rocks. A continuous stress deforms rocks in a sequential manner, initially producing strain or *elastic deformation* proportional to the magnitude of stress. Rocks, however, recover the linear strain up to a certain magnitude of applied stress. At the *elastic limit*, however, mineral lattices orient in the direction of stress and cause plastic deformation or permanent change in the shape of rocks. A continuing stress exceeding the *yield stress* eventually ruptures rocks.

DEFORMATION BEHAVIOR OF ROCKS: CONTROLLING FACTORS

The deformation behavior of rocks varies with the geologic attributes, pressure and temperature environment, and duration of applied stress. These controls of rock deformation, however, are understood only in test conditions for *isotropic* specimens (Hobbs et al., 1976; Billings, 1977). Therefore, experimental data are accepted as best approximations to the mechanisms of rock deformation at shallow and deep depths in the earth's crust.

Geologic Attributes

Rocks comprise brittle and ductile minerals. They are also anisotropic and isotropic in physical attributes. The rocks of *brittle minerals*, like sandstone, gneiss, and granite, rupture at the *elastic limit*, while shale, limestone, and schist, comprised of *ductile minerals*, undergo plastic deformation before rupture. An idealized stress-strain relationship for the deformation behavior of substances suggests that brittle minerals deform only slightly by about 1% before rupture, and ductile minerals continue to deform under stress before failure at the *ultimate strength* of rocks (Figure 2.13). Rocks of bedding separation, joints, and foliation are anisotropic, and of uniform geologic properties are isotropic. *Anisotropic rocks* disintegrate first at minor geologic structures of the type referred to above, which causes the stress to increase on intact portions of the rock. Hence, such rocks rapidly reach the *rupture limit*. By comparison, *isotropic rocks* like granite withstand a higher magnitude of stress before disintegration. *Pore water* is an additional variable of rock strength: it weakens intergranular cohesion, reduces the brittle character, and enhances the ductile properties of rocks.

Confining Pressure

Subcrustal stress increases rigidity, elastic limit, and ultimate strength of rocks (Billings, 1997). Thus, rocks are rigid and *brittle* at the earth's crust, and *ductile* within the subcrust where they deform into folded strata. Rocks at shallow depths, though, tend to be less rigid, where they open up faults and shear fractures in loading and unloading cycles of the seismic stress (Peltzer et al., 1999).

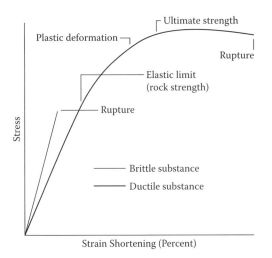

FIGURE 2.13   An idealized stress-strain diagram for elastic and plastic behavior of substances and conditions for brittle and ductile materials. (Source: Adapted from Figure 3.6 in Keller, E. A., *Environmental Geology*, 2nd ed. [Columbus, Ohio: Bell & Howell, 1979].)

Temperature

Rocks are brittle at lower and ductile at higher temperatures (Billings, 1997). Hence, higher temperatures deep beneath the crust affect rock strength in the manner of subcrustal stress and seismic stress.

Duration of Stress Application

Compressive stress of a magnitude lower than the rupture limit produces time-dependent primary, secondary, and tertiary creep in rocks (Hobbs et al., 1976; Billings, 1997). A few solids, called *Newtonian materials*, deform linearly in the manner of viscous fluids. Hence, rocks in natural settings are presumed to be Newtonian substances (Hobbs et al., 1976).

SEDIMENTS

Sediment aggregate is a complex of solid, water, and air phases, wherein the solid phase represents clastic fractions of certain size and mineral composition, and water and air phases occupy voids of the particulate matter. The deformation behavior of sediments depends on the properties of solid phase, pore water content, and duration of applied stress. A sediment aggregate is loosely called *soils*.

SOLID PHASE

The solid phase represents different-sized particles (Table 2.2). These fractions of certain mineral composition are also arranged in a particular manner within the aggregate mass. The above aspects of the solid phase are basic to the permeability, porosity, compressibility, and packing characteristics of the aggregate mass and, thereby, to its strength and deformation behavior under stress.

TABLE 2.2
Classification of Sediment by Size Class (Millimeters)

Type	Boulder	Cobble	Pebble	Granule	Sand	Silt	Clay
Size	>256	256	64	4	2	0.063	<0.002
		to	to	to	to	to	
		64	4	2	0.063	0.002	

Clastic sediments are massive and clay minerals by the standard of size, and inactive and active by the activity of mineral grains. *Massive minerals* are more than 0.002 mm in size, and *clay minerals* represent fractions of 0.002 mm or less in size. The activity of mineral grains is a quantitative measure of the *soil strength*, such that the soils of higher activity are relatively weak in strength. While massive minerals like quartz, feldspar, and calcite are *inactive*, the clay minerals are *active* in varying degrees. Among the clays, kaolinite minerals are least active, illite and chlorite groups of clay minerals are moderately active, and montmorillonite minerals are highly active (Kenney, 1984). Clay minerals also offer *sorption sites* for the *ionic species* in soil-water systems, which weakens the clay strength further. Observations suggest that clays lose strength, and flocculate by adsorbing sodium and calcium ions (Buckman and Brady, 1960). Hence, the clay strength varies with chemical composition and sorption properties of its constituent minerals.

WATER AND AIR PHASES

Soil water is attracted to the surface of dry sediments by the property of *adhesion*. Adhesion holds soil grains together by developing a film of water within interstitial voids, and enables the pore water to rise through the soil column as *capillary water*. The capillary water binds the moisture to mineral grains and, thereby, increases *cohesion* among grains. The cohesive force also varies directly with per unit number of contact points in the aggregate mass (Baver, 1956). Thus, clays with the largest number of per particle contact points are more cohesive than soils of other textures. The cohesive force is one of the two forces that resist shear; the other is the angle of internal friction.

Fine-grained cohesive soils lose strength when wet, and become plastic and liquid materials with increasing moisture content. The range of moisture content distinguishing the above deformation behavior of soils forms the basis of *Atterberg limits*, and derived plasticity and liquidity indices. The *plastic index* defines the amount of moisture in soils at which they are plastic and, hence, compressible. The *liquidity index* describes the nearness of soils to a liquid state. Atterberg limits provide limiting conditions of moisture content for the stability of slope-forming cohesive sediments (Chapter 4).

DEFORMATION OF SEDIMENTS

Sediment aggregate represents a complex arrangement of interlocked clastic grains held in place by the force of interlocking friction and the fluid force of intergranular

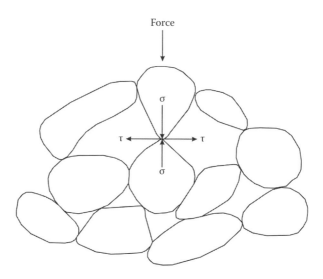

FIGURE 2.14   Resolution of stress forces for the solid phase of sediment aggregate.

cohesion. The force of *interlocking friction* develops at the contact of individual particles. Being only a fraction of the grain area, the individual contact points are of high normal stress (σ) along the vertical in the aggregate mass (Figure 2.14). This force produces shear stress (τ)causes an unstable array of particles to *slide* about their contact points and collapse into adjacent void spaces, producing a deformation at an angle to the horizontal. This angle is a measure of the maximum shear stress at failure, and called the *angle of internal friction*. The arrangement of interlocked grains in the sediment aggregate, however, is such that it does not permit one to keep track of individual stress forces in the system. Therefore, stress forces are evaluated in the same manner as for rocks, but on the assumption that the stress magnitude varies only slightly with distance from the contact points.

Loaded *wet soils* release pore pressure, which generates an upward *fluid force* equal to the difference between applied load and pore pressure in the system. Hence, *pore pressure* shares a part of the normal stress at the contact of grains. This is the *effective normal stress*, which also defines the strength of saturated soils. In test conditions, undrained saturated soils compress at continuous loading and reach a constant volume in time. This volume vis-à-vis the magnitude of stress is called *peak strength* of soils. The soils, however, continue to deform thereafter even at a lowered stress magnitude, and reach a static level of strain rate. This is the *residual strength* at which soils behave in the manner of dry aggregate. Experimental data suggest that soils of a larger difference between the peak and residual strength are more susceptible to slope failure (Lambe and Whitman, 1973).

STRESS–STRAIN RELATIONSHIPS

A mass of *dry sediments* deforms at all levels of applied stress, such that individual grains slide and roll into intergranular voids for initial stress of up to 15 MN cm^{-2}, and crush thereafter at compressive stress between 15 and 30 MN cm^{-2} (Embleton and Whalley, 1980). The crushed grains of finer composition, however, produce a tightly packed aggregate resistant to deformation. Hence, the sediment aggregate fractures at a stress magnitude exceeding 30 MN cm^{-2} (M = 10^6). The deformation behavior of loaded *saturated soils*, however, varies with the pore water content. The applied stress expels *pore water* and increases the stress absorbing *pore pressure* in soils. In a general term, the strain rate of moisture-bearing sediments depends on the magnitude of effective stress and void ratio in the sediment aggregate.

SUMMARY

Geothermal heat, the principal source of energy for global tectonic processes, is derived largely from the decay of radioactive elements in the lithospheric crust. The earth loses endogenic heat at the average rate of 2.4 × 10^{20} cal year^{-1} from continents and ocean basins, but the rate of heat flux differs markedly for major geologic features of the continental landmass and ocean basins. For continents, the heat loss is highest from areas of recent volcanic and mountain building activity and lowest from the Precambrian shields. For oceans, the average heat loss is an order of magnitude higher from mid-oceanic ridges and lowest from deep sea trenches. This uneven distribution of the earth's internal heat is the cause of plate tectonics.

The concept of plate tectonics is based on the sea floor spreading hypothesis. This hypothesis visualizes that the continental crust, oceanic crust, or both that comprise the earth's lithosphere move with the asthenosphere in the direction of thermal convection in the upper mantle. The asthenosphere is a sufficiently thick layer of weak strength in the upper mantle. This lithosphere is broken into seven major and several minor segments called plates that move independent of each other in rate and sense of direction. They, therefore, diverge, subduct, collide, or slide past each other, evolving major geologic features of the earth.

The oceanic crust is built of the magma that continually moves up through deep rifts in mid-oceanic ridges and escapes outward at ridge crests. This process has evolved the mid-oceanic ridge system and topographic eminences of islands and seamounts. The seamounts, which have moved away from associated ridges, provide visible evidence of sea floor spreading. Further, a record of magmatic reversals in the spreading oceanic crust suggests that it is no more than 200 million years old at its leading edge.

Divergent oceanic plates at mid-oceanic ridges subduct beneath the lighter-density continental plates, slide deeper into the mantle material, distort structurally, and melt to become a part of the mantle material. The sliding plates disturb the mantle equilibrium locally and generate the seismic activity of the earth. Deep sea trenches, island arcs, volcanoes, batholiths, and mountain ranges along continental margins are the expression of the subduction process. Equal-density continental plates, however, collide and generate enormous stress in the subcrust. The collision between continental plates has produced the alpine system from sediments of the Tethys Sea that had existed between the converging plates. The colliding plates produce crustal deformation, suture zone, and thrust faults, and generate seismic activity in alpine regions. The sliding of continental plates past each other produces a transform fault boundary at the surface.

Continents are built of igneous, sedimentary and metamorphic groups of rocks. The igneous rocks have evolved from the solidification of magma at the surface and in the subcrust, respectively producing the extrusive and intrusive variety of igneous rocks. The denudation of igneous rocks provided the solid and solute load of sedimentary basins from which a variety of clastic and chemical sedimentary rocks evolved by lithification, diagenesis, and chemical processes. The igneous and sedimentary rocks also transformed into various grades of metamorphic rocks by local and regional processes of metamorphism. The denudation of metamorphic rocks similarly provided for the evolution of sedimentary rocks.

The earth materials deform by forces of own weight and applied load. The deformation behavior of surface and near-surface rocks depends on chemical composition of constituent minerals, minor geologic structures in the body of rock, and the temperature and pressure environment to which they were subjected. In general, surface rocks are brittle and subsurface rocks tend to be ductile in nature. Surface rocks deform in response to the difference of magnitude between major and minor stress in the system. Sediments are a complex system of solid, water, and air phases in the aggregate mass. An aggregate of dry sediments deforms at all levels of applied load. At low magnitude of stress, sediments slide at their contact points and roll into intergranular voids. With further compressive stress, the grains pulverize and make the aggregate compact. This aggregate finally ruptures at still higher magnitude of applied load. By comparison, pore water determines the deformation behavior of cohesive sediments. The sediments of this type become plastic and liquid substances with increasing moisture content in the system. The deformation of slope-forming rocks and sediments initiates mass movement activity at the surface of the earth.

REFERENCES

Anderson, R. N. 1986. *Marine geology*. New York: John Wiley & Sons.

Arnaud, N., Brunel, M., Canragrel, J.-M., and Tapponnier, P. 1993. High cooling and denudation rates at Kongur-Shan, Eastern Pamir (Xinjiang, China) revealed by 40 Ar/39 Ar Alkai feldspar thermochronology. *Tectonics* 12:1335–46.

Baver, L. D. 1956. *Soil physics*. 3rd ed. New York: John Wiley & Sons.

Bevis, M., Taylor, F. W., Schutz, B. E., Recy, J., Iscas, B. L., Helu, S., Singh, R., Kendrick, E., Stowell, J., Taylor, B., and Calmant, S. 1995. Geodetic observations of very rapid convergence and back-arc extension at the Tonga arc. *Nature* 374:249–51.

Billings, M. P. 1997. *Structural geology*. New Delhi: Prentice Hall.

Bisacre, M., Carlisle, R., Robertson, D., and Ruck, J., eds. 1975–1984. *The illustrated encyclopedia of the earth's resources*. London: Marshall Cavendish Books Ltd.

Buckman, H. O., and Brady, N. C. 1960. *The nature and properties of soils*. New York: Macmillan.

Coleman, M., and Hodges, K. 1995. Evidence for Tibetan Plateau uplift before 14 Myr ago from a new minimum age for east-west extension. *Nature* 374:49–52.

Dewey, J. F. 1972. Plate tectonics. *Sci. Am.* 226:56–68.

Dietz, R. S. 1961. Continent and ocean basin evolution by spreading of the sea floor. *Nature* 190:854.

Dietz, R. S. 1972. Geosynclines, mountains and continent-building. *Sci. Am.* 226:31–38.

Dietz, R. S., and Holden, J. C. 1970. The breakup of Pangaea. *Sci. Am.* 223:30–41.

Ehlers, E. G., and Blatt, H. 1997. *Petrology*. New Delhi: CBS.

Embleton, C., and Whalley, B. 1980. Energy, forces, resistances and responses. In *Process in geomorphology*, ed. C. Embleton and J. Thornes, 11–38. New Delhi: Arnold-Heinemann.

Ernst, W. G. 1969. *Earth materials*. Englewood Cliffs, NJ: Prentice Hall.

Ewing, M., and Ewing, J. 1977. Marine geology. In *McGraw-Hill encyclopedia of ocean and atmospheric sciences*, ed. S. P. Parker, 232–41. New York: McGraw-Hill.

Gansser, A. 1964. *The geology of the Himalayas*. New York: Interscience Publishers.

Garrison, T. 2001. *Essentials of Oceanography*. 3rd ed. Thompson Brooks.

Goudie, A. S. 1993. *The nature of the environment*. 3rd ed. Oxford: Blackwell.

Hobbs, B. E., Means, W. D., and Williams, P. E. 1976. *An outline of structural geology*. New York: John Wiley & Sons.

Hurley, P. M. 1968. The confirmation of continental drift. *Sci. Am.* 218:53–64.

Keller, E. A. 1979. *Environmental geology*. 2nd ed. Columbus, Ohio: Bell & Howell.

Kenney, C. 1984. Properties and behaviour of soils relevant to slope instability. In *Slope instability*, ed. D. Brunsden and D. B. Prior, 27–65. Chichester: John Wiley & Sons.

Krumbein, W. C., and Sloss, L. L. 1963. *Stratigraphy and sedimentation*. 2nd ed. San Francisco, W. H. Freeman.

Lambe, T. W., and Whitman, R. V. 1973. *Soils mechanics*. New Delhi: Wiley Eastern.

Lay, T. 1995. Slab burial grounds. *Nature* 374:115–16.

McKenzie, D. P., and Richter, F. 1976. Convection currents in the earth's mantle. *Sci. Am.* 235:72–89.

Miller, E. L., and Ryan, W. B. F. 1977. Oceans. In *McGraw-Hill encyclopedia of ocean and atmospheric sciences*, ed. S. P. Parker, 354–62. New York: McGraw-Hill.

Molnar, P., and Tapponnier, P. 1975. Cenozoic tectonics of Asia: Effects of a continental collision. *Science* 189:419–26.

Molnar, P., and Tapponnier, P. 1978. Active tectonics of Tibet. *J. Geophys. Res.* 83:5361–75.

Niemela, J. J., Skrbek, L., Sreenivasan, K. R., and Donnelly, R. J. 2000. Turbulent convection at very high Rayleigh numbers. *Nature* 404:837–40.

Peltzer, G., Crampé, F., and King, G. 1999. Evidence of nonlinear elasticity of the crust from the mw 7.6 Manyi (Tibet) earthquake. *Science* 286:272–76.

Perfit, M. R. 1999. Molten rocks in motion. *Nature* 402:245–46.

Press, F., and Siever, R. 1986. *Earth*. 4th ed. New York: W. H. Freeman.

Ramsay, J. G. 1967. *Folding and fracturing of rocks*. New York: McGraw-Hill.

Richter, F. M. 1977. On the driving mechanism of plate tectonics. *Tectonophysics* 38:61–88.

Sankaran, A. V. 1997. The India-Asia collision warps and thaws Tibet's bowels. *Curr. Sci.* 72:700–1.

Searle, M. 1995. The rise and fall of Tibet. *Nature* 374:17–18.

Sparks, B. W. 1975. *Rocks and relief*. London: Longman.

Spiegelman, M., and Reynolds, J. R. 1999. Combined dynamic and geochemical evidence for convergent meltflow beneath the East Pacific Rise. *Nature* 402:282–85.

Stacey, F. D. 1992. *Physics of the earth*. 3rd ed. Kenmore, Queensland, Australia: Brookfield Press.

Strahler, A. N. 1969. *Physical geography*. 3rd ed. New York: John Wiley & Sons.

Tarbuck, E. J., and Lutgens, F. K. 1999. *Earth: An introduction to physical geology*. 6th ed. Englewood Cliffs, NJ: Prentice Hall.

Tapponnier, P., Armijo, R., Avouac, J. P., and Liu, Q. 1992. Subduction crustal folding, and slip partitioning along the edge of Tibet. In *Kashgar International Symposium on the Karakoram and Kunlun Mountains*, Kashgar, June 4–9, 1992, p. 8.

Tapponnier, P., and Molnar, P. 1976. Slip-line field theory and large-scale continental tectonics. *Nature* 264:319–24.

Taylor, B., Goodlife, A., Martinez, F., and Hey, R. 1995. Continental rifting and initial sea-floor spreading in the Woodlark Basin. *Nature* 374:534–37.

Toksöz, M. N. 1975. The subduction of the lithosphere. *Sci. Am.* 233:89–98.

Tyrrell, G. W. 1958. *The principles of petrology.* London: Methuen.

Van der Hilst, R. 1995. Complex morphology of subducted lithosphere in the mantle beneath the Tonga Trench. *Nature* 374:154–57.

3 Weathering

Weathering is *in situ* disintegration and decomposition of rocks into a residuum of original matter and alteration products in equilibrium with the stress of the environment. For convenience, weathering is classified into broad groups of mechanical, chemical, and biological domains with the understanding that many process activities are debated and known in theory only (Table 3.1). Weathering prepares rocks for *denudation* and contributes to the formation of *soils*.

MECHANICAL WEATHERING

Surface and subsurface rocks are in a state of tensile and compressive stress (Chapter 2). The surface rocks are also additionally exposed to the mechanical stress of temperature and moisture fluctuations in the environment. The tensile and compressive stress forms are commonly recognized as pressure release in surface rocks and compressive stress release in the subsurface rocks. The environment-related disintegration processes are broadly classified as thermal expansion and contraction of rocks, growth of crystals in rocks and sediments, hydration of minerals, crystallization of salts in rocks and porous media, and colloidal plucking. All of these domains of process activities disintegrate rocks and disrupt the earth materials in many different ways.

PRESSURE RELEASE/COMPRESSIVE STRESS

Erosional unloading of the landmass expands surface rocks, and compressive stress in the subcrust buckles up rocks at depth. The tendency of rocks to expand and squeeze opens up 0.5 m to several meters thick concentric joints, called *sheet fractures* or sheeting joints, roughly parallel to the topographic grain of massive rocks (Figure 3.1). The affected rocks split further and disintegrate at these joints, which become fewer and wider with depth and disappear at a depth of some 30 m from the surface. Sheet fractures are debated as the manifestation of unloading and compressive stress mechanisms.

Unloading Mechanism

Gilbert held that rocks compensate for the erosional loss of mass by expanding vertically through to the depth of the subcrust (Twidale et al., 1996). The elastic expansion of rocks releases *geodynamic stress* of the order of 1.5 to 8.0 kilobars with the attendant opening of sheet fractures perpendicular to the direction of pressure release (Ollier, 1979). The erosional unloading is identical in effect to that of dilation and instant bursting of prestressed intact rock in quarrying and tunneling operations (Ollier, 1977).

The unloading mechanism finds support from the presence of a 100 to 150 m deep intersecting set of older and newer generations of sheet fractures respectively

TABLE 3.1

Classification of Weathering Processes

Weathering Domain	Processes	General Characteristics
Mechanical	Pressure release/compressive stress Thermal expansion and contraction Growth of crystals Hydration Crystallization of salts Colloidal plucking	Reduction in particle size Increase in surface area No change in the composition of minerals Disruptive effects against impacted surfaces
Chemical	Oxidation and reduction Hydrolysis Carbonation Solution Sorption Biochemical, adsorption, and photochemical reactions	Complete change in the chemical properties of minerals Net increase in the volume of new products and residue Dissolution and precipitation of soluble rocks Formation of stable clay minerals Mobilization of hydrophobic contaminants Solubility and removal of a wide spectrum of metal cations Oxidation-reduction Dispersal Transformation and digestion of minerals and metals
Biological	Disintegration Chelation Bacterial Nanobacterial	Loosening of rocks and earth materials by organisms in the animal and plant kingdoms Isolation, displacement, dissolution of minerals, and formation of organometallic coordination compounds Transformation and decomposition of inorganic and organic compounds Methylation Mineral precipitation Conversion of igneous minerals to soils Corrosion of metals

parallel to the floor and flanks of the *glaciated valley* upstream of the thin-arch Vaiont Reservoir in the Italian Alps (Figure 3.2). Here, the older sheet fractures have evolved from the effect of *glacial erosion* in the Late Pleistocene and subsequent *isostatic rebound*, and the younger set of sheet fractures are attributed to a further loss of the landmass due to *stream erosion* in the Holocene (Kiersch, 1964). Being one of the external controls of *slope instability* (Chapter 4), the sheet fractures have caused inherently unstable slopes throughout the system, but more so at the shoulder of the dam. Consequently, the dam failed on October 9, 1963, by the increase of *pore pressure* in the limestone rock of the gorge.

FIGURE 3.1 Sheet fractures. Ucontitchie Hill, northwest Eyre Peninsula, South Australia. (Source: Photo courtesy Professor C. R. Twidale.)

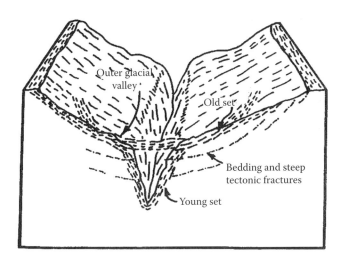

FIGURE 3.2 Sheet fractures due to offloading release in the Vaiont River Valley, Italy. (Source: Figure 4 in Kiersch, G. A., "Vaiont Reservoir Disaster," *Civil Eng.* 34 [1964]: 32–39. With permission from American Society of Civil Engineers.)

The unloading mechanism for sheet fractures, however, is debated. Twidale and others (1993, 1996) observe that the expansive tendency of rocks does not always find accommodation at the existing sheet fractures, and that a few sheet fractures in the system remain discordant to the surface, indicating that the sheet fractures possibly evolve by an alternative mechanism of compressive stress in the subcrust.

Compressive Stress Mechanism

Global tectonic forces (Chapter 2) generate enormous compressive stress in the *lithospheric crust*. In addition, petrologic processes of the solidification of intrusive igneous rocks, the crystallization of metamorphic rocks, and the lithification of sedimentary rocks also store energy in the form of compressive stress in the subcrust (Twidale et al., 1996). The compressive stress, in general, increases with depth. By one estimate, the compressive stress at a depth of 10 m from the surface is an order of magnitude higher than that due to the vertical loading of surface rocks.

Compressive stress in the subcrust buckles up rocks, releasing the energy of expansion as sheet fractures parallel to the surface of buried rocks (Twidale, 1973, 1981; Holzhausen, 1989; Twidale et al., 1996). The dome-shaped *bornhardts*, which possibly evolve in the subcrust and appear as an *exhumed feature* of the *multicyclic landscape*, are thus born with sheet fractures (Thomas, 1974; Dohrenwend et al., 1986). Experimental observations also suggest that a block of hard rock placed in a horizontal compressive stress field similarly buckles up and develops stress trajectories parallel to the surface and not through the *regolith* above it (Figure 3.3). Thus, sheet fractures evolve by the effect of compressive stress in the subcrust (Holzhausen, 1989). They are possibly initiated during the period of earth movements due to tectonic causes or in the period of rock formation within the subcrust (Twidale, 1981).

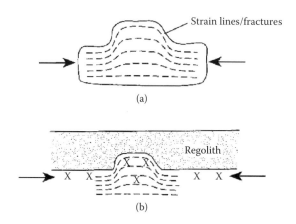

(a)

(b)

FIGURE 3.3   (a) Stress trajectories in partly unconfined block subject to horizontal compression (after Holzhausen, 1989). (b) Application of Holzhausen model to development of bornhardt with sheet fractures in differentially weathered terrain. (Source: Figure 18 in Twidale, C. R., Vidal Romani, J. R., Coruña, A., Campbell, E. M., and Centeno, J. D., "Sheet Fractures: Response to Erosional Offloading or to Tectonic Stress," *Z. Geomorph.* 106 [1996]: 1–24. With permission.)

Thermal Expansion and Contraction

Rocks comprise mineral suites, which differ in absorption, conductivity, and retention of heat. Hence, rock bodies are thought to unevenly expand and contract by the thermal stress of the environment and disintegrate at the boundary of mineral grains. Fresh-cut rock specimens subject over a long enough period of time to controlled cycles of extreme high and low temperatures, however, do not disintegrate, suggesting that air temperature variations are inconsequential to the disintegration of rocks (Ollier, 1979). The surface rocks, however, extensively disintegrate in desert environments of sufficient temporal thermal range, suggesting that the elements of weather, the exploitable weakness in rocks, and the surface moisture from dew possibly combine in a complex manner to produce the given effect in rocks that otherwise are *poor conductors of heat* (Smith, 1994). The moisture in contact with rocks possibly initiates temperature-controlled rhythmic *hydration* and *dehydration* of minerals, producing sufficient mechanical stress for disintegration along predetermined structural forms of weakness in rocks.

Growth of Ice Crystals

The freezing water increases 9% in volume at 0°C, and further to 13.5% at the lowest density of water at −22°C, producing a strong expansive force in the medium. The freezing of water with attendant growth of ice crystals, and deformation of ice within favorably placed cracks and joints of certain minimum size in the body of rock, generate a *cryostatic pressure* of 2100 kg cm^{-2} against the cavity space. The pressure of this magnitude exceeds many times the *tensile strength* of most rocks, ensuring widespread disintegration of the rock along its predetermined structural and lithologic forms of weakness (Davidson and Nye, 1985; Tharp, 1987). In comparative terms, the tensile strength of granite is only 70 kg cm^{-2} and that of sandstone is between 7 and 14 kg cm^{-2} (Ollier, 1979).

The ice pressure in the confined space of deep cavities within the body of rocks is relieved by *frost shattering*, which is the most intense form of disintegration on the surface of the earth. Environmental conditions for the growth of ice crystals and dynamics of frost shattering are discussed in Chapter 8. In the present, the intensity of frost shattering depends more on the frequency of *freeze-thaw cycles* than on the amplitude of temperature fluctuation about the freezing point (Harris, 1974). The growth of ice crystals in pore spaces of unconsolidated sediments, though, generates a strong uplift force to heave *frost-susceptible soils*. The source of moisture, and mechanisms of moisture migration for freezing in the soils of *cold region environments* are discussed in Chapter 8.

Hydration

A few rock-forming minerals hydrate by absorbing water molecule(s) in the mineral structure and dehydrate by losing the absorbed water from the mineral lattice. The hydrated minerals expand many times the volume of their anhydrous state (Table 3.2). In certain geomorphic environments, the *cyclicity* of hydration expansion and

TABLE 3.2
Increase in Volume of Some Minerals on Hydration

Anhydrous		Hydrous		Volume Change (%)
Name and Composition	Density	Name and Composition	Density	
Sodium carbonate (Na_2CO_3)	2.532	Natron ($Na_2CO_3.10H_2O$)	1.44	374.72
Thenardite (Na_2SO_4)	2.68	Mirabilite ($Na_2SO_4.10H_2O$)	1.464	315.00
Hydrophilite ($CaCl_2$)	2.15	Dihydrate chloride ($CaCl_2.2H_2O$)	0.835	241.10
Magnesium sulfate ($MgSO_4$)	2.66	Epsomite ($MgSO_4.7H_2O$)	1.68	233.20
Magnesium chloride ($MgCl_2$)	2.325	Bischofite ($MgCl_2.6H_2O$)	1.57	216.33
Anhydrite ($CaSO_4$)	2.61	Gypsum ($CaSO_4.2H_2O$)	2.32	42.27

Source: Weist, R. C., *Handbook of Chemistry and Physics* (Boca Raton, FL: CRC Press, 1973); Goudie, A. S., "Sodium Sulphate Weathering and the Disintegration of Mohenjo-Daro, Pakistan," *Earth Surface Processes* 2 (1977): 75–86.

dehydration contraction generates sufficient cumulative mechanical stress in rocks, by which they loosen and eventually disintegrate along the boundary of mineral grains.

CRYSTALLIZATION OF SALTS

Earth materials, biochemical activities, capillary suction, and airborne dust and salts release a variety of surface and near-surface salts, which hydrate and dehydrate in terrestrial environments and mobilize in the *ionic state*. The mobilized salts interact with natural features and man-made structures, producing the stress of hydration expansion and dehydration contraction. The magnitude of this stress, however, varies with the composition and concentration of ionic species, and the manner of crystallization at and within the impacted surfaces (Smith, 1994). The salts released into the environment also have the additional property of *thermal expansion*, which may be up to three times that of granite under similar conditions (Cooke and Smalley, 1968). Therefore, the volumetric expansion of salts precipitating within rock pores tends to stress the host rock, but its effects are not known in certain terms. The dynamics of salt migration and attendant effects, however, are widely understood for the salts of *capillary suction*.

Subsurface waters carry a variety of dissolved salts derived from the lithologic composition of *aquifers*. The salts migrate in *capillary suction* and, depending upon the thermal efficiency and depth to water table, crystallize as anhydrous *efflorescence* at the surface or precipitate as *hardpan* at a shallow depth from the surface. In semiarid environments, the surface salts mobilize at higher nighttime humidity, and interact with the man-made structures and facilities. Goudie (1977) observes that sodium sulfate alternating between the anhydrous thenardite and hydrous mirabilite states in contact with bricks has extensively crumbled the historic structure of Mohanjodaro, Pakistan. Other capillary salts, including nitrates and nitrites, have similarly caused extensive scaling, crumbling, and discoloration of the outer and inner building materials of the Islamic monuments of Khiva and Bukhara,

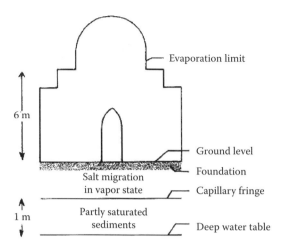

6 m

1 m

Evaporation limit

Ground level

Foundation

Salt migration
in vapor state

Capillary fringe

Partly saturated
sediments

Deep water table

FIGURE 3.4 Illustration of possible mechanism for two capillary zones in arid environments. (Source: Adapted from Figure 5 in Akiner, S., Cooke, R. U., and French, R. A., "Salt-Damage to Islamic Monuments in Uzbekistan," *Geogr. J.* 158 [1992]: 257–72.)

Uzbekistan, to a height not exceeding 6 m from the surface (Akiner et al., 1992). In the arid environment of this Central Asian republic, the capillary salts crystallize a little above the deep water table, rise above the capillary fringe in a *vapor state*, and precipitate a little below the surface as hardpan (Figure 3.4). The moisture and salts from this upper zone of salt accumulation seep through the bricks, move up the column, and laterally outward along the thermal gradient to the warmer exterior of the monuments. Hence, the mobilized salts produce greater disruption along the outer than the inner periphery of the building structure.

The salts also generate mechanical stress in the earth materials. The hydrated magnesium sulfate (epsomite) crumbles the debris of *rockfalls* and that of the recently exposed *moraines* in the Kashmir Himalayas (Goudie, 1984), the hydrated calcium sulfate (gypsum) possibly produces the basal concavity of *ventifacts* in the cold-dry environment of Antarctica (Whitney and Splettstoesser, 1982), brine crumbles granite, and salts in wave sprays cause *salt weathering* of rocks along the coasts of hot-humid environments.

COLLOIDAL PLUCKING

Colloids of clays and gelatine, and to some extent silica, represent a system halfway between *solution* and particle *suspension*. The mechanical effects of colloids are not known, but the wetting and drying of colloidal sols probably plucks particles from the outer skin of rock bodies.

CHEMICAL WEATHERING

Rock-forming minerals had formed at and within the earth's crust, and were stable at elevated temperature or pressure conditions, or both, obtained in the remote past of planet Earth (Chapter 2). These minerals are now unstable in present-day environmental conditions. They, therefore, decompose into mineral suites and products that are in *equilibrium* with the air and water phases of the *biosphere*. This tendency of the minerals to strive for equilibrium with the environment is called chemical weathering. Rock-forming minerals are susceptible to chemical alteration in different degrees and in the decomposition rate. Chemical weathering also occurs in a variety of ways.

The susceptibility of minerals to chemical alteration depends on the bond strength of the *atoms* of the same or different species called *molecules*, the packing of *cations*, and the number of bond links of different cations with the molecular structure of the crystal lattice (Ernst, 1969). The *bond strength* is a measure of the chemical stress or average energy required to dissociate one mole of the bond from the mineral structure. In comparative terms, the bond strength of silica-oxygen is highest (3110 to 3142 kcal mol^{-1}) and that of potassium-oxygen is lowest (299 kcal mol^{-1}). Therefore, the silica-rich rocks resist chemical alteration and the potassium-rich rocks are more susceptible to this form of weathering. *Packing* is the interlocking of cations with the mineral structure, such that the cations of weaker ionic bond strength escape the mineral structure and decompose more readily. *Bond links* tie diverse cations with the molecular structure of the mineral lattice. In general, the minerals of a large number of bond links are more susceptible to the chemical alteration than minerals with fewer and less diverse links.

Chemical weathering proceeds by evolving minerals of progressively higher molecular stability in the environment. This sequence of change in the mineral composition and structure may be illustrated with reference to the decomposition of *mafic* (dark-colored) and *felsic* (light-colored) groups of igneous minerals arranged in descending order of increasing mineral stability. Olivine and anorthite, the two unstable minerals in the mafic and felsic series of mineral stability, had crystallized at extremely high temperatures. Hence, they progressively decompose into the minerals of next higher stability until the end product of stable quartz is reached.

Mafic Minerals and Composition	Felsic Minerals and Composition
Olivine	Anorthite (calcic plagioclase)
$Ca (Al_2 Si_2 O_8)$	$(Mg, Fe)_2 Si O_4$
Auguite (pyroxene)	Albite (sodic plagioclase)
$Ca (Mg, Fe) Si_2 O_6$	$Na (Al Si_3 O_8)$
Hornblende (amphibole)	Orthoclase (potassium feldspar)
$Ca_2 (Mg, Fe)_6 Al Si_7 O_{22} (OH)_2$	$K (Al Si_3 O_8)$
Biotite (mica)	
$K (Mg, Fe)_3 (Al Si_3 O_{10}) (OH)_2$	
Muscovite	
$K Al_2 (Al Si_3 O_{10}) (OH)_2$	
Quartz	
$Si O_2$	

AGENTS OF CHEMICAL WEATHERING

Water, oxygen, and carbon dioxide are the principal agents of chemical weathering. *Water* is a catalyst, a solvent, and a reactant in all decomposition processes; *oxygen* participates in the decay of minerals rich in iron, manganese, and sulfur; and *carbon dioxide* in water dissolves the calcite-rich rocks. Chemical weathering processes are generally slow, irreversible, deceptively simple, and governed by local chemical equilibrium constants. They are described as oxidation-reduction, hydrolysis, carbonation, and solution processes.

In addition, a certain group of heterogeneous chemical activities involve interaction between the aqueous and solid phases of the chemical species in natural and contaminated environments. The *heterogeneous chemical reactions* are given as sorption processes on continents, and chemical, sorption, and photochemical processes in coastal environments.

OXIDATION AND REDUCTION

Combining of oxygen with *compounds* is oxidation, and the removal is reduction. Oxidation occurs by the loss and reduction by the gain of electron(s) from an *atom* or *ions*. In essence, a substance loses electron(s) and oxidizes only when another matter in its environment accepts the displaced electron(s) and is reduced. Therefore, oxidation accompanies reduction by an equal amount. In principle, most subaerial environments are oxidizing and shallow water bodies and aquifers carrying organic matter are reducing environments.

Oxidation

A loss of electron(s) from the outer shell configuration of minerals increases the *electric charge* on crystal structures by which the rock-forming minerals oxidize and decompose into hydrous compounds and hydrates of a higher state of stability. The mafic igneous minerals decompose into hydrated clay minerals, and felsic igneous minerals oxidize to peroxides and superoxides. The minerals containing iron, manganese and sulfide are particularly susceptible to oxidation as

$$2\ FeCO_3 + \tfrac{1}{2}\ O_2 + 2\ H_2O \rightarrow Fe_2O_3 + 2\ H_2CO_3$$

Siderite   Hematite   Carbonic acid

The oxidation of ferrous iron (Fe^{2+}) to ferric state (Fe^{3+}) produces hematite. The hematite is a stable mineral of iron and remains insoluble in natural conditions.

$$MnSiO_3 + \tfrac{1}{2}\ O_2 + 2\ H_2O \rightarrow MnO_2 + H_4SiO_4$$

Rhodonite   Pyrolusite   Silicic acid

Pyrolusite (or manganese oxide) provides for the *desert varnish* in aeolian environments.

$$PbS + 2 O_2 \rightarrow PbSO_4$$

Gelena   Anglesite

Anglesite causes *encrustation* of igneous rocks and dikes.

Reduction

Undecomposed and partly decomposed *organic matter* in water bodies is a strong reducing agent. The reducing property of organic matter results from the carbohydrates stored in the chlorophyll-containing plant tissues. The carbohydrates synthesize when sunlight combines with the dissolved carbon dioxide in water, and releases oxygen from the system as

$$6 CO_2 + 6 H_2O \rightarrow C_6H_{12}O_6 + 6 O_2\uparrow$$

Carbohydrate

The reduction of organic matter releases methane, hydrogen sulfide, and ferrous iron in the environment. These compounds and elements displace electrons and reduce few chemical species in the environment. The reducing properties of a part of the Gulf of Batabano in Cuba result from the presence of organic matter and sulfide content in bottom sediments (Plante et al., 1989), of shallow ocean basins to the depth of sunlight penetration in Indonesia to sulfate and nitrite content in bottom sediments (Rommets, 1988), and of a part of an aquifer in Denmark to the presence of organic matter and sulfide content in this water-bearing formation (Engesgaard and Kipp, 1992).

REDOX POTENTIAL

Oxidation or reduction of certain *chemical elements* is possible, but the ease with which these chemical species oxidize or reduce differs considerably from one element to another. The intensity of oxidation or oxidation potential of elements in solution is given by the electric potential of a single electrode of an element immersed in a solution of its ions (Hem, 1972). This index is expressed in volts and given by the symbol Eh. Many redox reactions, however, also involve the hydrogen ion activity or *solution pH*. The oxidation or reduction potential of a geologic environment in aqueous phase is called the redox potential (Krauskopf, 1967).

The Eh-pH diagrams, which represent the Eh on the X-axis and the pH on the Y-axis, find extensive geochemical applications, such as in determining the limits and potential of oxidizing and reducing environments, interpretation of the processes regulating the chemical composition of groundwater systems, and quantification of the conditions for precipitation and dissolution of chemical elements and compounds in natural environments. The chemical behavior of elements in solution depends on their molecular structure and the nature of chemical bonding. Thus, the chemical elements in two valance states change into a more stable oxidation or reduction state (Figure 3.5). The *ferric iron* (Fe^{3+}) is soluble at Eh +0.8 or more and a pH between 0 and 4, but remains insoluble over the rest of the Eh-pH field. By comparison, the

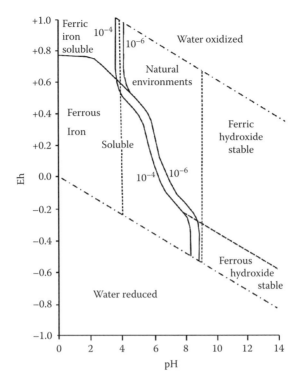

FIGURE 3.5 Solubility of iron as a function of Eh and pH. Labeled contour lines refer to activities of dissolved iron species of the indicated values: T = 25°C, total pressure = 1 atm. (Source: Figure 12.1 in Douglas, I., "Lithology, Landforms and Climate," in *Geomorphology and Climate*, ed. E. Derbyshire [London: John Wiley & Sons, 1976: 345–66]. With permission from John Wiley & Sons.)

ferrous iron (Fe^{2+}) is ordinarily soluble in water. Its solubility, however, increases with decreasing Eh and pH range.

Hydrolysis

A water molecule has the property of disassociation into hydrogen (H^+) and hydroxyl (OH^-) ions. The hydrogen ion imparts acidity to the aqueous solution and the hydroxyl ion participates in many weathering reactions. The rocks comprising silicate mineral groups are particularly susceptible to hydrolytic decomposition. This form of decomposition occurs by the diffusion of a hydrogen ion into the interior of a silicate mineral lattice, which liberates silica in the form of silicic acid and mobilizes alkalis as the ionic product of the desilication process as

$$2 \, KAlSi_3 \, O_8 + 2 \, H_2 CO_3 + 9 \, H_2 O \rightarrow Al_2 \, Si_2 \, O_5 \, (OH)_4 + 4 \, H_4 \, SiO_4 + 2 \, K^{2+} + 4 \, OH^-$$

Potash feldspar	Carbonic acid	Kaolinite	Silicic acid	Ions of potassium and hydroxyl

FIGURE 3.6   Spheroidal from of a boulder due to exfoliation. (Source: Figure 24.1 in Strahler, A. N., *Physical Geography*, 3rd ed. [New York: John Wiley & Sons, 1969]. With permission.)

$$Mg_2 SiO_4 + 4 H^+ + 4 OH^- \rightarrow H_4 SiO_4 + 2 Mg^{2+} + 4 OH^-$$

Olivine   Silicic   Ions of magnesium
                                                   acid   and hydroxyl

$$CaAl_2 Si_2 O_8 + 3 H_2 O + 2 CO_2 \rightarrow Al_2 Si_2 O_5 (OH)_4 + Ca^{2+} + 2 HCO_3^-$$

Anorthite   Kaolinite   Ions of calcium
                                                                  and bicarbonate

In arid regions of sufficient moisture, the hydrolytic weathering and granular disintegration of the igneous rocks causes small-scale *exfoliation*. The exfoliation attacks the interior of boulders from the sides, producing differentially weathered concentric layers and a core of undecomposed rock (Figure 3.6). This form of weathering is called *spheroidal weathering*. In hot-humid climates, however, the hydrolytic weathering acts horizontally and vertically through the joint system of igneous and metamorphic rocks, producing an unevenly decomposed rock to a certain depth in the subsoil. This process is known as *rock rotting*, and the weathering residue at the surface is called *saprolite*.

CARBONATION

Carbonation is the solvent action of carbonic acid—a hydrogen ion yielding solution of carbon dioxide in water—with *calcite-rich* chalk, limestone, and dolomite rocks. The reaction between carbonic acid and calcite may be given in a simple form as

$$CO_2 + H_2O \leftrightarrow H_2CO_3$$

Carbon dioxide   Carbonic acid

$$H_2CO_3 \leftrightarrow H^+ + HCO_3^-$$

Ions of hydrogen and bicarbonate

$$CaCO_3 + H^+ + HCO_3^- \leftrightarrow Ca^{2+} + 2 HCO_3^-$$

Calcite   Carbonic acid   Ions of calcium and
                              bicarbonate liberated

$$Ca^{2+} + 2\ HCO_3^- \rightarrow CO_2\uparrow + H_2O + CaCO_3\downarrow$$

Precipitation of calcium
carbonate

The acid-base reactions, given above, are extremely sensitive to the *solution pH*. The carbonic acid dissolves calcite in the acidic aqueous phase of pH less than 7 into bicarbonate and liberates the calcium ion. The calcium ion, however, reprecipitates as calcium carbonate in the alkali medium of a solution of pH more than 7. The dissolution and precipitation mechanisms of calcite are discussed in Chapter 10.

SOLUTION

Solutions are aqueous and hydrothermal in type. *Aqueous solutions* dissolve inorganic salts and soluble rock-forming minerals, and break down the molecular structure of certain minerals. A sediment formed of mineral precipitated from solution is called *chemical sediment. Hydrothermal solutions* comprise water and magmatic fluids, which originate deep in the lithospheric crust and rise under pressure through vents in the earth's crust. The magmatic fluids interact with the rock-forming minerals, and evolve alteration products and ore deposits. The hydrothermal solutions are also the principal source of energy for the evolution of life on the planet Earth. One biological view holds that the pyrite (or iron sulfide) from hydrothermal solutions builds up organic carbon molecules from carbon dioxide dissolved in water and synthesizes *amino acids*. The amino acids are the building block of deoxyribonucleic acid (*DNA*) and, hence, of life on the earth.

SORPTION REACTIONS ON LAND

Hydrophobic metal elements, decomposed and partly decomposed organic matter, suspended particles, and a variety of organisms coexist in natural and polluted waters of the continents. The organic matter in the system produces a variety of humic acids, of unique sorption properties which partition the hydrophobic contaminants into aqueous and sediment phases. The sediment phase binds with the humic acids, forming compounds of a higher molecular weight called polymers and complexes (Sorial et al., 1993). Such insoluble components pass out of their environment in the ionic state by absorption, adsorption, complexation, and competition processes riding on suitable mineral sites (Keoleian and Curl, 1989). The contextual meaning of the terms is illustrated in Figure 3.7 for clarity of understanding.

A number of *toxic* hydrophobic heavy metal contaminants also exist in natural environments. They adsorb onto the cell structure of suitable microorganisms, and are removed in the ionic state from the site of accumulation in soils. *Ganoderma lucidum*, which grows widely as a fungus in tropical forests, is a suitable *biosorbent* of copper present in the soils (Rao et al., 1993).

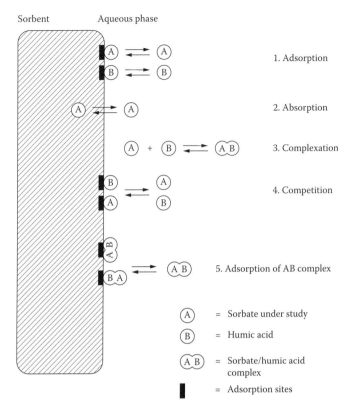

FIGURE 3.7   Possible mechanisms of interaction between a model sorbate, humic acid, and a natural sorbent. (Source: Figure 1 in Keoleian, G. A., and Curl, R. L., "Effects of Humic Acid on the Adsorption of Tetrachlorobiphenyl by Kaolinite," *Adv. Chem. Ser.* 219 [1989]: 231–50. With permission from American Chemical Society.)

BIOCHEMICAL, ADSORPTION, AND PHOTOCHEMICAL PROCESSES IN COASTAL ENVIRONMENTS

Chemical properties of coastal and estuarine waters are governed by the biological activity of marine organisms and bacteria in the environment as well as by the photochemical processes to the depth of sunlight penetration in the seawater. The *biological members* consume, precipitate, transform, and decompose a variety of minerals and toxic waste, and *photochemical processes* dissolve minerals, oxidize organic matter, and decompose complex organic molecules into their simpler structure.

The oxygen in seawater decomposes the *organic matter*, producing the life-supporting nutrients and chemical species for *photoautotrophic* marine life (Rommets, 1988), such as

$$(CH_2O)_{106} (HN_3)_{16} H_3PO_4 + 138 \; O_2 \rightarrow 106 \; CO_2 + 16 \; HNO_3 + H_3PO_4 + 122 \; H_2O$$

Organic matter in the system	Dissolved oxygen	Carbon dioxide	Nitric acid	Phosphoric acid	Water

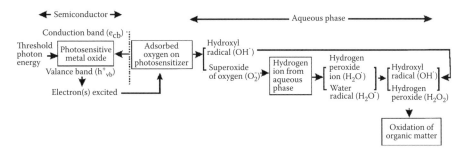

FIGURE 3.8 Mechanisms involved in generation of hydroxyl radicals by heterogeneous advanced oxidation. (Source: Adapted from Figure 1 in Notthakun, S., Crittenden, J. C., Hand, D. W., Perram, D. L., and Mullins, M. E., "Regeneration of Adsorbents Using Heterogeneous Advanced Oxygen," *J. Environ. Eng.* 119 [1993]: 695–714.)

Algae, fungi, and bacteria subsist on major and trace nutrients available in coastal waters, and release strong oxidizing hydrogen peroxide and unstable reduced metal ions in the environment through *metabolic processes*. The released entities affect the *nutrient cycle* and *redox potential* of the environment in many different ways (Morel et al., 1991). Marine organisms like phytoplankton, zooplankton, and benthos consume the *suspended sediments*, and a certain group of bacteria decompose *hydrocarbons* into a simpler molecular structure (Farley, 1990; Newman et al., 1990; Morel et al., 1991).

In estuarine environments, the heterogeneous processes of sorption and complexation coagulate light density *mud* from the suspended sediments moved in *tidal currents* (O'Melia, 1987; Morel and Gschwend, 1987; Morel et al., 1991; McBride, 1994). The floating mud is known to adsorb metal anions in the water column.

Photochemical processes to the depth of sunlight penetration in coastal waters oxidize the organic matter, break down a variety of organic compounds to simpler molecular structures, and dissolve toxic metal contaminants in the environment (Sulzberger, 1990). The photochemical processes are initiated when the *ultraviolet radiation* of certain *photon energy* adsorbs on photosensitive metal oxides, called *photosensitizers*, present in the environment and excites electrons from the valance band (h^+_{vb}). These electrons either migrate to the conduction band (e^-_{cb}) or return to the surface of photosensitive metal oxides (Notthakun et al., 1993). The electrons returning to the photosensitizers react with the adsorbed oxygen at the surface, and initiate self-sustaining chemical reactions (Figure 3.8). The interaction between electrons and oxygen initially produces a highly reactive hydroxyl radical (OH$\cdot$) and a superoxide of oxygen (O_2^-). The superoxide reacts with the H$^+$ in the medium and forms unstable hydroperoxide ion (HO_2^-) and water radical ($H_2O\cdot$), which combine to produce hydrogen peroxide (H_2O_2) and unstable free hydroxyl radical (OH$\cdot$) in the environment. The deficient electrons in the valence band similarly react with H_2O and OH$^-$, forming free OH$\cdot$. The free hydroxyl radical *oxidizes* the organic matter in the environment, and causes its destruction. Photochemical processes likewise oxidize oil spills to simpler soluble products of carboxylic acid, alcohol, ketones, and phenols.

BIOLOGICAL WEATHERING

Biological weathering refers to the impact of organisms on disintegration of rocks, biochemical displacement and dissolution of metallic and nonmetallic elements, bacterial- and nanobacterial-aided mobilization, and transformation and precipitation of elements and compounds in the *biosphere*. Hence, biological weathering overlaps the mechanical and chemical groups of activities in the environment.

DISINTEGRATION

A variety of organisms in animal and plant kingdoms rework sediments and disintegrate rocks. In general, the earthworms in cool humid environments, termites in tropical climates, and ants in arid regions rework the overburden, producing a matter of loose mineral soil. The lithophagic snails, which feed on the rock, return the ingested material in the form of loose mineral matter. Observations from a part of the Negev Desert, Israel, suggest that a variety of invertebrate snail species inhabiting the limestone terrain generate 69.5 to 110.4 g m^{-2} year^{-1} of the mineral soil, which is almost equal to the annual accumulation rate of the *windborne dust* in the region (Jones and Shachak, 1990). Besides the snail species *E. elbulus* feeding on limestone and associated lichen of the area release 24.2 mg of nitrogen m^{-2} year^{-1} in fecal deposits, which provide for the largest source of biomass production, succession of vegetation, and stability of landforms in the aeolian environment (Jones and Shachak, 1990).

Root wedging disintegrates rocks commonly in alpine environments, and the physical impact of falling trees and animal trampling universally loosens the soil grains. *Human activities* accelerate the process rates manifold and, thereby, affect the stability of rocks and sediments in many different ways (Chapter 12).

A few lichens disintegrate the host granite rock and liberate iron from suitable minerals. Observations also suggest that common species of lichen and moss secrete *humic acids*, which uniformly dissolve calcareous rocks beneath the root mat (Chapter 10).

CHELATION

Biochemical processes at the root zone produce a variety of organic acids. These acids claw the ionic species of nonmetallic and metallic soil nutrients between the two atoms of an organic molecule, forming organometallic coordination compounds called chelates (Lehman, 1963). *Chlorophyll*, which is the green coloring agent of vascular plants, is a chelate of magnesium clawed between four pyrrole rings (Gornitz, 1972). In general, chelation dissolves and displaces metallic and nonmetallic minerals, breaks down the mineral structure, and forms complexes and compounds that suppress the toxic effects in natural systems.

BACTERIAL ACTIVITY

Many types of organisms, including cyanobacteria, have been in existence since the remote past of the earth. *Cyanobacteria* are a phylum of photosynthetic bacteria of bluish pigment, which they use to capture sunlight for *photosynthesis*. Cyanobacteria were formerly called *blue-green algae*, even though it is now known that they are not algae. Over a thousand species of cyanobacteria exist in moist terrestrial environments, and in fresh and marine waters of the earth.

Cyanobacteria scavenge excess sodium and calcium content of the *salt-affected soils* (Singh, 1950; Apte and Alhari, 1994). These microbes are *autotrophs*. They take up nitrogen from the atmosphere, synthesize carbohydrates from the atmospheric carbon dioxide, store carbon dioxide in plant cells, and release humic acids and carbonic acid in the soils. The humic acids form *chelates* of sodium and calcium, and carbonic acid *dissolves* calcium carbonate nodules in the soils, freeing the system of its excess alkali content (Figure 3.9). Cyanobacteria also fix nitrogen in soils through the activity of *enzyme nitrogenase* in organisms.

Masses of cyanobacteria on the seabed have deposited abundant calcium carbonate in layers or domes. Allwood et al. (2006) present evidence for the cyanobacterial

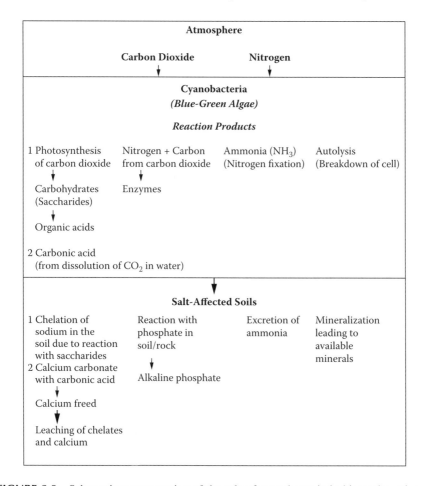

FIGURE 3.9 Schematic representation of the role of cyanobacteria in bioameleoration of salt-affected soils.

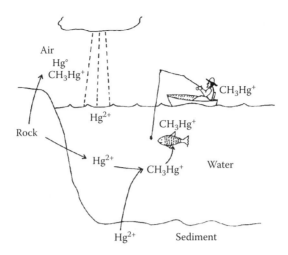

FIGURE 3.10 Mercury cycle in the biosphere. (Source: Figure 2 in Summers, A. O., and Silver, S., "Microbial Transformations of Metals," *Ann. Rev. Microbial.* 32 [1978]: 637–72. With permission.)

trapping and binding of a laminated sedimentary reef-like calcareous structure called *stromatolites* 3430 million years ago in the Pilbara craton, western Australia. These microbes of the Archaean era had thrived in ancient shallow marine *ecosystems* as an upward-migrating filamentous type of bacteria at the sediment-water interface. Stromatolites also occur as dark-colored deposits in the Sonbhadra region of the Indo-Gangetic Plain of India (2 Ma), but their evolutionary process and age are not known at the present time.

A large variety of microbes exist in clean and polluted environments. They transform nontoxic and toxic metallic minerals to new products and plant-available nutrient forms. *Conventional ecosystems* are the habitat of diverse species of bacteria, which are known to transform metal minerals on land and methylate organic and inorganic ionic species in water bodies (Summers and Silver, 1978). Certain bacteria on land, like *Thiobacillus* oxidize ferrous iron and *Sulfolubus*, reduce ferric iron to the ferrous state. These sulfate-reducing bacteria *corrode* buried iron pipelines, and a bacterium of the family Desulfobulaceae reduces iron in sedimentary environments (Holmes et al., 2004).

Certain *heavy metals* undergo geochemical and microbial transformation in water bodies, and become more toxic in nature. The methylation of mercury is the best-known example of geochemical and microbial transformation of heavy metals in aquatic systems (Summers and Silver, 1978). *Mercury*, which arrives into the water bodies from land and atmospheric sources, is transformed into a soluble mercurous state (Hg^{2+}) by as yet inadequately understood biochemical processes. This mercury methylates (Figure 3.10) possibly by the methylcobalamin-excreting bacteria present in the sludge waste of water bodies and also by the *bacterial flora* of the gills and guts of fish into soluble *methylmercury* (CH_3Hg^+). The methyl mercury is fifty to one hundred times more toxic than the mercurous form.

TABLE 3.3
Biodiversity of Extreme Environments

Extreme Conditions	Microorganism	Habitat	Extreme Growth Conditions	Metabolic Characteristics
High temperature	*Pyrococcus furiosus*	Geothermal marine sediments	100°C	Anaerobic, heterotroph
Cold temperature	*Bacillus TA41*	Antarctic seawater	4°C	Aerobic, heterotroph
High pressure	*Methanococcus janaschii*	Deep sea hydrothermal vent	25 atm 85°C	Growth and methane production stimulated by pressure
High pH	*Clostridium paradoxum*	Sewage sludge	pH 10.1 56°C	Anaerobic, heterotroph
Low pH	*Metallosphaera sedula*	Acid mine drainage	pH 2 75°C	Facultative, chemolithotroph
High salt	*Halobacterium halobium*	Hypersaline water	4–5 M NaCl	Aerobic, heterotroph

Source: Table 1 in Adams, M. W. W., Perler, F. B., and Kelly, R. M., "Extremozymes: Expanding the Limit of Biocatalysis," *Biotechnology* 13 (1995): 662–69. With permission from Nature Biotechnology.

EXTREMOZYMES

Certain microbes are adapted to uncompromising environments considered destructive to the life form (Adams et al., 1995; Danson and Hough, 1998). They are known as extremozymes. Extremozymes are classified by habitat and metabolic characteristics into six types (Table 3.3). Many extremozymes, however, had also existed in the remote geologic past of the earth as a primitive life form. The methanogenic microbes in *hydrothermal precipitates* of the Pilbara craton, Australia, are 3.5 billion years old and one such type (Ueno et al., 2006). Extremozymes, in general, are known to destroy hazardous waste, oxidize metal sulfides and elemental sulfur, and precipitate metals (Adams et al., 1995). Certain groups of bottom-dwelling marine bacteria oxidize or reduce manganese and form offshore ferromanganese nodules with up to 63% MnO_2 by weight (Summers and Silver, 1978).

NANOBACTERIA

Nanobacteria are intermediate in size between *bacteria* and *viruses*. They abound in conventional environments, and are equally adapted to extreme environments in the manner of extremozymes. Nanobacteria precipitate aragonite in the water medium (Folk, 1994), and are also known to dissolve aluminum and corrode metals (Folk, 1997; Folk and Lynch, 1997a, 1997b).

The precipitation mechanism of *aragonite* through the mediation of nanobacteria in a water column is based on the principle of *ionic adsorption* (Folk, 1994). This model argues that a positively charged calcium ion progressively adsorbs onto the

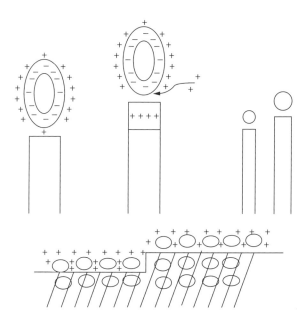

FIGURE 3.11   Conception of how nanobacteria precipitate carbonate. The illustration on the left shows a negatively charged cell wall of a nanobacterium that attracts the hull of Ca^{2+} ions from the solution. The illustration on the right shows deposition of $CaCO_3$ in the form of aragonite stalks beneath. The "bald spot" on the nanobacterium is soon replenished by the Ca^{2+} ions. This process is repeated over and over again. (Source: Figure 16 in Folk, R. L., "Interaction between Bacteria, Nanobacteria, and Mineral Precipitation in Hot Springs of Central Italy," *Géographie Physique Quaternaire* 48 [1994]: 233–46. With permission from Les Presses de L'université de Montréal.)

negatively charged bag-shaped outer shell of the nanobacteria, and neutralizes the charge on the nucleus. This process affects the *solubility product* of the molar concentration of calcium and carbonate ions in the immediate vicinity of nanobacteria, by which calcium sheds from the surface as aragonite stalks (Figure 3.11). The space vacated at the hull of nanobacteria similarly attracts the available calcium ion in the system, repeating the process of aragonite precipitation.

WEATHERING PROFILE

Prolonged rock weathering produces an incoherent mass of fragmentary and decomposed debris, in which the component of clastic fragments increases with depth to the intact rock. This is the weathering profile of a chemically undifferentiated product known as *regolith*. The profile attributes, in general, depend on climate and vary with the fissure frequency and mineral composition of rocks. The weathering profiles of *dry climates* and *frozen ground environments* are shallow and are comprised largely of the fragmentary material, and of the *taiga* and *humid tropical climates* are deep and rich in the decomposed mineral matter of oxides, hydroxides, and clay minerals (Strakhov, 1967). The rocks of higher *fissure frequency* and *chemically reactive minerals* carry deep weathering profiles of a higher proportion of decomposed than fragmentary products (Thomas, 1974).

Variations in the chemical composition and depth attributes of weathering profiles may be illustrated with respect to the weathering of granite terrain, which is characteristic of widespread *sheet fractures* and *deep joints* in the body of rock. In hot humid climates, the rock-forming minerals decompose along lateral sheet fractures and vertical joints to hydrated clay minerals. The intense weathering along predetermined narrow zones leaves a partly decomposed rock at the surface as *tors*, an undecomposed rock with *core stones* within the variable depth of the *regolith*, and surface debris that have moved from the area above (Figure 3.12). The minerals comprising the regolith are mobilized in the *ionic state* for evolving *soil profiles*.

SOIL PROFILE

A soil profile represents distinct *horizons* of certain minerals and is characteristic of certain texture, color, and depth attributes. These horizons are termed A- and B-horizons of true soils, C-horizon of a partly decomposed parent material, and D-horizon of an unweathered rock (Figure 3.13). The soil horizons evolve by the soil-forming processes, called *pedogenic regimes*, which function within the framework of climate, parent material, topography, and organic matter over a period of time. Hence, the nature and intensity of soil-forming processes changes with changes in the *symbiotic association* of abiotic and biotic elements of the landscape. Deforestation and forest fires, which alter the *biochemical environment* of the soils, evolve *degrade soils* at local and regional scales.

Climate had long been recognized as the dominant control of soil-forming processes. It is now, however, variously argued that the soils evolve in equilibrium with the *climate change* (Johnson et al., 1990), the role of climate is insignificant in pedogenesis (Nesbitt and Young, 1989), and climate only indirectly contributes to the soil dynamics (McFadden, 1988; Pope et al., 1995). The soils, accordingly, are an immature, steady-state, or complex attribute of the environmental conditions through which they have passed to arrive at the present state of development. *Immature soils* change with the change of environment, *steady-state soils* are in equilibrium with the environment, and *complex soils* evolve around more than one control of the soil-forming processes.

PEDOGENIC REGIMES

Podzolization, laterization, calcification, gleization, and salinization are major pedogenic regimes (Figure 3.14). They provide a framework for the classification of soils into groups and their principal varieties that have a worldwide distribution under similar geographic conditions.

PODZOLIZATION

Podzolization is the pedogenic regime of light-colored podzol soils of cool humid climates of higher latitude and altitude zones. The cool climate supports thick

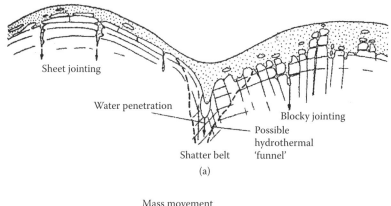

Sheet jointing

Water penetration

Blocky jointing

Possible
hydrothermal
'funnel'

Shatter belt

(a)

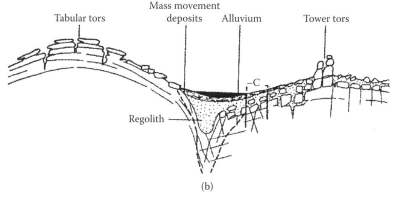

Mass movement
deposits   Alluvium

Tabular tors   Tower tors

—C

Regolith

(b)

FIGURE 3.12   Generalized pattern of weathering in a granite rock. (a) Regolith evolves from the joint-controlled movement of water that causes widespread chemical weathering at the surface and localized alteration of granite-forming minerals at depth. Very deep regolith possibly reflects hydrothermal alteration of the rock-forming minerals. In general, uneven intensity of chemical weathering leaves detached weathered boulders or core stones within the depth of regolith. (b) The stripping of regolith exhumes partly decomposed rocks at the surface. Tors are an expression of subsurface rotting in closely-jointed granites and spheroidal modification of larger granitic blocks. Mass movement of the regolith accumulates debris, some times called head, at the foot of slope or within local rock depressions. A section (C) through the regolith of colder environments suggests that it generally comprises frost shattered granite blocks, regolith affected by mass movement and greater amount of unweathered rock through the depth of the weathering profile. (Source: Adapted from Figure 14.1 in Thomas, M. F., "Criteria for the Recognition of Climatically Induced Variations in Granite Landforms," in *Geomorphology and Climate*, ed. E. Derbyshire [London: John Wiley & Sons, 1976], 411–45.)

coniferous forests, and inhibits bacterial decomposition of the organic litter. The organic matter produces humic acids, which leach bases, colloids, and oxides of aluminum and iron from the A-horizon and, thereby, enrich this horizon with the insoluble ash-gray residue largely of silica. The leached sesquioxides of iron and aluminum accumulate in the B-horizon, giving it a dark color and dense structure. The ionic alkali products, however, pass through the C-horizon and eventually out of the system as the *dissolved load* of streams.

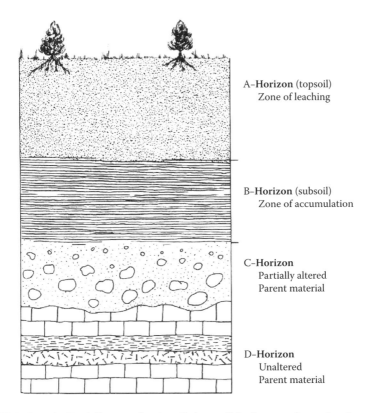

A–**Horizon** (topsoil)
Zone of leaching

B–**Horizon** (subsoil)
Zone of accumulation

C–**Horizon**
Partially altered
Parent material

D–**Horizon**
Unaltered
Parent material

FIGURE 3.13 Idealized diagram showing distinct soil horizons and associated zones. Soil horizons evolve from the mobilization of soluble elements like sodium, potassium, calcium, chlorides, and sulfates, which either are leached as soil solution or are utilized by plants for metabolic growth. In a general term, the uppermost part of the soil profile, called A horizon, carries decomposed or partly decomposed organic matter, and a content of lighter inorganic material depleted of soluble minerals. This zone, called the zone of eluviation, is most affected by geomorphic agents and human activities. By comparison, the B horizon has much less organic matter, more fragments of the parent material and accumulation of the products of soil solution percolating from the A horizon. The material is re-synthesized by chemical processes into secondary clay minerals. The B horizon is also called the zone of illuviation. The C horizon is more compact and comprises decomposed parent material called regolith. This material rests over unaltered parent material of the soil. This part of the soil profile is called D horizon. (Source: Figure 3.16 in Keller, E. A., *Environmental Geology* [Columbus, Ohio: Bell & Howell Co., 1979]. With permission from Bell & Howell Co.)

LATERIZATION

Laterization is the pedogenic regime of laterite soils that are common to equatorial, tropical wet-dry, and humid tropical climates. The laterite soils make thick sequences of iron and aluminum-rich varieties beneath woodlands. The *iron-rich laterite* develops on the acidic and *aluminum-rich laterite* on the basic parent material.

The iron-rich laterites contain 30 to 80% iron oxide, opposed to the 2 to 10% available in the parent material. Hence, the development of iron-rich lateritic soils

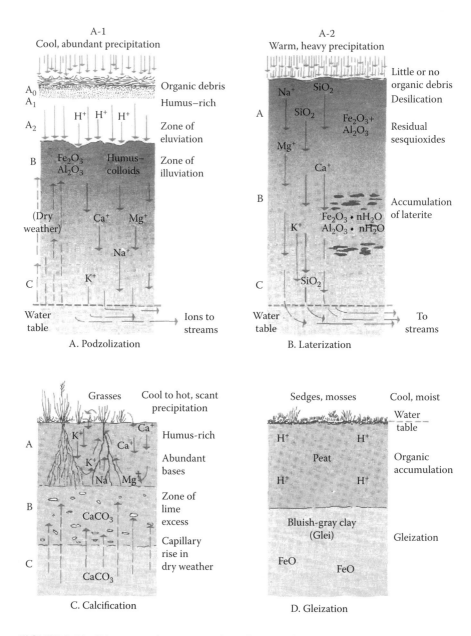

FIGURE 3.14 Diagrammatic representation of soil profile development under four pedogenic regimes. (Source: Figure 18.9 in Strahler, A. N., *Physical Geography*, 3rd ed. [New York: John Wiley & Sons, 1969]. With permission.)

requires weathering of the source material many times the thickness of the soil profile or an equally improbable lateral migration of iron from sources outside the site of laterite formation (McNeil, 1964). Therefore, not only the source of iron but also its mobilization and precipitation mechanisms are debated for the laterite soils.

Several hypotheses have been proposed for the high iron content of laterite soils (Thomas, 1974). They variously envisage solubility, oxidation, and coagulation of iron in the soil profile by chelation and other chemical processes. The *chemical selection hypothesis*, which postulates that laterite profiles enrich with iron and alkali content by the processes of desilication, mobilization of bases in the ionic state, and accumulation of iron (Fe_2O_3) and aluminum (Al_2O_3) in the A-horizon and of their hydrates ($Fe_2O_3.nH_2O$, $Al_2O_3.nH_2O$) in the B-horizon, is by far the best among proposed hypotheses (Bunting, 1969). *Desilication* possibly occurs in a strong alkali environment, forming *gibbsite* ($Al(OH_3)$) and *boehmite* ($Al\,O\,(OH)$) as the principal constituents of *bauxite deposits*. The alkali environment essential for desilication, however, cannot mobilize calcium, potassium, and magnesium in the ionic state. Therefore, alternative hypotheses are required for the evolution of laterite soils.

CALCIFICATION

Calcification is the pedogenic regime of calcium- and organic matter–rich soils of moisture-deficient mid-latitude and semiarid tropical grassland regions, in which the calcium and magnesium content of the soils does not pass into solution and is retained in the A-horizon. Besides, the capillary precipitate of *caliche* in the B-horizon also strengthens the alkali character of these soils.

GLEIZATION

Gleization is the pedogenic regime of excessively moist sticky clays, called the glei soils, of the *peat* in tundra and cool continental climates. The peat comprises undecomposed and partly decomposed organic matter, which *reduces* the iron content within it to provide a blue, gray, or olive color to the glei soils (Bunting, 1969).

SALINIZATION

Salinization is the pedogenic process of slow-draining desert surfaces. The evaporation of water precipitates a variety of chemical sediments in the soils of this environment.

Geomorphic Significance

Soils evolve over a long period of time. Hence, they preserve a signature of the climate change in the chemical and mineral composition of respective profiles (Hall and Michand, 1988). *Buried soils* that are isolated from the effects of present climate by subsequent sedimentation are thus excellent marker horizons for interpretation of the environment change. They are called *paleosols*. The inherent soil properties enable realistic *land use decisions*, a few aspects of which are discussed in Chapter 12.

RELATED DEPOSITS

The products of chemical weathering are mobilized in the ionic state in *capillary suction*. They precipitate at and within soils, and within the interstices of near-surface rocks as a resistant crust called *duricrust*. The duricrust is variously composed of the oxides of aluminum (alcrete), silica (silcrete), iron (ferricrete), and calcium (calcrete).

The *alcrete* or bauxite is the lateritic aluminum. It best develops in fast-draining tropical and subtropical areas of 1200 to 1500 mm mean annual rainfall. *Bauxite* is extensive in Africa and Australia, where it commonly occurs in association with *ferricrete* (Gentilli, 1968). *Silcrete* is up to 99% silica by weight (Dixon, 1994). It develops in areas of about 250 mm mean annual rainfall as primary and secondary duricrust. The silcrete of *primary origin* is the product of the precipitation of mineral matter in sheet flow and that in stagnant waters. The silcrete of *secondary origin* represents the desilication of laterite soils. The ferricrete is more or less a cemented soil horizon of iron oxide in areas of 1000 to 1200 mm mean annual rainfall. The ferricrete in a part of North Africa is 37% Fe_2O_3, 31% MnO_2, 9% Al_2O_3, and 8% SiO_2 in chemical composition (Fairbridge, 1968). The *calcrete* is comprised largely of $CaCO_3$, and is pedogenic and nonpedogenic in origin (Goudie, 1994). The calcrete of *pedogenic origin* evolves by calcification in areas of 400 to 600 mm mean annual rainfall (Dixon, 1994), and that of *nonpedogenic* origin precipitates in the environment of karst as *tufa*.

SUMMARY

Weathering is *in situ* disintegration and decomposition of rocks to products in equilibrium with the physical, chemical, and biological stress of the environment. For convenience, weathering is classified into mechanical, chemical, and biological domains to explain how and why rocks irreversibly alter to new products in the earth's environment.

Rocks disintegrate by the mechanical stress of elastic expansion at the surface and compressive stress in the subcrust, thermal expansion and contraction of surface rocks, hydration-dehydration of the rock-forming minerals, growth of ice crystals in rock cavities, and crystallization of salts in contact with the rock surfaces. The dynamics of many mechanical processes of rock disintegration, however, is debated.

Water, oxygen, and carbon dioxide are the principal agents of chemical weathering. Rocks decompose by the weakening of bond energy holding atoms and molecules to the mineral structure. The rock-forming minerals break down by oxidation-reduction, hydrolysis, carbonation, and solution processes to more stable products in equilibrium with the environment. A certain group of activities, called heterogeneous chemical processes, are governed by the interaction between humic acids and mineral species. This form of interaction displaces hydrophobic minerals as complexes and compounds in the chemical environment. The sea-dwelling organisms, bacteria, and photochemical processes to the depth of sunlight penetration in the seawater also initiate biochemical and chemical processes. The marine organisms and bacteria consume, precipitate, and decompose minerals and organic matter in the environment. The photochemical processes produce a strong oxidizing hydroxyl radical, which oxidizes oil spills to simpler products and destroys the organic matter in the environment.

Biological weathering is the impact of organisms on disintegration and biochemical alteration of the earth materials. A variety of organisms in the biosphere rework sediments, digest rocks, and release loose mineral soil rich in nitrogen content. Plants generate humic acids, which dissolve minerals, extract soluble nutrients from the parent material, form organometallic coordination compounds, and suppress undesirable properties of the natural systems. A few bacterial species of unconventional or extreme environments precipitate calcium carbonate at the seabed, mediate in the formation of reef structures in shallow coastal waters, oxidize and reduce metal ions, enrich low-grade ores, and of conventional environments, methylate mercury to high toxic levels. The bacteria of extreme environments are known to oxidize, precipitate, transform, and destroy few metals and compounds. Nanobacteria also dissolve, corrode, and precipitate certain groups of metals, convert the minerals of igneous rocks to soils, and precipitate aragonite by adsorbing the calcium ion.

Weathering over a long period of time evolves a weathering profile of disintegrated and decomposed mineral matter to a certain depth of the rock. The composition and thickness of the weathering profile depend on the climatic regime, reactivity of the rock-forming minerals, and fissure frequency in the body of rock. The weathering profile evolves distinctive soil groups by the processes of podzolization, laterization, calcification, gleization, and salination. Of these, the pedogenic regime of laterite soils is most debated. The ionic products of chemical weathering return to the surface in capillary suction, and precipitate in the soils and near-surface rocks as duricrusts of the oxides of aluminum, silica, iron and calcium. Their global distribution closely follows the pattern of mean annual rainfall.

REFERENCES

Adams, M. W. W., Perler, F. B., and Kelly, R. M. 1995. Extremozymes: Expanding the limit of biocatalysis. *Biotechnology* 13:662–69.

Akiner, S., Cooke, R. U., and French, R. A. 1992. Salt-damage to Islamic monuments in Uzbekistan. *Geogr. J.* 158:257–72.

Allwood, A. C., Walter, M. R., Kamber, B. S., Marshall, C. P., and Birch, I. W. 2006. Stromatolite reef from the Early Archaean era of Australia. *Nature* 441:714–18.

Apte, S. K., and Alhari, A. 1994. Role of alkali cations (K+ and Na+) in cyanobacterial nitrogen fixation and adaptation to salinity and osmatic stress. *Ind. J. Biochem. Biophys.* 31:267–79.

Bunting, B. T. 1969. *The geography of soil.* London: Hutchinson & Co. Ltd.

Cooke, R. U., and Smalley, I. J. 1968. Salt weathering in deserts. *Nature* 220:1226–27.

Danson, M. J., and Hough, D. W. 1998. Structure, function and stability of enzymes from the Archaea. *Trends Microbiol.* 6:307–14.

Davidson, G. P., and Nye, J. F. 1985. A photoelastic study of ice pressure in rock cracks. *Cold Regions Sci. Technol.* 11:141–53.

Dixon, J. C. 1994. Duricrusts. In *Geomorphology of desert environments*, ed. A. D. Abrahams and A. J. Parson, 82–105. London: Chapman & Hall.

Dohrenwend, J. C., Wells, S. C., McFadden, L. D., and Turrin, B. D. 1986. Pediment dome evolution in the eastern Mojave Desert, California. In *International geomorphology*, ed. V. Gardner, 1047–62. Chichester: John Wiley & Sons.

Douglas, I. 1976. Lithology, landforms and climate. In *Geomorphology and climate*, ed. E. Derbyshire, 345–66. London: John Wiley & Sons.

Engesgaard, P., and Kipp, K. L. 1992. A geochemical transport model for redox controlled movement of mineral fronts in groundwater flow systems: A case of nitrate removal by oxidation of pyrite. *Water Resources Res.* 28:2829–43.

Ernst, W. G. 1969. *Earth materials.* Englewood Cliffs, NJ: Prentice Hall.

Fairbridge, R. W. 1968. Induration. In *The encyclopedia of geomorphology*, ed. R. W. Fairbridge, 552–56. New York: Reinhold Book Corp.

Farley, K. J. 1990. Predicting organic accumulation in sediments near marine outfall. *J. Environ. Eng.* 116:144–65.

Folk, R. L. 1994. Interaction between bacteria, nannobacteria, and mineral precipitation in hot springs of central Italy. *Géographie Physique Quaternaire* 48:233–46.

Folk, R. L. 1997. Nannobacteria: Surely not figments, but what under heaven are they? *Natural Science*, vol. 1, article 3.

Folk, R. L., and Lynch, F. L. 1997a. The possible role of nannobacteria (dwarf bacteria) in clay-mineral diagenesis and the importance of careful sample preparation in high-magnification SEM study. *J. Sediment Res.* 67:583–89.

Folk, R. L., and Lynch, F. L. 1997b. Nannobacteria are alive on earth as well as mars. In International Society for Optical Engineering (SPIE), 3111, pp. 406–19.

Gentilli, J. 1968. Duricrust. In *The encyclopedia of geomorphology*, ed. R. W. Fairbridge, 296–97. New York: Reinhold Book Corp.

Gornitz, V. 1972. Chelation. In *The encyclopedia of geochemistry and environmental sciences*, ed. R. W. Fairbridge, 149–52. Stroudsberg: Dowden, Hutchinson & Ross.

Goudie, A. S. 1977. Sodium sulphate weathering and the disintegration of Mohenjo-Daro, Pakistan. *Earth Surface Processes* 2:75–86.

Goudie, A. S. 1984. Salt efflorescences and salt weathering in the Hunza Valley, Karakoram Mountains, Pakistan. In *International Karakoram Project,* ed. K. J. Miller, 607–15. Vol. II. Cambridge: Cambridge University Press.

Goudie, A. S. 1994. Duricrust. In *The encyclopaedic dictionary of physical geography*, ed. A. Goudie, I. G. Simmons, B. W. Atkinson, and J. K. Gregory, 140–42. 2nd ed. Oxford: Blackwell.

Hall, R. D., and Michand, D. 1988. The use of hornblende etching clast weathering and soils to date alpine glacial and periglacial deposits: A study from southwestern Montana. *Bull. Geol. Soc. Am.* 100:458–67.

Harris, C. 1974. Autumn, winter and spring soil temperatures in Okstindan, northern Norway. *J. Glaciol.* 13:521–33.

Hem, J. D. 1972. Oxidation and reduction. In *The encyclopedia of geochemistry and environmental sciences*, ed. R. W. Fairbridge, 839–43. Stroudsberg: Dowden, Hutchinson & Ross.

Holmes, D. E., Bond, D. R., and Lovely, D. R. 2004. Electron transfer by *Desulfobulus propionicus* to Fe (III) and graphite eletrods. *Appl. Environ. Microbiol.* vol. 70 (2) 1234–37.

Holzhausen, G. R. 1989. Origin of sheet structure. 1. Morphology and boundary conditions. *Eng. Geol.* 27:225–78.

Johnson, D., Keller, E., and Rockwell, T. 1990. Dynamic pedogenesis: New views on some key soil concepts and a model for interpreting Quaternary soils. *Quat. Res.* 33:306–19.

Jones, C. G., and Shachak, M. 1990. Fertilization of the desert soil by rock-eating snails. *Nature* 346:839–41.

Keller, E. A. 1979. *Environmental geology.* Columbus: Bell & Howell Co.

Keoleian, G. A., and Curl, R. L. 1989. Effects of humic acid on the adsorption of tetrachlorobiphenyl by kaolinite. *Adv. Chem. Ser.* 219:231–50.

Kiersch, G. A. 1964. Vaiont Reservoir disaster. *Civil Eng.* 34:32–39.

Krauskopf, K. B. 1967. *Introduction to geochemistry.* Tokyo: Kogakusha Co. Ltd.

Lehman, D. S. 1963. Some principles of chelation chemistry. *Soil Sci. Soc. Am. Proc.* 27:167–70.

McBride, M. B. 1994. *Environmental chemistry of soils.* New York: Oxford University Press.

McFadden, L. D. 1988. Climatic influences on rates and processes of soil development in Quaternary deposits of southern California. In *Paleosols and weathering through geologic time,* ed. J. Reinhardt and W. Sigleo, 153–77. Geological Society of America Special Paper 216. Geological Society of America.

McNeil, M. 1964. Lateritic soils. *Sci. Am.* 211:96–102.

Morel, F. M. M., Dzombak, D. A., and Price N. M. 1991. Heterogeneous reactions in coastal waters. *Phys. Chem.* 9:165–80.

Morel, F. M. M., and Gschwend P. G. 1987. The role of colloids in partitioning of solutes in natural waters. In *Aquatic surface chemistry,* ed. W. Stumm, 405–22. New York: John Wiley & Sons.

Nesbitt, H. W., and Young G. M. 1989. Formation and diagenesis of weathering profiles. *J. Geol.* 97:129–47.

Newman, K. A., Morel, F. M. M., and Stolzenbach, K. D. 1990. Settling and coagulation characteristics of fluorescent particles determined by flow cytometry and fluorometry. *Environ. Sci. Technol.* 24:506–13.

Notthakun, S., Crittenden, J. C., Hand, D. W., Perram, D. L., and Mullins, M. E. 1993. Regeneration of adsorbents using heterogeneous advanced oxygen. *J. Environ. Eng.* 119:695–714.

Ollier, C. D. 1977. Applications of weathering studies. In *Applied geomorphology,* ed. J. R. Hails, 9–49. Amsterdam: Elsevier Scientific Publishing Co.

Ollier, C. D. 1979. *Weathering.* London: Longman.

O'Melia, C. R. 1987. Particle-particle interaction. In *Aquatic surface chemistry,* ed. W. Stumm, 385–403. New York: John Wiley & Sons.

Plante, R., Alcolado, P. M., Martinez-Iglesias, J. C., and Ibarzabal, D. 1989. Redox potential in water and sediments of the Gulf of Batabanó, Cuba. *Estuarine Coastal Shelf Sci.* 28:173–84.

Pope, G. A., Dorn, R. I., and Dixon, J. C. 1995. A new conceptual model for understanding geographical variations in weathering. *Ann. Assoc. Am. Geogr.* 85:38–64.

Rao, C. R. N., Iyenger, L., and Venkobachar, C. 1993. Sorption of copper(II) from aqueous phase by waste biomass. *J. Environ. Eng.* 119:369–77.

Rommets, I. W. 1988. The carbon dioxide system; its behaviour in decomposition process in east Indonesian basins. *Netherlands J. Sea. Res.* 22:383–93.

Singh, R. N. 1950. Reclamation of *usar* lands in India through blue-green algae. *Nature* 165:325–26.

Smith, B. J. 1994. Weathering processes and forms. In *Geomorphology of desert environments,* ed. A. D. Abrahams and A. J. Parsons, 39–63. London: Chapman & Hall.

Strahler, A. N. 1969. *Physical geography.* 3rd ed. New York: John Wiley & Sons.

Strakhov, N. M. 1967. *Principles of lithogenesis,* trans. J. P. Fitzimmons. Vol. 1. Edinburgh: Oliver and Boyd.

Sulzberger, B. 1990. Photoredox reactions at hydrous metal oxide surfaces: A surface coordination chemistry approach. In *Aquatic chemical kinetics,* ed. W. Stumm, 401–29. New York: John Wiley & Sons.

Summers, A. O., and Silver, S. 1978. Microbial transformations of metals. *Ann. Rev. Microbial.* 32:637–72.

Tharp, T. M. 1987. Conditions for crack propagation by frost wedging. *Bull. Geol. Soc. Am.* 99:94–102.

Thomas, M. F. 1974. *Tropical geomorphology.* Delhi: Macmillan.

Thomas, M. F. 1976. Criteria for the recognition of climatically induced variations in granite landforms. In *Geomorphology and climate,* ed. E. Derbyshire, 411–45. London: John Wiley & Sons.

Twidale, C. R. 1973. On the origin of sheet jointings. *Rock Mechanics* 5:163–87.

Twidale, C. R. 1981. Granitic inselbergs: Domed, block-strewn and castellated. *Geogr. J.* 147:54–71.

Twidale, C. R., Vidal Romani, J. R., and Campbell, E. M. 1993. A-tents from the granites, near Mt. Magnet, Western Australia. *Rev. Géomorph. Dynam.* 42:97–103.

Twidale, C. R., Vidal Romani, J. R., Coruña, A., Campbell, E. M., and Centeno, J. D. 1996. Sheet fractures: Response to erosional offloading or to tectonic stress. *Z. Geomorph.* 106:1–24.

Ueno, Y., Yamada, K., Yoshida, N., Maruyama, S., and Isozaki, Y. 2006. Evidence from fluid inclusions for microbial methanogenesis in the Early Archaean era. *Nature* 440:516–19.

Weist, R. C. 1973. *Handbook of chemistry and physics.* Boca Raton, FL: CRC Press.

Whitney, M. I., and Splettstoesser, J. F. 1982. Ventifacts and their formation: Darwin Mountains, Antarctica. In *Proceedings of the International Society of Soil Science: Aridic Soils and Geomorphic Processes*, Jerusalem, Israel, March 29–April 4. pp. 175–94.

Yaalon, D. H. (ed.) 1982. *Aridic soils and geomorphic processes.* Catena Supplement 1, 103–115.

4 Mass Movement

Mass movement refers to the activity of downslope movement of competent and incompetent earth materials alike when they become unstable under *stress*. This instability, called *slope failure*, can be affected in different degrees and types of movement that depend on certain aspects of geology, climate, topography, hydrology, and other environmental stress conditions, including human activities. Slope failure can be surficial or deep-seated along shear planes, and localized or regional in nature. It can likewise be imperceptibly slow to rapid, involving the transfer of microscopic fractions through to mega-sized detritus down the slope, and subsidence. The movement can similarly be independent of or assisted by moisture and air as mediums of transport in the deforming mass. Seismic and volcanic activities are other stress-producing events that directly impact on slope failure at local and regional scales. The stress of diverse human activities on the environment also variously manifests in many different forms of the instability of the earth materials on slopes. The mass movement activity thus can have many causes. It is also complex and diverse in nature. The mass movement term is often interchangeably used with *mass wasting*; however, that additionally connotes the failure of a unit mass of slope-forming material as in creep or slide.

CAUSES

Every slope experiences a gravity-controlled *shear stress*, the magnitude of which increases with slope inclination and unit weight of slope-forming material. In general, at-a-site slope instability depends on the balance of opposing forces of shear stress and shear resistance of the slope-forming earth materials. The slope becomes unstable and deforms when the shear stress due to external causes, internal causes, or both exceeds the *shear resistance* of the earth materials (Terzaghi, 1950; Hansen, 1984). *External controls* of shear stress refer to a variety of natural causes and anthropogenic effects as land cover and land use change that increase the magnitude of stress in the earth materials due to changes in slope geometry, moisture conditions, pore pressure, and loading and unloading of the earth materials in periods of seismic stress. *Internal controls* of shear stress like weathering, seepage erosion, and progressive failure are inherent to the nature of the earth materials. The external and internal variables of shear stress provide a framework for the classification of mass movement activity into types. However, possible permutations and combinations increasing the magnitude of shear stress on slope-forming earth materials are so numerous that the application of even a selected few among them yields complicated mass movement classifications of little practical use.

Shear resistance of the earth materials depends on the strength attributes of rocks and sediments discussed in Chapters 2 and 3. The *rock strength* depends on the type and arrangement of mineral constituents of the rock, fissure frequency in the body of rock, climatic regime, and depth placement of the body of rock relative to the earth's surface. The *sediment strength* varies with interlocking friction, cohesion among grains, pore pressure, and angle of internal friction in the sediment aggregate.

Dynamics of Mass Movement Activity

Mass movement is an extremely diverse and complex activity that, among others, involves downslope transfer of clastic fragments and surface sediments, plastic deformation of near-surface sediments, failure of cohesive sediments and rock masses along shear planes, and liquefaction of saturated silt and sand-sized particles. These movements occur at a certain threshold stress, producing the slowest form of creep failure through to the most rapid avalanche deformation. Most forms of slope instability develop by planar or translational and rotational failure at shear (or rupture) planes. The principles governing deformation of loose clastic fragments on slopes, and of the slope-forming clastic sediments, discussed below, also generally apply to the failure of slope-supporting rocks.

Isolated Rock Particles

An isolated rock particle at the surface is subject to driving and resisting forces respectively of *shear stress* down the slope and *normal stress* perpendicular to the slope incline and through the center of gravity of the particle mass (Figure 4.1(a)). For the particle mass, m, and slope inclination, θ, the shear stress is mg sin θ and normal stress mg cos θ. The particle becomes unstable when the magnitude of shear stress exceeds that of the normal stress, such as by the increase of slope inclination or of the particle mass. As slope inclination cannot be altered under normal circumstances, the movement occurs by the increase of particle mass.

Planar Failure of Cohesive Sediments

Substantially thick cohesive sediments of certain physical properties and disposition dislocate at a shallow depth along *planar surfaces of failure*, producing deformation roughly parallel to the general slope. Planar (or translational) failure is typical of those *frictional materials*, which rapidly gain shear strength with depth and overlie sediments of heterogeneous size composition, and of *cohesive sediments* overlying hard materials at depth (Hutchinson, 1968). The failure is initiated at a threshold shear stress that depends on the perpendicular force of the unit weight of sediment, x, above the potential failure surface at depth, d, from the surface (Figure 4.1(b)). This force, given as dx, resolves into components of shear stress and normal stress. The shear stress along the failure surface inclined at an angle θ to the horizontal is dx sin θ cos θ, and the normal stress is given as dx cos^2 θ. The cohesive sediments fail by planar dislocation when the component of shear stress just exceeds that of the normal stress at the shear plane.

Rotational Failure of Cohesive Sediments

Sufficiently thick cohesive sediments are particularly susceptible to dislocation and slump failure along *spoon-shaped rupture planes* that evolve at depth in the subsoil

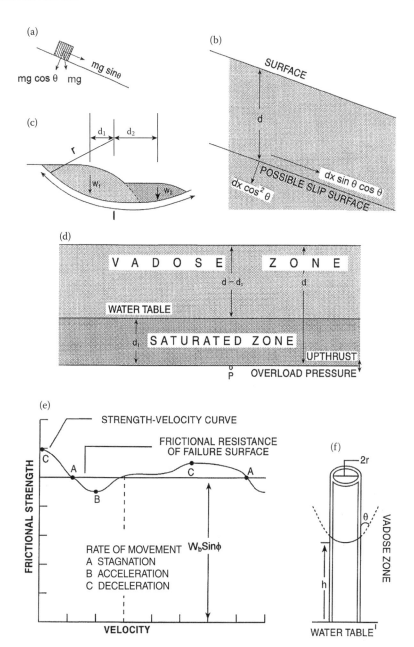

FIGURE 4.1 Analysis of slope instability. (a) Forces acting on a particle resting on a sloping surface. (b) Forces acting at a shallow planar surface of failure inclined parallel to the ground slope. (c) Forces acting on a spoon-shaped surface of failure in cohesive materials. (d) Upthrust of hydraulic pressure beneath water table. (e) Forces resulting from pore pressure and frictional strength of sediments in planar landslides. (f) Forces due to capillary tension in porous materials.

(Figure 4.1(c)). The gravity force causing rotation of the failed mass about the curvature of a shear plane is called moment. In slump failure, the *driving moment* is the product of the weight of sediment tending to produce failure and horizontal distance from the center of gravity of the mass. The *resisting moment* depends on the weight of the sediment resisting failure, the radius of curvature, and the length of the arc along the failure surface. The equilibrium forces for rotational failure are given as

$$W_1 d_1 = W_2 d_2 + s \, r \, l$$

$$\text{Driving moment} \quad \text{Resisting moment}$$

in which W_1 and W_2 are weights of the sediment promoting and resisting failure, d_1 and d_2 are respective distances from the center of gravity of the sediment tending to produce and resist deformation, s is the shear resistance of the material, r is the radius of arc along the rupture plane, and l is the length of arc. Rotational failure typically occurs along undercut banks and valleys trenched in cohesive sediments. The lithologic groups that lose shear strength when wet likewise yield to this form of failure.

Role of Pore Water in Slope Failure

With increasing soil moisture content, the cohesive sediments lose strength and deform as plastic and liquid substances (Chapter 2). In partly saturated soils, the pore water exerts an *upthrust force* that opposes the force of gravity holding the sediment aggregate onto the slope. The interstitial water is largely derived from the surface source of water. However, water table fluctuations in porous media also contribute significantly to the pore water content in mineral soils (Chapter 2). Hence, equilibrium forces for the stability of sediment aggregate lying parallel to the water table surface also require evaluation (Figure 4.1(d)). At a point P just below the water table, the overload pressure equals depth (d) times the unit weight of sediment (γ): as γd. The upthrust force equals $\gamma d (d - d_1) + (\gamma_s + \gamma_w) d_1$, where γ_s is the bulk density of sediment in the phreatic zone, γ_w is the unit weight of pore water, and $d - d_1$ is the thickness of the vadose zone. The sediment mass becomes unstable when the upthrust force exceeds the component of gravity force.

Pore water changes due to water table fluctuations weaken the *frictional resistance* of sediments. Therefore, critical temporal variations in the soil water content cause the cohesive material to pass through nonrigid and rigid states and deform at a variable rate (Figure 4.1(e)). The sediment aggregate is nonrigid at high and rigid at significantly low pore pressure in the system. In test conditions, the deformation rate varies with the strength of cohesive sediments and sliding velocity as (Davis et al., 1993)

$$s = c + W_b + \cos \varphi' \, [\tan \varphi' + f(v)]$$

in which s is the shear strength and c the cohesive strength of sediments, W_b is the effective weight of failed mass, φ' is the angle of internal friction, and $f(v)$ is the sliding velocity of deformed mass. By the above equation, the deformed mass becomes stagnant for equal magnitude of the shear strength and shear resistance of the material in soil-water systems, accelerates when the shear strength of the material weakens by the increase of pore pressure, and decelerates when the frictional resistance of the mass increases by the lowering of pore pressure in the system.

The adhesive attraction of pore water for dry sediment surfaces causes the water to rise through the soil column as *capillary water*. The capillary water generates a *hydraulic force* in partly saturated soil-water systems, the magnitude of which varies inversely with the diameter of the capillary tube (Figure 4.1(f)). The capillary water however, increases *cohesion* among grains and, thus, the effective weight of soils by an amount equal to T cos θ 2 π r. The capillary column also exerts a downward force that equals $\gamma_w \pi r^2 h$, and augments the force of gravity holding sediments in place. In these expressions, T is the surface tension, θ is the angle of meniscus with the capillary, π is the diameter and r the radius of the capillary tube, γ_w is the unit weight of soil and h is the height of capillary rise. The hydraulic force of capillary water thus has a moderating affect on the failure of cohesive sediments (Carson and Kirkby, 1972).

CLASSIFICATION OF MASS MOVEMENT ACTIVITY

Mass movements are a complex earth surface activity and are loosely called *landslides*. Landslides are characteristic of morphologic and dimensional variability, style and manner of movement, and the state of activity and manner of its distribution in the failed mass (Cruden et al., 1994). Hence, several descriptive and genetic classifications of mass movement phenomena have been proposed in literature. The descriptive classifications combine one or the other aspect of primary and secondary variables of significance to slope stability to differentiate and express the hierarchy of mass movement phenomena. The mass movement activity has also been explained in terms of internal processes of slope failure.

Most *descriptive schemes* project the rate of movement and coherence of the failed mass as primary variables, and the material size, moisture content, and associated rupture plane as secondary variables characterizing the type of slope instability. Sharpe (1938) pioneered a classification of mass movement phenomena. He proposed two major classes of flow and slide failures by the rate of movement, and recognized their subtypes by the amount of ice/water content and material composition in the failed mass. Bolt et al. (1975) also used the rate of movement to differentiate falls, slides, and flows, and distinguished their subclasses by the type of material and behavior of deformed mass. Záruba and Mencl (1969) considered the coherence and type of material as primary variables, and the nature of shear plane in the failed mass as a secondary variable of significance to various forms of mass movement activity.

Carson and Kirkby (1972) differentiated heave, slide, and flow phenomena by the *internal relationship* of stress or shear force and strain in the deformed mass. The shear force sharply drops to zero once dislocation occurs in slides, but remains uniformly distributed in the flow-type failures. Hence, shear planes are characteristic of slides and not of flows. Heave is diagnostic of *frost-susceptible soils*, which lose shear strength by the growth of segregated ice. These major forms of mass movement are further classified into seven varieties by the rate of movement and moisture content in the failed mass (Figure 4.2). In general, heaves are dry, slides require little moisture, and flows develop at moderate to high moisture content in cohesive earth materials.

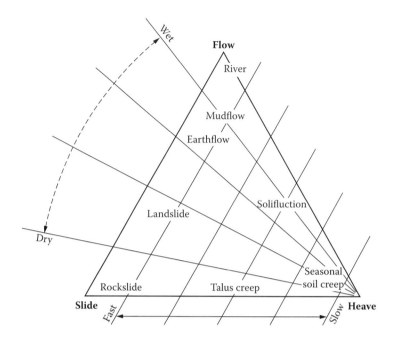

FIGURE 4.2   Classification of mass movement processes. (Source: Figure 5.2 in Carson M. A., and Kirkby, M. J., *Hillslope Form and Processes* [Cambridge, Massachusetts: Cambridge University Press, 1972]. With permission from Cambridge University Press.)

The *liquidity index*, discussed in passing in Chapter 2, provides yet another classification of slope instability. This index is a quantitative measure of the moisture content actually present in the failed mass to its liquid limit. The liquidity index is more than 1 for debris flow, mudflow, and earthflow activity and less than 1 for slump, earth slide, and mudslide failures (Figure 4.3). Hence, higher moisture content is required to initiate flow than slump and slide deformations (Carson, 1976).

Varnes (1978) recognized the type of movement as a distinguishing trait of free fall through to complex movements of the earth materials, in which subtypes of slope instability are distinguished by the material size (Table 4.1 and Figure 4.4). This classification additionally *quantifies* the rate of slope instability, providing meaning to an otherwise personal assessment of the deformation rate for each type of failure (Figure 4.5).

Slope instability places restrictions on land use decisions in many different ways (Crozier, 1984). The frequency and magnitude of slope failure affects the *suitability status* of land for certain proposed uses, and *landslide hazards* pose a risk to the population and economy of impact areas. Hence, several *purpose-oriented classifications*, quantifying the type and magnitude of slope instability and perceived risk, have been proposed. The application of a few such prognostic classifications is discussed in Chapter 12.

Most classifications recognize the rate of movement as a characteristic primary variable of slow through to rapid forms of slope failure. In the present context, therefore, mass movements are also recognized as slow and rapid forms of the earth

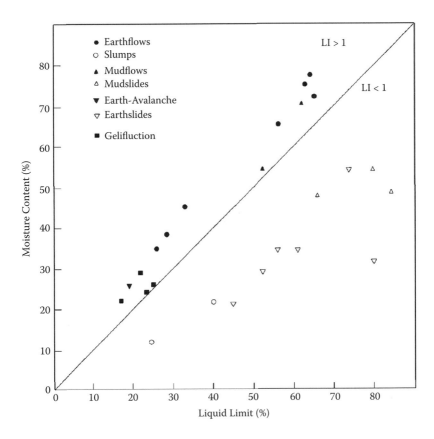

FIGURE 4.3 Types of mass movements in relation to liquidity index, based on diverse source of data. (Source: Figure 4.4 in Carson, M. A., "Mass-Wasting, Slope Development and Climate," in *Geomorphology and Climate*, ed. E. Derbyshire [London: John Wiley & Sons, 1976], 101–36. With permission from John Wiley & Sons.)

materials. The two major types of slope failure are distinguished further into sub-types by the material composition, moisture content, and presence or absence of shear planes in the failed mass.

SLOW MOVEMENTS

Creep is an imperceptibly slow movement of soil and rock fragments in diverse geo-morphic environments. The nature of creep as a surface or surface and subsurface activity, and as a continuous or discrete downslope displacement of soil and rock particles, is debated.

SOIL CREEP

Soil creep of *surface particles* varies with slope steepness, mechanical properties of the soil aggregate, temporal changes in soil temperature and moisture status, freeze-

TABLE 4.1
Varnes's Landslide Classification of 1978

Type of Movement	Type of Material		
		Engineering Soils	
	Bedrock	Predominantly Coarse	Predominantly Fine
Falls	Rockfall	Debris fall	Earth fall
Topples	Rock topple	Debris topple	Earth topple
Rotational	Rock slump	Debris slump	Earth slump
Few units	Rock block slide	Debris block slide	Earth block slide
Slides Translational			
Many units	Rock slide	Debris slide	Earth slide
Lateral spreads	Rock spread	Debris spread	Earth spread
Flows	Rock flow (Deep creep)	Debris flow (Soil creep)	Earth low
Complex	Combination of two or more principal types of movement		

Source: Figure 1.2 in Hansen, M. J., "Strategies for Classification of Landslides," in *Slope Instability*, ed. D. Brunsden and D. B. Prior (Chichester, UK: John Wiley & Sons, 1984), 101–36. With permission from John Wiley & Sons.

thaw rhythm, and other environmental stress conditions. Kirkby (1967) determined that the rate of soil creep varies directly with slope inclination as

$$S = A \, T \sin \theta$$

in which S is the rate of creep movement, A is a constant for soil types, T is the period of measurement, and θ is the slope inclination.

However, observations suggest that soil creep is also a subsurface activity through to the surface depth of 20 cm in Scotland (Kirkby, 1967), up to 1 m of the surface depth in Malaysia (Eyles and Ho, 1970), and to the depth of root zone in the rain forests of Puerto Rico (Lewis, 1974). Field data on soil creep from two sites in Puerto Rico suggest the creep rate of about 5 mm year[-1] near the surface and between 1.5 and 3.5 mm year[-1] at a depth of some 30 cm from the surface (Figure 4.6). The observed depth variation in the rate of soil creep possibly suggests *thermal expansion and contraction* of soil particles at the surface and *weakening of cohesion* among grains just below the root zone (Lewis, 1974).

Several theoretical models have been proposed to understand the dynamics of the surface creep of soil particles. They are based on the *rate process theory*, which predicts that new chemical products evolve when colliding atoms or molecules break the threshold of energy barrier in thermochemical reactions (Feda, 1989; Kuhn and

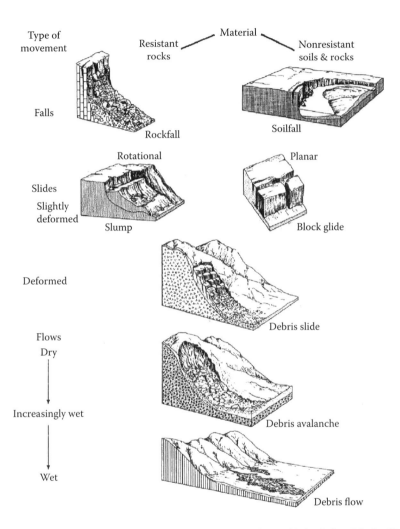

FIGURE 4.4 Common forms of landslides. (Source: Figure 23 in Selby, M. J., *Slopes and Slope Processes*, New Zealand Geographical Society Publication 1 [New Zealand Geographical Society, 1970]. With permission from New Zealand Geographical Society.)

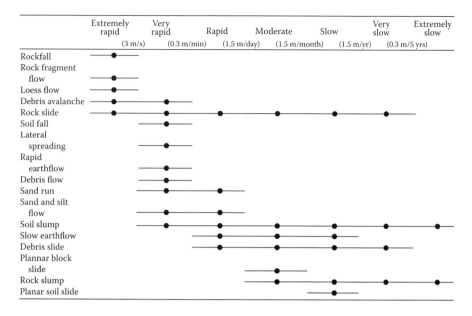

FIGURE 4.5   Approximate landslide movement rates after Varnes (1958). (Source: Table 4.7 in Crozier, M. J., "Field Assessment of Slope Instability. Landslips in Sensitive Clays," in *Slope Instability*, ed. D. Brunsden and D. B. Prior [Chichester, UK: John Wiley & Sons, 1984], 103–45. With permission from John Wiley & Sons.)

Mitchell, 1993). This energy barrier is called *activation energy* and very nearly depends on the ambient temperature and mass of the chemical systems.

Kuhn and Mitchell (1993) predict that creep failure results from *interparticle sliding* of soil grains held together by the force of *interlocking friction*. The particles become unstable and creep at a certain threshold of *thermal activation energy* breaching the barrier of *friction force* at the contact of grains, such that the activation energy and creep rate depend directly on the air temperature amplitude. The friction force varies with per particle ratio of normal and shear stress at the contact of grains (see Figure 2.14). The simulated rates of soil creep given by the rate process theory (Kuhn and Mitchell, 1993) are in general agreement with the observed creep rates that vary from less than 1 mm to about 2.5 cm year^{-1} in different geomorphic environments (Kirkby, 1967; Lewis, 1974).

Soils are a system of solid, air, and water phases, wherein the unstable array of grains collapses under the stress of applied load into adjacent voids, producing primary, secondary, and tertiary forms of creep (Figure 4.7). *Primary creep* results from initial sliding and rolling of particles at the contact of grains. With further loading of the system, the grains arrange in the direction of movement, become strain-free, and slide at a higher rate of deformation called *secondary creep*. At still higher magnitudes of the applied load, however, the interlocking friction increases and arrests the rate of creep deformation due to case hardening. This is the phenomenon of *tertiary creep*. The compressed aggregate ruptures thereafter at a higher magnitude of applied load. The experimental friction data also suggest that the primary

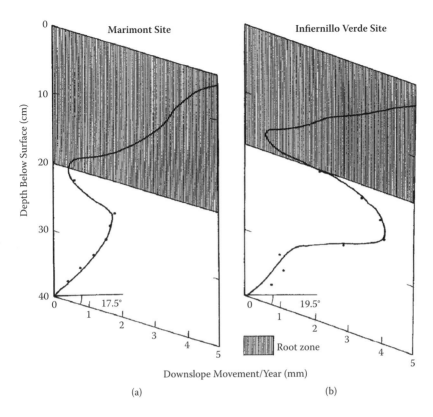

FIGURE 4.6 Velocity distribution of soil movement by creep for Marimont site (a) and Infiernillo Verle site (b) at El Yunque National Forest, Puerto Rico. (Source: Figure 1 in Lewis, L. A., "Slow Movement of Earth under Tropical Rain Forest Conditions," *Geology* 1 [1974]: 9–10. With permission from Geological Society of America.)

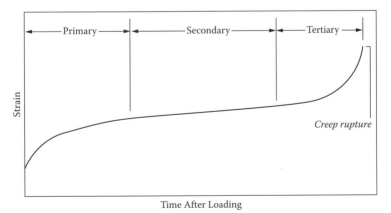

FIGURE 4.7 Forms of creep: primary, secondary and tertiary. (Source: Figure 1 in Kuhn, M. R., and Mitchell, J. K., "New Perspectives on Soil Creep," *J. Geotech. Eng.* 119 [1993]: 507–24. With permission from American Society of Civil Engineers.)

and secondary forms of creep deformation are governed by the magnitude of friction force at the contact of grains (Kuhn and Mitchell, 1993).

ROCK CREEP

Rock creep refers to the activity of surface creep of rock fragments. This form of movement is related to the environmental stress of thermal expansion and contraction of clastic fragments on slopes. Jointed rocks and massive rocks affected by sheet fractures, which invariably produce abundant loose rock fragments, are ideal sites for this form of mass movement activity.

RAPID MOVEMENTS

Flows, slides, falls, subsidence, and avalanche are rapid movements of slope failure. The flow failures are intergranular deformation of cohesive sediments and clastic fragments; slides are caused by the failure of rock, clastic debris, and cohesive sediments along shear planes; falls describe the free fall of a mass of rock or sediments from joints behind cliff faces; local-scale subsidence occurs by the liquefaction of wet silt and sand-sized sediments; and avalanches accompany rapid disintegration and deformation of rocks and debris along shear planes.

FLOWS

Flow-type failures comprise a continuously deforming mass of cohesive sediments and coarse rock fragments over steep slopes of an intact material. They owe the mobility to high moisture content in the failed mass. The flows are classified by the particle size and water content into types of solifluction, debris flow, earthflow, and mudflow failures.

Solifluction

Solifluction is the slowest type of flow failure of *saturated* cohesive sediments in almost all topographic and climatic provinces. This form of failure in the *active layer* of permafrost is called *gelifluction*. The dynamics of moisture migration to the active layer and morphologic aspects of gelifluction are discussed in Chapter 8. In the present, repetitive flow of excessively moist cohesive sediments evolves overriding solifluction/gelefluction lobes even on gentle slopes. Solifluction lobes are as well developed in the soils of the *seasonally frozen* environment of the Central Himalayas (Figure 4.8) as they are in the mineral soils of the *perennially frozen* subarctic environment.

Debris Flow, Earthflow, and Mudflow

Debris flow, earth flow, and mudflow activities are *planar failure* of the earth materials of certain grain size and moisture content in the failed mass. Debris flows are largely the deformation of large-sized rock fragments with little moisture in the failed mass. By comparison, earthflow and mudflow activities comprise progressively finer fractions and higher moisture content in deformed cohesive sediments.

FIGURE 4.8 Solifluction on a grass-covered sticky clay surface of 24° slope near Keylong in the Lahul area of the northwestern Himalayas, India.

Debris Flow

The debris flow is a highly destructive geomorphic activity throughout the world but excels in arid regions and mountainous environments. The failure is triggered by heavy rain and rapid increase of *pore pressure* in a mass of fragmentary debris over slopes. Hence, the incidence of debris flows generally coincides with the rainy season in France (Van Steijn, 1991) and China (Baoping, 1994), spring snowmelt in Yukon, Canada (Johnson, 1984a) and the Karakoram Mountains of the northwestern Himalayas (Owen, 1991), and with the rain and snowmelt periods in the mountainous terrain of central Japan (Marui, 1994). Debris flows also result from other causes. The burst in the early 1970s of a moraine-dammed Klattsine Creek, British Columbia, Canada, released a massive volume of sand and gravel downstream. This deposit is identified as a debris flow.

Debris flows move rapidly through channels, but more aggressive failures among them overtop channel banks and spread laterally outward. One opinion holds that high mobility of the failed mass obtains from excessive pore pressure in the fragmentary mass that keeps the debris afloat during transport (Pierson, 1981; Johnson and Rodine, 1984). The other view, however, doubts the existence of sufficiently high pore pressure in a continuously deforming mass, and predicts the role of viscous and inertial forces in providing high mobility to the debris flow activity (Sassa, 1985; Davies, 1988; Mainali and Rajaratnam, 1994). The *viscous flow model* predicts that the shear stress (τ) in a deforming mass, largely of finer rock fragments, varies with the yield stress due to grain movement and viscous forces due to the interstitial water as (Mainali and Rajaratnam, 1994)

$$\tau = \tau_y + \mu \times d_u/d_y$$

where τ_y is the yield stress due to grain movement, μ is the dynamic viscosity of the fluid phase, and d_u/d_y is the vertical velocity gradient in the deforming mass.

The term $\mu \times d_u/d_y$ then describes the viscous force due to the interstitial water. The *granular flow model* evaluates the relative significance of inertial and viscous forces in the deformation behavior of debris flows of large-sized rock fragments by a non-dimensional number, N, as (Mainali and Rajaratnam, 1994)

$$N = (\lambda^{1/2} \, \rho_s \, D^2)/(\mu \times d_u/d_y)$$

where λ is the linear concentration of particles in the flow, ρ_s is the particle density, D is the particle diameter, and the term $\mu \times d_u/d_y$ refers to viscous forces of the fluid phase. In this model, inertial forces dominate the flow for $N > 450$ and viscous forces for $N < 40$.

Earthflow

The earthflow activity is common to *quick clays*. These clays rapidly lose consistency and shear strength when wet, and *retrogressively fail* at the lower slope segment. The instability extends slice by slice upslope to more than ten times the initial slope height, producing a pear-shaped area of the failed surface (Mitchell and Markell, 1974; Bentley and Smalley, 1984). The *shear strength* of clays, however, not only depends on the moisture content, but also varies directly with the inclination of failure surface (Su and Liau, 1999). Therefore, the slope steepness also affects the earthflow activity. Observations suggest that the *montmorillonite clay* of Medeira Island off the coast of Portugal fails in the rainy season, deforming at a rate proportional to the amount of rainfall (Figure 4.9). The earthflow activity is widespread in the St. Lawrence lowlands of eastern Canada. A 1991 large retrogressive failure in the montmorillonite Leda clay had caused fairly extensive loss of life and property in St. Jean-Vianney, Quebec. The *pyroclastic clay* in a part of seasonally cold mountainous central Japan, however, similarly deforms by the *upthrust force* of segregated ice (Nozaki et al., 1994).

Mudflows

Mudflows are initiated by the deformation of saturated sediments in small-sized catchments. Hence, they are typically channel-bound *viscous failure* of fine-grained sediments with up to 50% moisture content in the failed mass. Mudflows are a global phenomenon, but the source of moisture required for the activity varies with the geomorphic environment. The mudflow activity in *periglacial regions* coincides with the melting of ice in ice-cored moraines (Johnson, 1984a), in *arid regions* with the thunderstorm activity, and in *temperate regions* with the thaw of spring snow. In southern England, mudflows occur in the overconsolidated clay weakened by summer drying and shrinkage, which produces clay of weak strength. Mudflows of tropical Madagascar, called *lavakas*, are the result of extensive deforestation of hillslopes in the last century (Helfert and Wood, 1986). Mudflows involving volcanic ash and other ejecta on the slopes of active and dormant volcanoes are called *lahars*. They are initiated when the material becomes unstable by saturation from rain. Lahars are common to Japan, the Philippines, Indonesia, Costa Rica, and other countries where active and dormant volcanoes exist.

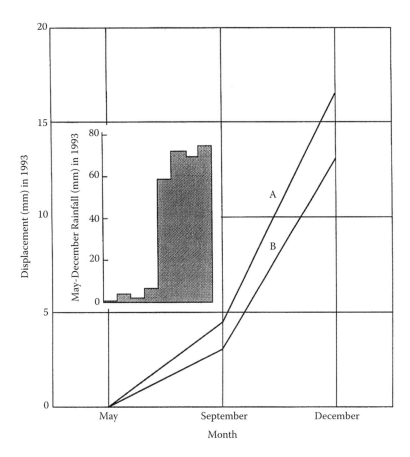

FIGURE 4.9   Relationship between the mean monthly rainfall and earthflow activity at Machico, Medeira Island. (Source: Figure 8 in Rodrigues, D. D. M., and Ayala-Carcedo, F. J., "Landslides in Machico Area on Madeira Island," in *Proceedings of the 7th International Association of English Geologist*, Lisboa, Portugal, September 5–9, 1994, Vol. 3, pp. 1495–500. With permission from A. A. Balkema Publishers.)

Mudflows move at a variable rate that depends on the moisture content, composition and size of the failed mass, and the channel gradient. Leopold et al. (1964) observed that a shallow mudflow at Paricutín in Mexico had moved at the rate of 61 cm s^{-1} on a 2° slope, and large mudflows of the recent past in the mountains of southern California traveled at a maximum speed of 4 m s^{-1} through 6° steep channels.

FALLS

Free fall of rock fragments develops from the disintegration and abrupt collapse of rocks along sufficiently deep and wide surface joints at cliff faces (Carson and Kirkby, 1972; Whalley, 1984). This activity is common to the *alpine environment*, where it is initiated by specific geologic conditions and several other causes that widen joints and produce rock fragments. Jointed rocks aligned to the slope near

Khuni Nala site on Jammu-Srinagar Highway 1A in Kashmir Himalayas, and along the surface transport network in Nilgiri Hills in peninsular India, are particularly prone to the free-fall failure of rock fragments (Krishnaswamy, 1980). Rockfalls are similarly a perpetual problem along the transport routes through the mountainous terrain in Canada and eastern California. Widespread rockfall activity in Yukon Mountains, Canada, is the result of rocks weakened by the passage of Pleistocene ice (Johnson, 1984b), and the one in Karakoram Mountains of northwest Himalayas is related to extensively sheared, jointed, and frost-shattered rocks (Owen, 1991). A high-magnitude rockfall activity of 1998, which wiped out a population of over six hundred in a part of the Central Himalayan region of the northwest Himalayas, was possibly caused either by the instability of extensively sheared and intensely shattered rocks along faults due to the *neotectonic movements* (Valdia, 1998) or by the saturation of debris due to the effect of a cloudburst of the day (Pant and Luirei, 1999).

Rockfall activity commences at enlarged joints in rocks. Joints enlarge by several processes, of which *frost shattering* is the most potent process of deepening and widening of joints in rocks. Thus, the subarctic region with the largest number of *freeze-thaw cycles* (Chapter 8) is uniquely placed for the highest incidence of rockfall activity. The annual peak of rockfall activity in Norway (Hutchinson, 1968) follows the period of freeze-thaw and water pressure increase in rock joints. Bjerrum and Jørstad (1968) similarly observed that the highest frequency of rockfall activity in the *chalk area* of Kent, England, coincides with the duration of air frost and moisture surplus periods (Figure 4.10). The moisture in frost-loosened joints is known to weaken the rock strength further (Hutchinson, 1971; Brunsden, 1980).

SLIDES

The failed mass in slides breaks in few or many units, and travels by planar or rotational movement at shear planes and beyond. *Planar slides* are usually confined to rock masses, and *rotational slides* commonly occur in unconsolidated sediments. Slides are blockglides, slumps, debris slides, liquefaction, and rock avalanches by the material composition, deformation process, dislocation depth, and nature of rupture planes.

Blockglide

Blockglide or slab failure is the slippage of a part of massive rock along a bedding separation. This form of failure commonly occurs at the interface of strata of contrasting lithologic composition. The failure is initiated at a vertical tension crack in the top stratum that subsequently enlarges to a joint of certain critical depth behind the cliff face and generates sufficient downslope *lateral stress* in the rock. Consequently, a portion of the top stratum between the joint and cliff face topples at the bedding separation. Carson and Kirkby (1972) predict that the crack depth for failure is a function of the shear resistance of cliff-forming material and cliff height above the bedding plane separation as

$$Z = 2\ c/\gamma \tan (45 + \varphi/2) - H'_c$$

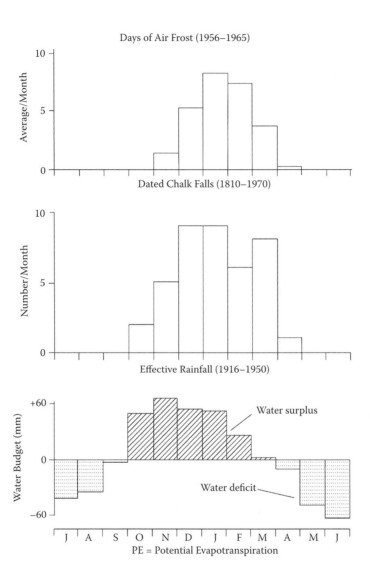

FIGURE 4.10   The relationship between rockfall, days of air frost, and effective rainfall for chalk cliffs on the Kent coast. (Source: Figure 5.14 in Brunsden, D., "Mass Movement," in *Process in Geomorphology*, ed. C. Embleton and J. Thornes [New Delhi: Arnold-Heinemann, 1980], 130–86. With permission from Edward Arnold Ltd.)

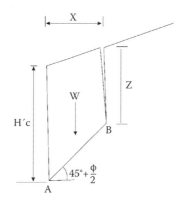

FIGURE 4.11 Stability analysis of a slope subject to slab failure. AB represents the potential sliding surface; its length is $x/(\cos 45 + \varphi/2)$; weight of block is $W = 1/2x\ (H'_c + Z)\gamma$; downslope component of weight of block = $W \sin (45 + \varphi/2)$; component of weight normal to AB = $W \cos (45 + \varphi/2)$. (Source: Figure 6.4 in Carson M. A., and Kirkby, M. J., *Hillslope Form and Processes* [Cambridge: Cambridge University Press, 1972]. With permission from Cambridge University Press.)

in which Z is the crack depth at failure, c, γ, and φ are respectively the intergranular cohesion, unit weight of sediment, and angle of internal friction, and H'_c is the cliff height above bedding separation (Figure 4.11). Slab failure thus depends on the *residual strength* of materials at a threshold depth of the tension crack in rocks.

Slab failure of the Threatening Rock in Chaco Canyon National Monument, in the United States, is perhaps the only documented slope instability of this type, providing the cause, period of initiation, and rate of deformation for the jointed porous sandstone overlying a shale bed. The custodian of the monument had regularly monitored the monthly displacement of rock beginning in November 1935 until it finally toppled over the substratum in January 1941. The failure coincided with the widening and deepening of a joint immediately behind the sandstone cliff to 4 m at the surface and 1 m into the rock. Incidentally, the instability of rock was also recognized by native Indians, who had wedged pine logs at the base of sandstone to keep it from moving. These logs are between 1004 and 1057 ^{14}C years old (Schumm and Chorley, 1964). Analysis of the movement data suggests that the sandstone block first began to detach from the main body 2300 years ago, and continued to deform at an exponential rate till failure (Figure 4.12). The seasonal displacement pattern of the blockglide further suggests that most of the deformation of the "braced-up rock" occurred in the wet winter period of higher moisture content in the substratum.

Slump Failure

Cohesive sediments slump by *rotational failure* along shallow and deep-seated shear planes. *Shallow slumps* develop adjacent to streams that either undercut banks or rapidly deepen valleys, and *deep-seated slumps* evolve along streams that trench to a critical depth in cohesive sediments (Carson and Kirkby, 1972; Lambe and Whitman, 1973).

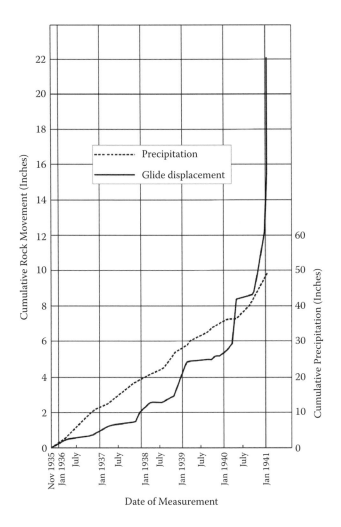

FIGURE 4.12   Cumulative movement of Threatening Rock (full line) and cumulative pre-cipitation (broken line) plotted against date of measurement or time. (Source: Figure 2 in Schumm, S. A., and Chorley, R. J., "The Fall of Threatening Rock," *Am. J. Sci.* 262 [1964]: 1041–54. With permission from Yale University Press.)

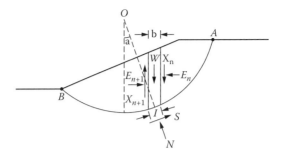

FIGURE 4.13   Bishop's method for the stability analysis of deep-seated slumps. E_n, E_{n+1}, resultants of the total horizontal forces between the slices; X_n, X_{n+1}, resultants of the total vertical forces between the slices. (Source: Figure 7.12 in Carson M. A., and Kirkby, M. J., *Hillslope Form and Processes* [Cambridge: Cambridge University Press, 1972]. With permission from Cambridge University Press.)

Shallow Slumps

Shallow slump failures develop along *curved tension cracks* of highest lateral shear in moist cohesive sediments on steep slopes. These cracks subsequently open spoon-shaped shear planes along which the material deforms, initiating successive slump failures riding one over the other and evolving a hummocky topography downslope. The Nashiri slide on Jammu-Srinagar Highway 1A in Kashmir Himalayas is a shallow slump in sandstone and shale lithology of poor shear strength when wet (Krishnaswamy, 1980). The dynamics of rotational failure was discussed earlier in this chapter.

Deep-Seated Slumps

Deep-seated rotational failures are initiated at a certain threshold of shear stress exceeding the strength of cohesive materials. The stress and resisting forces for deep-seated slumps are analyzed by Bishop's 1955 graphical method of slices, wherein the failed mass at the depth of a shear plane is segmented into individual slices of known width and weight as to perfectly fit a part of the planar shape (Figure 4.13). In this method, the resisting horizontal forces (E_n, E_n + 1) and the driving vertical forces acting through the mass (X_n, X_n + 1) are analyzed for each slice and summed for the entire slump failure (Bishop, 1955; Carson and Kirkby, 1972; Lambe and Whitman, 1973). The *shear strength* of sediment mass is given as

$$s = \sum_{B}^{A} [(c'1(W / \cos a - u1)\tan\phi'\} / (1 + \tan a \tan\phi' / F_s\}]$$

in which s is the shear strength of cohesive sediment, c′ is the effective stress, l is the length of slice, W is the weight of slice, u is the pore pressure, a is the angle of arc from the center of the circle, φ′ is the apparent angle of internal friction, and F_s is the term for shear strength not available for resisting the weight of slice. The *driving force* is similarly given as

$$T = \sum_{B}^{A} W \sin a$$

The ratio between the shear strength and shear stress is called *safety factor.* Bishop (1955) determined the safety factors for two materials studied to be 1.14 and 1.84.

Rockslide, Debris Slide, Liquefaction

Rockslides and debris slides are *shallow planar dislocations* to a depth generally less than one-tenth of the distance moved by the failed mass (Hutchinson, 1968). Hence, the failed mass does not travel far from the source of origin. Liquefaction is a localized phenomenon of *lateral spread* of moist silt and sand particles beneath cohesive sediments.

Rockslides

Rockslides are generally fairly rapid shallow planar dislocations of rock fragments detached along joints, bedding plane separations, and faults by natural and cultural processes. A rockslide in Madeira Island, 900 km off the coast of Portugal in Atlantic Ocean, was initiated by the development of a *wave-cut notch* at the foot of a *sea cliff* (Rodrigues and Ayala-Carcedo, 1994). Rockslides in the Karakoram Mountains of the northwestern Himalayas reoccur in intensely sheared and fractured rocks that suddenly fail even at a moderate *seismic stress* and deform rather rapidly due to the presence of lubricant *kaolinite* in the fragmentary mass (Owen, 1991). A 1963 massive rockslide that caused the failure of Vaiont Reservoir in the Italian Alps was initiated by the *pore pressure increase* in limestone rocks (Kiersch, 1964). The Goldau rockslide of 1806 in the Swiss Alps occurred on a bedding plane in stratified sandstone roughly parallel to the valley side, and moved some $15 \times 10^6 \, \mathrm{m}^3$ of the rock downslope (Terzaghi, 1950).

Cultural processes also initiate rockslides. Cut excavation and fill sequence for a surface transport network in mountainous terrains induces slope instability, of which rockslide failure is one. Shaft mining generates instability in surface rocks, which similarly deform as shallow planar failure of the material (Cunningham, 1988). Porous rocks adjacent to large dams in tectonically active Himalayan terrain lose strength by the increase of pore pressure and occasionally fail in the form of rockslides (Krishnaswamy, 1980).

Debris Slides

Debris slides typically occur on 15 to 40° steep slopes, in which the failed material behaves as a more or less cohesionless mass. Extremely rapid debris slides often result from sudden heavy rainfall, particularly in the tropics (Hutchinson, 1968). Also, sufficiently thick surface debris, which rapidly lose *cohesive strength* by the increase of pore pressure, fail as debris slides (Carson and Kirkby, 1972). The *weathering residue* on Nilgiri Hills in the tropical climate of southern India is also prone to this type of deformation (Krishnaswamy, 1980).

Liquefaction

Liquefaction is the process of transformation of saturated granular material from a solid to liquid state by the effect of intense ground shaking in periods of earthquake activity (Keller, 1979). The loading and unloading cycles of earthquake events generate enormous stress in sufficiently thick local bodies of saturated silts and sands underlying compact strata. This stress increases the pore pressure in the system, and decreases the sediment volume by rearranging grains. The increase of pore pressure in the system accompanies complete loss of the shear strength of materials by which they fail by lateral spread even on gentle to nearly flat surfaces, producing cracks and fissures in the top stratum and differential settling of the ground. The Bhuj, India, earthquake of January 26, 2001, similarly generated an enormous stress in the substratum, causing liquefaction of sand and silt sediments (Rydelek and Tuttle, 2004). The cyclic stress of loading and unloading cycles also reduces the volume of substratum sediments which produces *ground subsidence*. Observations suggest that the ground locally subsided 0.6 m by the compaction of sediments in the 1906 San Francisco earthquake, and by 1.8 m during the 1964 Alaska earthquake (Goudie, 1993). Liquefaction, in general, produces three types of failure: lateral spread on gentle to moderate slopes, sinking of the ground, and flow slides (or earthflows) on moderate slopes. The St. Lawrence lowlands experience flow slides because of the presence of noncohesive Leda clay. A 1993 massive flow slide at Lemieux, Ontario, had moved into and blocked the South Nation River for 3 days.

Studies on the dynamics of lateral spread suggest that the *cyclic stress* of loading and unloading cycles rearranges silt and sand-sized particles, expels pore water, and increases pore pressure in the substratum by which silt and sand-sized particles pass in suspension and behave like a *viscous fluid*. The impermeable top stratum, however, interferes with the escape of pore water, which accumulates either as a film of water or as a *water gap* at the interface of strata. The water gap generates an *uplift force* that sets the overlying mass of sediment afloat, leading to zero stress and severely weakened sediment strength (Figueroa et al., 1994; Fiegel and Kutter, 1994; Kokusho, 1999). Hence, the top stratum falls through the interface and disintegrates laterally outward from the thicker portion of the failed mass (Figure 4.14), displacing pore water vertically as a relatively high velocity flow that even cracks open the surface at its thinner segments (Figure 4.15). Thus, liquefaction is also regarded a localized phenomenon of unsteady flow of sand and pore fluid (Spence and Guymer, 1997).

The disintegration and lateral displacement of liquefied mass varies with the magnitude of cyclic stress and thickness of saturated body of silt and sand fractions. Studies from the United States and Japan suggest that this displacement varies directly with the thickness of the liquefied layer and surface slope as (Bartlett and Youd, 1995)

$$D_h = 0.75 \ T^{0.50} \ \theta^{0.33}$$

in which D_h is lateral displacement in meters, T is the thickness of liquefied layer, and θ is the slope steepness in percent.

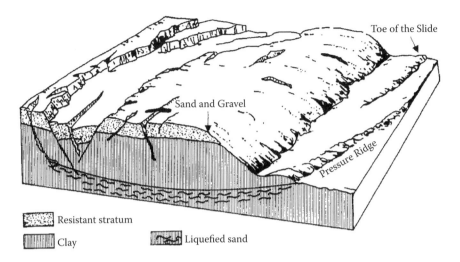

Resistant stratum

Clay   Liquefied sand

FIGURE 4.14   A failure by lateral spread of resistant material on liquefied clay. (Source: Figure 26 in Selby, M. J., *Slopes and Slope Processes*, New Zealand Geographical Society Publication 1 [New Zealand Geographical Society, 1970]. With permission from New Zealand Geographical Society.)

AVALANCHES

Avalanches are catastrophic failure of a large mass of debris and of bedrock disintegration products along shear planes. *Debris avalanches* are a shallow failure of moist clastic fragments and volcanic ejecta on preexisting slide tracts on steep slopes. The failure is triggered by earthquake shocks, violent volcanic activity, or increase of moisture content in debris during or following a period of high-intensity rainfall. In general, debris avalanches comprise of some 80 to 85% of debris content and 15 to 20% of moisture content by weight. Debris avalanches develop an arcuate head on steep slopes and leave a single or multiple overriding lobes of failed mass at the base of the chute. Glacial deposits in the Vancouver area of western Canada are prone to this form of failure during autumn and winter rains (Eisbacher and Clague, 1981). The Nilgiri Hills in the tropical rain forest climate of south India are also beset with the problem of debris avalanches of two events per square kilometer (Krishnaswamy, 1980). Depressurization of active volcanoes prior to eruption causes *flank failure* and generates fragmentary products (Voight and Elsworth, 1997). The magmatic debris deform by excess pore pressure that also provides rapid mobility to the failed mass down the chute slope (Crandell et al., 1984; Frances et al., 1985; Voight and Elsworth, 1997). Nearly 300 ka to 360 ka old debris avalanche lower down the Mount Shasta volcano in California is the largest known Quaternary landslide on earth (Crandell et al., 1984). This failure was possibly caused by an earthquake or volcanic activity with moisture of an external or internal source. The debris avalanche on Socompa volcano in Chile, however, was triggered by a volcanic eruption accompanied by a violent lateral blast 10000 to 500 years ago (Francis et al., 1985).

Most rock avalanches in landslide-prone central southern Alps of New Zealand are triggered by large earthquakes and only a few by storms (Whitehouse and

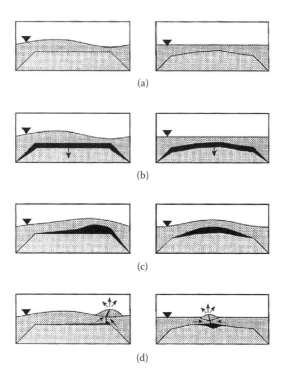

FIGURE 4.15 Mechanism of liquefaction for layered soil deposits where less permeable overburden overlies liquefiable soil: (a) initial profile, (b) shaking and liquefaction, (c) bulging, and (d) rupture. (Source: Figure 1 in Fiegel, G. L., and Kutter, B. L., "Liquefaction Mechanism for Layered Soils," *J. Geotech. Eng.* 120 [1994]: 737–55. With permission from American Society of Civil Engineers.)

Griffiths, 1983). *Rock avalanches* are highly mobile failure of bedrock fragments along deep-seated shear planes. Terzaghi (1962) theorized the mechanism for the splitting of massive rock bodies, and consequent production of a large volume of rock fragments from the intact rock. He concluded that the failure is determined not by the strength of rocks but by the pattern of continuous and discontinuous joints in the body of rock. Deep-seated rock avalanches develop at the boundary of continuous and discontinuous joints in rocks.

Theoretically, rocks tend to shear along joints, but this tendency is held in check by effective cohesion and stress-related shearing resistance within portions of intact rock surrounding joints. *Effective cohesion* is the cohesion of intact rock times the effective joint area, where the *effective joint area* is the ratio of the area in joints to the total area of the section. Changes in slope geometry, increase of pore pressure in the system, or opening of sheet structures in the rock mass increases the effective joint area. This activity opens fresh joints in the intact rock, splits the rock by progressive failure, and produces a densely packed cohesionless aggregate of angular blocks on slopes. Eventually, a wedge-shaped shear plane inclined between 30 and 90° is established between the intact and split rock (Figure 4.16).

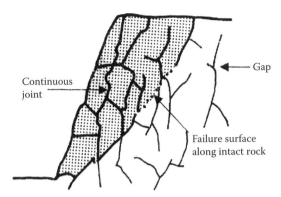

FIGURE 4.16   The concept of effective joint area and the development of a failure surface along the intact rock. (Source: Figure 6.7 in Carson M. A., and Kirkby, M. J., *Hillslope Form and Processes* [Cambridge: Cambridge University Press, 1972]. With permission from Cambridge University Press.)

Rock avalanches comprise a highly porous and thick mass of shattered debris, which pulverizes during movement and travels far from the source area even without the presence of moisture in the failed mass (Plafker and Erickson, 1978). This form of failure is highly mobile, gaining speeds in excess of 150 km/h even on relatively low gradients. The high mobility of rock avalanches is attributed to slide, flow, momentum transfer, and acoustic fluidization mechanisms.

The *sliding mechanism* (Shreve, 1968) envisages air lubrication of the deforming mass over its bed. It is held that the air is compressed and pressure-heated beneath a freely-falling mass of rockfalls. The failed mass riding over the cushion of this heated buoyant air travels at high speeds without much frictional contact with the ground. It ceases movement when the trapped air escapes the rubble. The *flow mechanism* (Hsu, 1975) attributes rapid deformation of rock avalanches to the interstitial finer debris and rock dust particles dispersed within voids of larger rock fragments. These sediments reduce the effective normal pressure of the failed mass on the surface and, thereby, the frictional contact with the ground. In flow failure, the deformation occurs by cohesionless grain flow in which stress is transferred by grains through collision. The *flow mechanism* of Kent (1966) is based on the precept of fluidization of the detached mass of rock from a cliff or a steep slope. This form of flow deformation accompanies collapse of dry clastic fragments. The debris fluidizes when its interstitial air reduces the internal friction between grains, and hold the deforming mass slightly above the ground. The fluidization provides high mobility to the flowing or streaming body of highly fragmented debris. The *momentum transfer mechanism* of Eisbacher (1979) is based on the distribution pattern of clastic fragments, such that the largest broken slabs and nearly two-thirds of the failed mass lie at the proximal rock ramps and the remaining mass becomes progressively fragmentary and finer toward the distal end of the avalanche. The overall distribution of deformed mass by size class suggests that clastic fragments filter through open spaces of incoherent mass by kinetic sieving or the momentum transfer mechanism. Melosh (Strom, 1994) explains the crushed and stratified basal debris in the rock avalanches of Pamir and

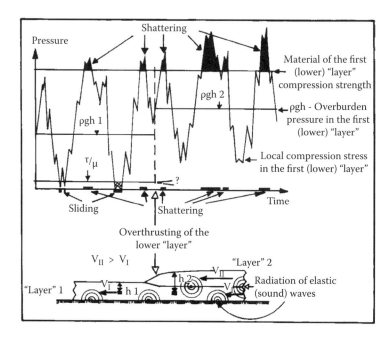

FIGURE 4.17 Acoustic fluidization model for rock avalanches. The debris in the lower layer of the deformed mass shatter by the effect of local compressive stress in the material due to pressure of sound vibrations. Rock fragments shatter and pulverize when pressure fluctuation exceeds the overburden pressure, and slide where local pressure is below the ratio between shear stress (τ) and coefficient of friction (μ). (Source: Figure 6 in Strom, A. L., "Mechanism of Stratification and Abnormal Crushing of Rockslide Deposits," in *Proceedings of the 7th International Association of English Geologist*, Lisboa, Portugal, September 5–9, 1994, Vol. 3, pp. 1287–95. With permission from A. A. Balkema Publishers.)

Tian Shan mountains of Central Asia by the *acoustic fluidization mechanism*. The model proposes that the deforming mass generates pockets of high-frequency sound vibrations, which produce intense local stress to pulverize even large-sized rock fragments. The extent of pulverization, manner of stratification, and movement of rock fragments depend upon the energy of sound vibrations at the base of a deforming mass and dissipation of energy through the failure surface (Figure 4.17). The model predicts that shear stress (τ) and the coefficient of friction (μ) determine the behavior of rock avalanches, such that clastic fragments shatter at local stress exceeding τ/μ, and the mass slides over the bed at local stress less than τ/μ.

SUMMARY

Mass movement or slope failure is caused by the instability of slope-forming earth materials due to causes internal or external, or both, to them. Rocks and sediments fail when the magnitude of shear stress due to changes in slope geometry, loading and unloading of the earth materials in periods of seismic stress, and weathering activity exceeds the shear resistance of the earth materials. The shear resistance of rocks is governed by the mineral composition, fissure frequency, and disposition relative to the earth's surface, and of sediments by the interlocking friction, cohesion, pore pressure, and angle of internal friction among grains.

Mass movements exhibit a great variety of processes and morphologic forms. Hence, their comprehensive classification is hardly possible. Mass movements are generally classified by the dynamics and rate of movement as slow to rapid failures, and subsidence. Their subtypes are recognized by the type of material and other attributes.

The creep of soil and rock fragments is the slowest form of mass movement activity. Soil creep is a surface phenomenon but has also been observed to a depth of up to 1 m from the surface. Surface creep is probably a manifestation of thermal stress exceeding the friction force at the contact of individual soil grains, and near surface creep follows from the weakening of cohesion among grains.

Rapid mass movements are planar and rotational in style. They are classified by the dynamics of failure into flow, slide, fall, and avalanche phenomena. Subsidence is a type of failure in which saturated silt and sand grains pass into a liquid state by the effect of cyclic seismic stress. Solifluction, debris flow, earthflow, and mudflow are major forms of the flow-type failure. Solifluction is the slowest flow-type deformation of saturated cohesive sediments in almost all climatic provinces, and the process of altiplanation in frozen ground environments. Debris flow, earthflow, and mudflow failures are distinguished by a progressively increasing moisture content and a decreasing sediment size in the failed mass. Debris flows are initiated by the increase of pore pressure in surface debris, which deform by plastic and granular flow mechanisms. Wet clays of low consistency fail by the activity called earthflow, in which deformation retrogrades to more than ten times the initial slope height. Mudflows are viscous failure of cohesive sediments at pore water content of up to 50% by weight.

Free fall of rock fragments originates at wide and deep joints behind cliff faces. The frequency of rockfalls is higher in areas of sheared and shattered rocks, and where jointed rocks are aligned to the slope. This activity is particularly pronounced in cold region environments of freeze-thaw cycles.

Slides are shallow and deep-seated planar and rotational failure of the earth materials. They are blockglides, slumps, debris slides, liquefaction, and rock avalanches by the process of deformation, depth and type of dislocation, and material composition. Blockglides are a form of slope failure on bedded strata of differing lithologic composition. They are caused by the enlargement of a surface joint to some critical depth behind the cliff face, producing large enough lateral stress to topple a part of the top stratum at the interface of strata. Slumps are shallow and deep failure of cohesive sediments along spoon-shaped rupture planes. Shallow slumps are initiated along radial tension cracks at the surface, which later evolve into shear planes. Deep-seated slumps develop along streams that trench deeply into the valley floor.

Rockslides and debris slides are shallow planar failure of an incoherent mass of rock fragments and weathering residue on slopes, respectively, which become unstable by the increase of pore pressure. Liquefaction is a localized subsidence of thick saturated silt and sand bodies underlying an impermeable stratum. The substratum becomes unstable by the seismic stress of loading and unloading cycles in earthquake events, expelling the pore water from moist sand. The expelled water generates a strong uplift force, causing the top stratum to fall through the interface and disintegrate by lateral spread.

Debris and rock avalanches are catastrophic failures of the earth materials along shear planes. Debris avalanches are shallow dislocations of saturated surface debris and volcanic ejecta on slopes. Rock avalanches are massive and deep-seated failures of disintegrated rock along a wedge-shaped rupture plane separating continuous and discontinuous joints. The high rate of avalanche deformation is attributed to slide, flow, momentum transfer, and acoustic fluidization mechanisms.

REFERENCES

Baoping, W. 1994. The features of landslides, rockfalls and debris flow disasters in China. In *Proceedings of the 7th International Association of English Geologist*, Lisboa, Portugal, September 5–9, 1994, Vol. 3, pp. 1459–64.

Bartlett, S. F., and Youd, Y. L. 1995. Empirical prediction of liquefaction induced lateral spread. *J. Geotech. Eng.* 121:316–29.

Bentley, S. P., and Smalley, I. J. 1984. Landslips in sensitive clays. In *Slope instability*, ed. D. Brunsden and D. B. Prior, 457–90. Chichester: John Wiley & Sons.

Bishop, A. W. 1955. The use of slip circle in the stability analysis of slopes. *Géotechnique* 5:7–17.

Bjerrum, L., and Jørstad, F. 1968. Stability of rock slopes in Norway. *Norwegian Geotech. Inst. Publ.* 79:1–11.

Bolt, B. A., Horn, W. L., Macdonald, G. A., and Scott, R. F. 1975. *Geological hazards*. Berlin: Springer-Verlag.

Brunsden, D. 1980. Mass movement. In *Process in geomorphology*, ed. C. Embleton and J. Thornes, 130–86. New Delhi: Arnold-Heinemann.

Carson, M. A. 1976. Mass-wasting, slope development and climate. In *Geomorphology and climate*, ed. E. Derbyshire, 101–36. London: John Wiley & Sons.

Carson M. A., and Kirkby, M. J. 1972. *Hillslope form and processes*. Cambridge: Cambridge University Press.

Crandell, D. R., Miller, C. D., Glicken, H. X., Christiansen, R. L., and Newhall, G. C. 1984. Catastrophic debris avalanche from ancestral Mount Shasta Volcano, California. *Geology* 12:143–46.

Crozier, M. J. 1984. Field assessment of slope instability. Landslips in sensitive clays. In *Slope instability*, ed. D. Brunsden and D. B. Prior, 103–45. Chichester: John Wiley & Sons.

Cruden, D. M., Krauter, E., Beltran, L., Lefebvre, G., Ter-Stepanian, G. I., and Zhang, Z. Y. 1994. Describing landslides in several languages: The multilingual landslide glossary. In *Proceedings of the 7th International Association of English Geologist*, Lisboa, Portugal, September 5–9, 1994, Vol. 3, pp. 1325–34.

Cunningham, D. M. 1988. A rockfall avalanche in sandstone landscape, Nattai North, NSW. *Austr. Geogr.* 19:221–29.

Davies, R. H. 1988. Debris flows surges—A laboratory investigation. In *Mitteilugen der Versuchsanstalt fur Wasser ban, Hydrologie und Glaziologie*, Zurich, Switzerland, p. 96.

Davis, R. O., Desai, C. S., and Smith, N. R. 1993. Stability of motions of translational landslides. *J. Geotech. Eng.* 119:420–32.

Eisbacher, G. H. 1979. Cliff collapse and rock avalanches (sturzstroms) in Mackenzie Mountains, northwestern Canada. *Can. Geotech. J.* 16:309–34.

Eisbacher, G. H., and Clague, J. J. 1981. Urban landslides in the vicinity of Vancouver, British Columbia, with special reference to the December 1979 rainstorm. *Can. Geotech. J.* 18:205–16.

Eyles, R. J., and Ho, R. 1970. Soil creep on a tropical slope. *J. Tropical Geogr.* 31:42.

Feda, J. 1989. Interpretation of creep of soils by rate process theory. *Géotechnique* 39:667–77.

Fiegel, G. L., and Kutter, B. L. 1994. Liquefaction mechanism for layered soils. *J. Geotech. Eng.* 120:737–55.

Figueroa, J. L., Saada, S., Liang, L., and Dahisaria, N. M. 1994. Evaluation of soil liquefaction by energy principles. *J. Geotech. Eng.* 120:1554–69.

Francis, P. W., Gardeweg, M., Ramirez, C. F., and Rothery, D. A. 1985. Catastrophic debris avalanche deposits of Scopa Volcano, northern Chile. *Geology* 13:600–3.

Goudie, A. 1993. *The nature of the environment.* 3rd ed. Oxford: Blackwell Publishers.

Hansen, M. J. 1984. Strategies for classification of landslides. In *Slope instability*, ed. D. Brunsden and D. B. Prior, 101–36. Chichester: John Wiley & Sons.

Helfert, M. R., and Wood, C. A. 1986. Shuttle photos show Madagascar erosion. *Geotimes*, March 4–5.

Hsu, K. J. 1975. Catastrophic debris streams (sturzstroms) generated by rockfalls. *Bull. Geol. Soc. Am.* 86:129–40.

Hutchinson, J. N. 1968. Mass movement. In *The encyclopedia of geomorphology*, ed. R. W. Fairbridge, 688–96. New York: Reinhold Book Corp.

Hutchinson, J. N. 1971. Field and laboratory studies of a fall in Upper Chalk Cliffs at Jess Bay, Isle of Thanet. In *Roscoe Memorial Symposium*, University of Cambridge, March 29–31, 1971.

Johnson, A. M., and Rodine, J. R. 1984. Debris flow. In *Slope instability*, ed. D. Brunsden and D. B. Prior, 257–361. Chichester: John Wiley & Sons.

Johnson, P. G. 1984a. Paraglacial conditions of instability and mass movement—A discussion. *Z. Geomorph.* 28:235–50.

Johnson, P. G. 1984b. Rock glacier formation in high magnitude low-frequency slope processes in the southwest Yukon. *Ann. Assoc. Am. Geogr.* 74:408–19.

Keller, E. A. 1979. *Environmental geology.* 2nd ed. Columbus, OH: Charles E. Merill Publ. Co.

Kent, P. E. 1966. The transport mechanism in catastrophic rock falls. *J. Geol.* 74:79–83.

Kiersch, G. A. 1964. Vaiont Reservoir disaster. *Civil Eng.* 34:32–39.

Kirkby, M. J. 1967. Measurement and theory of soil creep. *J. Geol.* 75:359–78.

Kokusho, T. 1999. Water film in liquefied sand and its effect on lateral spread. *J. Geotech. Geoenv. Eng.* 125:817–26.

Krishnaswamy, V. S. 1980. State-of-the-art report. In *Proceedings of the International Seminar on Landslides: Geological Aspects and Seismological Relations of Landslides and Other Types of Mass Movements*, New Delhi, April 7–11, 1980, Vol. 3.

Kuhn, M. R., and Mitchell, J. K. 1993. New perspectives on soil creep. *J. Geotech. Eng.* 119:507–24.

Lambe, T. W., and Whitman, R. V. 1973. *Soil mechanics.* New Delhi: Wiley Eastern Ltd.

Leopold, L. B., Wolman, M. G., and Miller, J. P. 1964. *Fluvial processes in geomorphology.* San Francisco: W. H. Freeman and Co.

Lewis, L. A. 1974. Slow movement of earth under tropical rain forest conditions. *Geology* 1:9–10.

Mainali, A., and Rajaratnam, N. 1994. Experimental studies of debris flow. *J. Hydraul. Eng.* 120:104–23.

Marui, H., 1994. Geotechnical study on the debris flows in the Urakawa river basin. In *Proceedings of the 7th International Association of English Geologist*, Lisboa, Portugal, September 5–9, 1994, Vol. 3, pp. 1623–28.

Mitchell, J. R., and Markell, A. R. 1974. Flowsliding in sensitive soils. *Can. Geotech. J.* 11:11–31.

Nozaki, T., Miyazawa, M., Uchida, M., and Okada, Y., 1994. Earthflow induced by snow load in Nagano Prefecture, central Japan. In *Proceedings of the 7th International Association of English Geologist*, Lisboa, Portugal, September 5–9, 1994, Vol. 3, pp. 1431–39.

Owen, L. A. 1991. Mass movement deposits in the Karakoram Mountains: Their sedimentary characteristics, recognition and role in Karakoram landform evolution. *Z. Geomorph.* 35:401–24.

Pant, P. D., and Luirei, K. 1999. Malpa rockfalls of 18 August 1998 in the northeastern Kumaun Himalaya. *J. Geol. Soc. India* 54:415–20.

Pierson, T. C. 1981. Dominant particle support mechanism in debris flow at Mt. Thomas, New Zealand, and implications for flow mobility. *Sedimentology* 28:49–60.

Plafker, G., and Erickson, G. E. 1978. Navados Huascarán avalanches, Peru. In *Rockslides and avalanches,* ed. Voight, 227–314. Vol. 1. Amsterdam: Elsevier.

Rodrigues, D. D. M., and Ayala-Carcedo, F. J. 1994. Landslides in Machico area on Madeira Island. In *Proceedings of the 7th International Association of English Geologist,* Lisboa, Portugal, September 5–9, 1994, Vol. 3, pp. 1495–500.

Rydelek, P. A., and Tuttle, M. 2004. Explosive craters and soil liquefaction. *Nature* 427:115–16.

Sassa, K. 1985. The mechanics of debris flows. In *Proceedings of the XI International Conference on Soil Mechanics and Foundation Engineering,* San Francisco, pp. 1173–76.

Schumm, S. A., and Chorley, R. J. 1964. The fall of Threatening Rock. *Am. J. Sci.* 262:1041–54.

Selby, M. J. 1970. *Slopes and slope processes.* New Zealand Geographical Society Publication 1. New Zealand Geographical Society.

Sharpe, C. F. S. 1938. *Landslides and related phenomena.* New York: Columbia University Press.

Shreve, R. L. 1968. Leakage and fluidization in air-layer lubricated avalanches. *Bull. Geol. Soc. Am.* 79:653–58.

Spence, K. J., and Guymer, I. 1997. Small-scale laboratory flowslides. *Géotechnique* 47:915–32.

Strom, A. L. 1994. Mechanism of stratification and abnormal crushing of rockslide deposits. In *Proceedings of the 7th International Association of English Geologist,* Lisboa, Portugal, September 5–9, 1994, Vol. 3, pp. 1287–95.

Su, S. F., and Liao, H. J. 1999. Effect of anisotropy on undrained slope stability in clays. *Géotechnique* 49:215–30.

Terzaghi, K. 1950. Mechanism of landslides. In *Application of geology for engineering practice,* ed. S. Paige, 83–123. New York: Geological Society of America.

Terzaghi, K. 1962. Stability of steep slopes on hard weathered rock. *Géotechnique* 12:251–63, 269–70.

Valdia, K. S. 1998. Catastrophic landslides in Uttranchal, Central Himalaya. *J. Geol. Soc. India* 52:483–86.

Van Steijn, H. 1991. Frequency of hillslope debris flows in a part of French Alps. *Bull. Geomorph.* 19:83–90.

Varnes, J. D. 1978. Slope movement types and processes. In *Landslide analysis and control,* ed. R. L. Schuster and R. J. Krizek, 11–33. Special Report 176. Washington, DC: National Academy of Sciences.

Voight, B., and Elsworth, D. 1997. Failure of volcanic slopes. *Géotechnique* 47:1–31.

Whalley, W. B. 1984. Rock falls. In *Slope instability,* ed. D. Brunsden and D. B. Prior, 217–56. Chichester: John Wiley & Sons.

Whitehouse, I. E. and Griffiths, G. A. 1983. Frequency and hazard of large rock avalanches in the central southern Alps, New Zealand. *Geology* 11:331–34.

Záruba, Q., and Mencl, V. 1969. *Landslides and their controls.* Prague: Academia and Elsevier.

5 Fluvial Processes

Streams are the most dynamic component of fluvial landscapes and landscape change. The mutual interaction between the flow and sediments changes the streambed continuously and, thereby, influences the resistance to flow and sediment transport characteristics per unit time of the channel cross section. This frictional resistance between water and mobile roughness boundary causes channel erosion and transport of sediments in the fluvial system. Hydraulic engineers have developed empirical and theoretical models for understanding the behavior of frictionally dominated flow and the morphologic activity of streams. Fluvial geomorphologists, however, classify and describe fluvial forms, emphasizing the interaction among interdependent hydraulic variables of spatially organized drainage systems. These and other aspects of fluvial processes are discussed in several texts on the subject (Leopold et al., 1964; Morisawa, 1968; Garde and Ranga Raju, 1977; Bogárdi, 1978; Richards, 1982; Graf, 1998).

TYPES OF FLOW

Fluids deform continuously as long as the *fluid stress* exceeds the *viscosity* or internal resistance of fluid molecules to deformation, producing laminar and turbulent flows with transitions in between. *Laminar flow* occurs at high viscous forces accompanying low velocity and small depth of flow, such that neighboring fluid layers move smoothly past each other without mixing. The flow becomes unstable or turbulent when velocity increases in the flow of a fixed size. *Turbulent flow* comprises unstable eddy motions, which rapidly dissipate kinetic energy in the flow, thoroughly distribute the shear stress in the medium, and increase the resistance to flow.

FLOW DISCRIMINATION

Dimensionless Reynolds number and Froude number are measures of the type and general characteristics of the fluid flow. The *Reynolds number* (R_e) differentiates laminar and turbulent flows by the velocity and flow depth, and viscosity and density of the fluid as

$$R_e = (v\ d)/(\mu/\rho)$$

Driving force/Resisting force

in which v is the mean velocity of flow, d is the average depth of flow, μ is the viscosity, and ρ is the density of water. In general, R_e up to 500 characterizes *laminar flow*, R_e more than 2000 represents *turbulent flow*, and R_e between 500 and 2000 defines *transition* from laminar to turbulent flow (Figure 5.1).

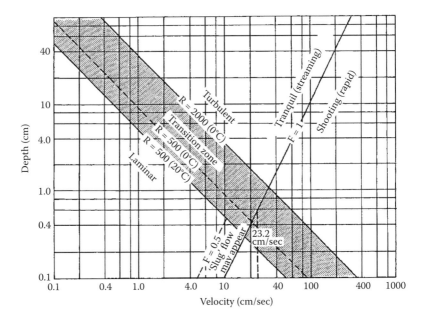

FIGURE 5.1   Regimes of flow as function of velocity and flow depth in broad open channels. The transition zone between laminar and turbulent flows is defined by the Reynolds number. The value of unity for the Froude number separates tranquil from shooting flows. (Source: Figure 1 in Sundborg, Å., "The Dynamics of Flowing Water," *Geogr. Annlr.* XXXVIII [1956]: 133–64. With permission.)

The *Froude number* (F) is a measure of gravity effects on the fluid flow as

$$F = v/\sqrt{g\,d}$$

The Froude criterion distinguishes tranquil (streaming) from shooting (rapid) flow. The flow is *tranquil* for F < 1, *shooting* for F > 1, and in *transition* from tranquil to shooting at F values between approximately 0.7 and 1.0.

STREAM LOAD

Weathering, erosion, and mass movement activities within drainage basins generate clastic and solute loads for stream transport. A *clastic load* is heterogeneous in composition from clay through to gravel-sized fractions. A *solute load* is derived from the basin lithology. It is comprised of inorganic salts and compounds that move as the dissolved load of drainage systems. The volume of stream load, however, is supply limited. It is primarily determined by the drainage area, geologic and topographic attributes of terrain, and climatic regime.

CLASTIC LOAD

The clastic load is arbitrarily classified as bedload, suspended load, and wash load of drainage systems. *Bedload* is the coarser fraction of bed material load that travels as a contact and saltation load at or near the bed of streams. The *contact load* rolls or slides in a series of discontinuous movements, and the *saltation load* travels by leap and bounce movement at the bed of streams. In practice, however, the two forms of bedload are difficult to distinguish. The manner of bedload transport also differs significantly in ephemeral and perennial streams. The bedload of *ephemeral streams* migrates by local scour of sand and gravel-bedded channels in flash floods (Foley, 1978; Schick et al., 1987) and of *perennial streams* moves at all stages of channel discharge (Reid and Frostick, 1994).

Suspended load is the finer fraction of stream load held in suspension by the upward momentum of turbulent eddies imposed on the main channel flow. Hence, the suspended load may be considered an advanced stage of bedload transport not in contact with the bed for sufficient time. The suspended load arrives into drainage systems from the *overland flow* during storms. Besides, the turbulence-dislodged sediments from within the voids of larger bed material clasts also add insignificantly to the total volume of suspended load. The suspended load is expressed in parts per million by weight (ppm), giving its concentration in the flow. The *concentration* of suspended load is highest at the *turbulent boundary layer* close to the channel bed and least toward the water surface level. In general, the concentration of suspended load varies directly with the runoff volume. As discussed later in the section on hydraulic geometry of streams, the suspended load dampens the *resistance to flow*. The suspended load is also a measure of the total volume of clastic load moved by streams.

Wash load is comprised of colloidal clay particles that possess a *settling velocity* of 10^{-6} m s^{-1} or less (Reid and Frostick, 1994). Hence, it is always held in suspension irrespective of the channel condition.

Solute Load

Chemical composition of the solute or dissolved load of streams depends on basin lithology. In general, the concentration of dissolved salts in channel flow varies directly with the denudation rate of terrain and inversely with the magnitude of flow. Leopold et al. (1964) observed that the measured dissolved load of streams across the United States varies with the *climatic regime*, averaging 20% of the total stream load. The dissolved load is 9% of the total in arid regions, which progressively increases to 37% of the stream load in humid environments.

STREAM FLOW

Stream flow is initiated when the shear stress of channel flow exceeds the friction between water and channel boundaries. In open channels, *shear stress* (τ) is the component of gravity times weight of water that provides a certain driving force for the flow of water. The flow of water evolves a highly organized drainage system and regulates the morphologic activity of stream erosion and deposition of sediments.

The *resistance to flow* from the bed and bank materials, channel bends, and bank protuberances affects the flow velocity in the vertical and across channels. This resisting force or *shearing stress* per unit of the channel boundary is proportional to hydraulic radius and slope. The slope, called *energy gradient*, represents potential energy loss through turbulence and friction in the flowing water, and is usually given by the *water surface slope*. The resistance to flow per unit area of the streambed is defined as τ/v^2, where v is the mean velocity of flow.

FLOW VELOCITY

In open channels, the flow velocity varies with water depth, bed slope, hydraulic geometry, bed roughness, and viscosity of water. The flow velocity is zero at the streambed. It increases toward the water surface level in a logarithmic fashion, suggesting that local velocity equals the mean velocity of flow at a certain depth of the water column. For most streams, this proportion is 0.6 of the depth of channel flow from the bed. The rate of change of velocity in the *vertical* is governed by the manner of mixing of fluid layers in the water column and bed roughness elements. In general, the velocity gradient is steeper for turbulent flow and larger roughness elements. The velocity distribution *across channels*, however, varies with channel pattern. In *straight channels*, the highest velocity is toward the center of streams, and in *meandering streams* the path of highest velocity swings between channel banks.

Hydraulic engineers determine the flow velocity by empirical Chézy and Manning formulae. The Chézy formula expresses the flow velocity as

$$v = C \sqrt{R S}$$

in which v is the mean flow velocity, C is a constant for factors contributing to the resistance force, R is the hydraulic radius nearly equal to the depth of flow, and S is the channel gradient. The Manning formula quantifies the mean velocity of flow for channels of uniform cross section and slope as

$$v = 1.49 \ (R^{0.67} \ S^{0.50})/n$$

where n is the bed roughness factor. Experimental values of n vary from 0.025 for alluvial channels to 0.03 for straight-channel segments without pools and riffles to 0.05 for gravel and boulder-bedded mountain streams.

SEDIMENT TRANSPORT

In laboratory channels of uniform cross section and bed material size, the bed particles are set in motion at a flow velocity that just exceeds the forces resisting the movement of grains. This is the *bed shear velocity* for entraining particles of a given size. The deforming bed, however, continuously changes the bed configuration and thereby influences the mobile roughness boundaries, channel shape, and water and material flow. Hence, the movement of sediments by flowing water presents a complex problem for resolution.

The bed material of alluvial streams is admixed and heterogeneous in size composition in the vertical and across channels. Hence, hydraulic conditions for the incipient motion of sediments vary with individual streams. Coarser sediments resist shear mainly due to the *submerged weight*, and finer silt and clay fractions resist movement by the effect of *cohesion* among grains. Streams also carry the bulk of sediment charge as *suspended load*, which dampens the resistance to flow and affects channel shape. The sediment transport rate, thus, reflects the inherent variability of bed configuration and sediment texture in the fluid flow system (Gomez, 1991; Powell et al., 1999). Hence, the physical laws governing sediment transport even under ideal conditions can only be conservatively applied to streams. Given the constraints, several theoretical models in terms of critical tractive force and semiempirical formulae based on competent velocity, and lift and drag forces, have been proposed for predicting hydrologic conditions at which the bed particles are just set in motion. All derivations, however, are based on the movement of coarse-grained sediments for which the incipient condition of grain motion is clearly discernible at the streambed.

CRITICAL TRACTIVE FORCE

Bedload transport is initiated when the fluid force (stress) of channel flow reaches a certain magnitude and the resulting shear stress, called critical tractive force, pushes the loose bed particles of a given size downstream. The *fluid stress* on a channel bed, as in Du Boys' (1879) and Chézy's equations, is given by the water depth times bed slope. Due to channel conditions, however, the bedload exhibits unsteady and nonuniform transport behavior. Therefore, several theoretical models predicting the critical tractive force for a variety of sediment sizes and channel flow conditions have been proposed. They are discussed in several texts on the subject (Garde and Ranga Raju, 1977; Bogárdi, 1978; Richards, 1982; Graf, 1998).

COMPETENT VELOCITY

The bed particles of a given size are entrained in the fluid flow at a certain critical mean bed velocity called the competent or erosion velocity. The competent velocity, however, is difficult to evaluate precisely as the bed particles are a heterogeneous admixture of different sizes, and a few are sufficiently cohesive and packed in nature. Sundborg (1956) determined that the competent velocity is almost proportional to the square root of particle size diameter.

Sundborg (1967) presented experimental data on the instantaneous state of sediment transport of specified size for flow velocity 1 m above the channel bed (Figure 5.2). These data suggest that particles more than 0.02 mm in size travel as the bedload, and finer than 0.02 mm as the suspended load of streams, and that the boundary between two modes of sediment transport remains sufficiently diffused due to channel conditions. Further, the competent velocity is lower for sand than silt and gravel-sized particles and higher for particles more and less than 0.5 mm in size composition. The particles once entrained in the flow, however, continue to be moved at a velocity lower than the critical. The data further suggest that bed particles cease movement, and are deposited, at a flow velocity less than two-thirds of the competent velocity.

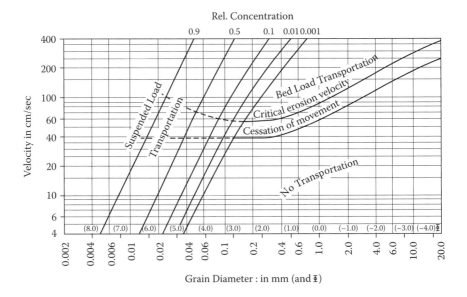

FIGURE 5.2 Relation between flow velocity, grain size of a uniform 2.65 density composition, and status of sediment in fluid flow. The velocities are 1.0 m above the channel bed. (Source: Figure 1 in Sundborg, Å., "Some Aspects on Fluvial Sediments and Fluvial Morphology. I. General Views, and Graphic Methods," *Geogr. Annlr.* 49-A [1967]: 333–43. With permission.)

LIFT AND DRAG FORCES

On a deformable bed of streams, the fluid force on particles at rest resolves into components of lift and drag forces respectively perpendicular and parallel to the channel flow. The flow velocity is zero at the bottom of particles and slightly higher at the top, producing a relatively higher pressure beneath than at the top of grains. This pressure difference enables the bed particles to lift from the bed, displace in the direction of flow, and land at the bed by describing a *saltation trajectory*. The difference of pressure on up- and downstream faces of the particles also introduces a drag force in the system that pushes the bed particles downstream. In general, the lift force is more significant to the entrainment of bed particles than the drag force.

REGIMES OF FLOW

Interaction between the fluid force of water and the shear resistance of bed-forming materials in *alluvial channels* sets grains in motion. The movement of bed particles changes bed configuration and water surface characteristics, evolving a sequence of downstream traveling bed forms, in which the rate of formation increases with the increasing rate of sediment discharge until the growing bed form reaches equilibrium with the flow velocity and sediment transport rate (ASCE, 1966; Coleman and Melville, 1994; Swamee and Ojha, 1994; McLean et al., 1999; Karim, 1999). The range of flows, which produce similar bed configurations, resistance to flow, and sediment transport rate are called regimes of flow (Sundborg, 1956; Simons and Richardson, 1966).

Resistance to flow is an appropriate quantitative measure of the regimes of flow given as lower and upper, with transition in between. The resistance to flow is conveniently estimated from the hydraulic variables of depth, velocity, slope, and bed material size, and expressed by the dimensionless Darcy-Wiesbach resistance coefficient (Simons and Richardson, 1966). In flumes, the *Darcy-Wiesbach resistance coefficient* varies from 0.052 to 0.16 for the lower and 0.02 to 0.07 for the upper flow regime.

Lower Flow Regime

The lower flow regime is typical of tranquil flow and large resistance to flow. These channel conditions evolve ripples and dunes as characteristic bed forms (Figure 5.3(a)–(c)). The ripples are up to 0.06 m high and 0.6 m long bed forms in phase with the water surface level. With increase in flow velocity, shear stress, and rate of sediment transport, however, the ripple bed transforms into up to 0.3 m high and 0.6 m long dunes in flumes, and up to several tens of meters long dunes in

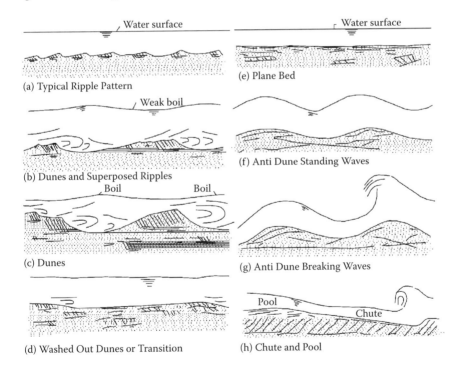

(a) Typical Ripple Pattern

(b) Dunes and Superposed Ripples

(c) Dunes

(d) Washed Out Dunes or Transition

(e) Plane Bed

(f) Anti Dune Standing Waves

(g) Anti Dune Breaking Waves

(h) Chute and Pool

FIGURE 5.3 Idealized diagram of the forms of bed roughness in alluvial channels. (Source: Figure 3 in Simons, D. B., and Richardson, E. V., *Resistance to Flow in Alluvial Channels*, U.S. Geological Survey Professional Paper 422J [U.S. Geological Survey, 1966]. With permission from U.S. Geological Survey.)

alluvial streams. The dune bed increases instability in the flow system by which the bed and water surface remain out of phase with each other.

Transition

With further increase in flow velocity and sediment transport rate, the resistance to flow decreases rather than increases by the effect of increased suspended load in the system. Hence, the resistance to flow becomes incompatible for the bed configuration that flattens the dunes and increases their length. Such bed forms are called washed-out dunes out of phase and plane bed in phase with the water surface slope (Figure 5.3(d),(e)).

Upper Flow Regime

The resistance to flow is small, but the rate of sediment transport is large in the upper flow regime, producing antidune and chute-and-pool bed forms in phase with the water surface slope (Figure 5.3(f)–(h)). The resistance to flow results from grain roughness with grains moving, wave formation and subsidence, and rapid energy dissipation in breaking waves that make a large amount of finer bed material sizes available for transport. Antidunes are several meters in wavelength. They evolve from upstream migration of finer particles, and collapse and regrouping of grains, in series of short trains. The chute-and-pool bed evolves beneath a long chute of accelerating flow, hydraulic jump at the end of the chute, and a long pool of tranquil but accelerating flow downstream. The upper flow regime is characteristic of alluvial streams of steep gradient.

HYDRAULIC GEOMETRY

Channel variables of width, depth, velocity, and suspended load vary systematically with changes in discharge. Leopold and Maddock (1953) demonstrated that channel variables are in a state of *quasi-equilibrium* with variations in discharge, and are related to discharge by simple power functions. Hydraulic geometry is a graphical description of the interdependent channel variables and analysis of the mutual adjustment of channel variables to discharge in channels developed in part by earlier flows.

Channel variables of width, depth, velocity, and suspended load are related to discharge by power functions as (Leopold and Maddock, 1953)

$$w = a \, Q^b$$

$$d = c \, Q^f$$

$$v = k \, Q^m$$

$$L = p \, Q^j$$

where w, d, v, and L are respectively the water surface channel width, average channel depth, mean flow velocity, and suspended load of streams, and Q is discharge. The letters a, c, k, and p are coefficients, and b, f, m, and j are exponents in the above equations. By the *flow equation*, channel width, depth, and velocity are related to channel discharge as

$$Q = w \times d \times v$$

By substitution,

$$Q = a\,Q^b \times c\,Q^f \times k\,Q^m$$

such that $a \times c \times k = 1$ and $b + f + m = 1$.

The interdependent channel variables adjust among themselves in many different ways to changes in discharge by individually sharing a part of the change, such that they perform the least work in the adjustment process (Chapter 1). The *principle of least work* emphasizes minimum variance in adjustment and equitable distribution of work among hydraulic variables mutually adjusting to the varying magnitude of at-a-station discharge and equal frequency of discharge in downstream direction. The relation of suspended load to other hydraulic variables is discussed in another section of this chapter.

AT-A-STATION HYDRAULIC GEOMETRY

At-a-station channel width, depth, and velocity variations with discharge are independent of basin characteristics, and unique to each system (Table 5.1). The manner of adjustment of individual channel variables to discharge is illustrated in Figure 5.4 for clarity of understanding.

At-a-station exponents describing the hydraulic geometry of streams are somewhat consistent in the rate of change of width, depth, and velocity with the magnitude of channel flow. The rate of change, in general, is governed by channel roughness, bed

TABLE 5.1
At-a-Station Hydraulic Geometry

Stream	Database	At-a-Station					
		Intercepts			Exponents		
		a	c	k	b	f	m
Streams of Great Plains and Southwest USA[1]	20	26	.15	.37	.26	.40	.34
Brandywine Creek, Pennsylvania[2]	7	54	.23	.10	.04	.41	.55
White River, Mt. Rainier, Washington[3]	112	4	.22	1.1	.38	.44	.27
Bollin-Dean, northwest Cheshire[4]	12				.114	.402	.483

Sources: 1, Leopold and Maddock (1953); 2, Wolman (1955); 3, Fahnestock (1963); 4, Knighton (1974).

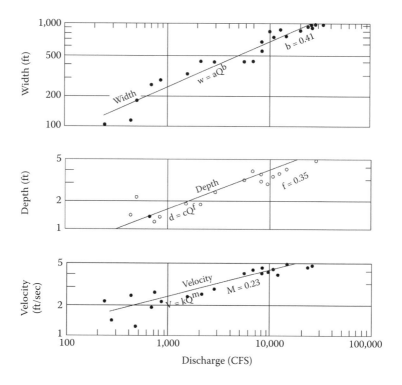

FIGURE 5.4   Typical relations among width, depth, velocity, and discharge, Cheyenne River, near Eagle Butte, South Dakota. (Source: Figure 3 in Leopold, L. B., and Maddock, T., Jr., *The Hydraulic Geometry of Streams and Some Physiographic Implications*, U.S. Geological Survey Professional Paper 252 [U.S. Geological Survey, 1953]. With permission from U.S. Geological Survey.)

and bank material size, channel pattern, and internal inconsistencies of adjustment (Knighton, 1975). Hence, the proportion of increase in the exponents is expected to vary in streams even if channels experience the same at-a-station magnitude of discharge. A comparison of the exponents for the data in Table 5.1 suggests that channel depth increases at a somewhat faster rate than the water surface width (f > b). Hence, channels of a higher rate of depth increase experience a smaller rate of velocity increase with discharge. The Brandywine and Bollin-Dean systems, however, experience a small change of width (b) and a large change of velocity increase (m) with discharge. These streams are typically narrow and developed in cohesive sediments.

DOWNSTREAM HYDRAULIC GEOMETRY

Downstream hydraulic geometry compares the rate of change of channel variables along the length of streams for *equal frequency* of discharge. A comparison of exponents for at-a-station and downstream segments suggests that channel width increases rapidly, depth normally does not change by as much, and velocity increases only at a moderate rate (Table 5.2). The downstream increase in channel width is governed

TABLE 5.2
Comparison of At-a-Station and Downstream Exponents

Stream	Database	At-a-Station Exponents			Downstream Exponents		
		b	f	m	b	f	m
Streams of Great Plains and Southwest USA[1]	20	.26	.40	.34	.5	.4	.1
Brandywine Creek, Pennsylvania[2]	7	.04	.41	.55	.57	.40	.03
White River, Mt. Rainier, Washington[3]	112	.38	.33	.27	.38	.33	.27
Bollin-Dean, northwest Cheshire[4]	12	.114	.402	.483	.46	.16	.38

Sources: 1, Leopold and Maddock (1953); 2, Wolman (1955); 3, Fahnestock (1963); 4, Knighton (1974).

by the increase in discharge, while the velocity increase with depth illustrates compensation for the declining channel gradient in the same direction. The White River data, however, suggest that channels in noncohesive gravels rapidly adjust to changes in discharge, and maintain the same shape over a wide range of flow conditions.

SUSPENDED LOAD AND CHANNEL SHAPE

Suspended load is a hydraulic variable that affects the shape of alluvial channels through adjustments in velocity to discharge and depth to discharge. A typical relation of suspended load to *at-a-station* discharge, given by the power function $L = p Q^j$ and presented in Figure 5.5, suggests that the suspended load usually increases more rapidly with discharge as the slope of best-fit line (j) is greater than unity. In a *downstream direction*, however, the concentration of suspended load decreases somewhat for the given frequency of discharge. This aspect is explained by a progressive downstream area increase in gentler slopes of the drainage basins.

The effect of suspended load on *channel shape* is best exemplified at a constant discharge. Leopold and Maddock (1953) demonstrated that decrease of width at constant velocity or increase of velocity at constant width increases the volume of suspended sediment at constant discharge. The channel shape also adjusts in a definite pattern with variations in the suspended load in flood spasms. The hydraulic data on temporal variation in channel width, depth, velocity, and suspended load for a flood event in San Juan River, Utah, demonstrate the manner of channel adjustment to changes in the concentration of suspended load of the system (Figure 5.6). These data suggest that channel width increased at a slight rate up to the peak discharge and retracted the same path in the falling stage of the flood event. The depth, however, uniformly increased with discharge up to 5000 cusec (141.5 cumec), but increased more rapidly thereafter as discharge increased to a peak of 60,000 cusec (1698 cumec). The channel at 5000 cusec was 3.2 ft (0.97 m) in the rising and 4.4 ft (1.34 m) deep in the falling stage of a similar magnitude of discharge. A corresponding adjustment also occurred in the flow velocity: it was 8.6 fps (234.6 cm s[-1]) in

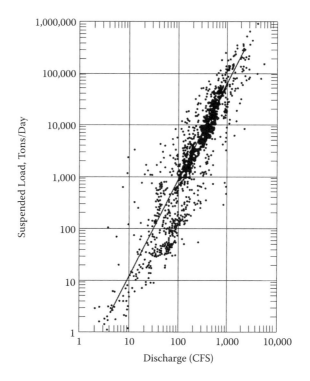

FIGURE 5.5 Relation of suspended sediment load to discharge, Powder River at Arvada, Wyoming. (Source: Figure 13 in Leopold, L. B., and Maddock, T., Jr., *The Hydraulic Geometry of Streams and Some Physiographic Implications*, U.S. Geological Survey Professional Paper 252 [U.S. Geological Survey, 1953]. With permission from U.S. Geological Survey.)

the rising flood and 8.0 fps (218.2 cm s^{-1}) in the falling stage of discharge at 5000 cusec. The concentration of suspended sediment, however, varied throughout the flood event. The suspended load increased at a rapid rate for initial increase in discharge from 600 to 5000 cusec, the rate of increase declined until the peak discharge was reached, and it continued to decline rapidly throughout the receding flood event. This variation in the concentration of suspended load is related to fill and scour of the channel bed during the flood spasm. The bed is silted in the period of higher concentration of suspended sediment load in the rising flood and scoured in the period of declining concentration of sediment load in the falling stage of the flood spasm. These observations suggest that the suspended load dampens *resistance to flow* and increases the flow velocity with a remarkable effect on channel shape (ASCE, 1998). Following the *Manning equation*, referred to in an earlier section on resistance to flow, increase in flow velocity and decrease of channel depth require an increase of channel slope or a decrease of roughness, or both.

Sediment and Channel Slope

Channel bed elevation declines downstream, providing a channel gradient or *thalweg* that for mature streams is concave up in profile (Figure 5.7). The fall

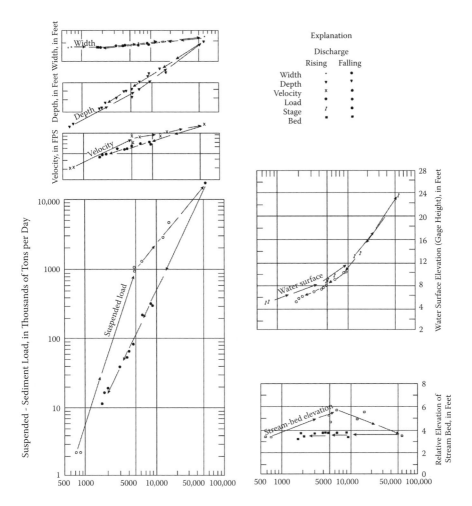

FIGURE 5.6 Changes in width, depth, velocity, water surface elevation, and streambed elevation with discharge during the 1941 flood in San Juan River near Bluff, Utah. (Source: Figure 21 in Leopold, L. B., and Maddock, T., Jr., *The Hydraulic Geometry of Streams and Some Physiographic Implications*, U.S. Geological Survey Professional Paper 252 [U.S. Geological Survey, 1953]. With permission from U.S. Geological Survey.)

in bed elevation downstream is apparently related to decrease in the bed material size in the same direction. This type of adjustment requires a decrease in downchannel flow velocity, which is contrary to the observed increase in velocity downstream. The downstream decrease in bed slope also reflects a decrease in channel roughness due to reduction in the particle size downstream and, hence, resistance to flow. Therefore, the overall concavity of stream profile is viewed as an internal adjustment among hydraulic variables to discharge in ways that provide for an efficient slope of sediment transport in equilibrium with increased channel width, depth, and velocity in the downstream direction (Leopold and Maddock, 1953).

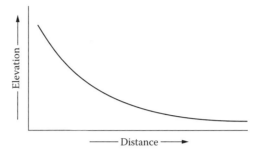

FIGURE 5.7 An idealized longitudinal profile of a mature stream.

Other relations for the profile concavity have also been suggested. Langbein and Leopold (1964) proposed that the channel slope is an inverse function of the mean annual discharge as

$$S \propto Q_m^{-z}$$

where S is the channel slope, Q_m is the mean annual discharge, and z is the exponent that varies between 0.5 and 1.0. The channel slope is a nearly straight sloping line for $z = -0.5$ and excessively concave up for $z = -1.0$. Hack (1957) predicted that downstream flattening of channel slope is related to the decrease of bed material size and increase of drainage basin area as

$$S = 18 \, (D_m/A_d)^{0.6}$$

in which S is the channel slope, D_m is the bed material size, and A_d is the drainage basin area.

CHANNEL PATTERNS

Processes of adjustment among hydraulic variables of channel slope, sediment load, and discharge in long stretches of alluvial streams evolve large-sized bed forms called channel patterns (Reid and Frostick, 1994). Observations suggest that channel patterns are commonly single and multithread in plan view. *Single channels* exhibit a great variety of form, from straight through to fully developed meanders, and *multithread channels* appear in braided and anastomosed (anabranched) patterns. These patterns can change gradually or suddenly along streams. In general, channel patterns depict spatially varying conditions of *stream power* and *sediment resistance to transport* and, thus, the processes of sediment transport and deposition within channels (Richards et al., 1993).

CLASSIFICATION

Leopold and Wolman (1957) observed that braided, meandering, and straight-channel patterns stand out prominently in most streams (Figure 5.8), and demonstrated

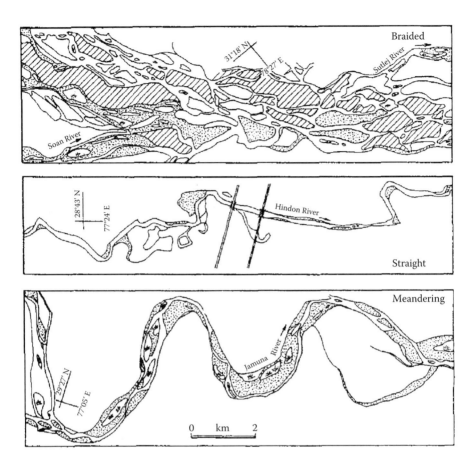

FIGURE 5.8 Typical channel patterns in plan view: braided, meandering, and straight.

that the relationship between channel slope (S) and bankful discharge (Q_b) discriminates braided and meandering channels by a power function of the type

$$S = 0.6 \ Q_b^{-0.44}$$

suggesting that channel slope per unit of discharge is steeper for braided than meandering streams (Figure 5.9). Schumm (1963) proposed an empirical classification of channel patterns by a *sinuosity index*, giving straight and meandering channels with transitions in between. Chitale (1970) suggested a somewhat similar classification by a *tortuosity ratio*, recognizing single and multithread channels as the principal patterns. Brice and Blogett (1978) and Chang (1985) observed that *thresholds* (see Chapter 1) of channel gradient or discharge, or both, distinguish braided and meandering patterns adequately well. Schumm (1981) demonstrated that channel patterns change from straight to meandering and from meandering to straight at two geomorphic thresholds of valley slope. Schumm and Khan (1971) showed that flume channels change from straight to meandering at a certain threshold of sediment load in the

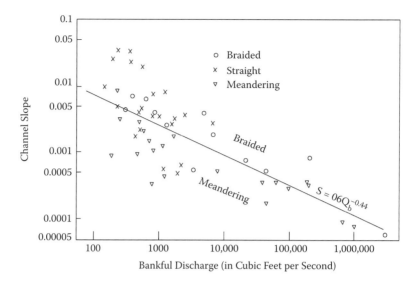

FIGURE 5.9   Slope and bankful discharge: critical distinction of meandering from braiding. (Source: Figure 14 in Leopold, L. B., and Wolman, M. G., *River Channel Patterns; Braided, Meandering and Straight*, U.S. Geological Survey Professional Paper 282B [U.S. Geological Survey, 1957]. With permission from U.S. Geological Survey.)

system. Shen and Schumm (1981) and Millar and Quick (1998) observe that channels change pattern at thresholds governed by hydraulic variables of width–depth ratio, channel gradient, and sediment transport rate. Braided channels are characteristic of a larger width–depth ratio, steeper channel gradient, and higher sediment load than meandering channels (Figure 5.10). Bridge (1993) recognized that *dimensionless hydraulic indices* of flow intensity and channel form index also distinguish braided and meandering patterns over a wide range of discharge conditions. *Flow intensity* is the ratio of shear velocity to shear velocity at the threshold of bedload movement, and *channel form index* is the ratio between the mean flow depth to bed material size and the channel width to bed material size. *Bank stability* is similarly a crucial variable in the evolution of channel patterns, producing bank-height-constrained or braided, and bank-shear-constrained or meandering channel patterns (ASCE, 1998; Millar and Quick, 1998). *Bank-height-constrained* braided channels evolve by the mass wasting of noncohesive banks, and *bank-shear-constrained* meandering channels develop by the erosion of cohesive banks. Despite numerous classifications, braided, meandering, and straight-channel patterns depicting multithread and single channels are common to streams the world over.

Braided Channels

Braided channels are single and multithread in type. *Single-channel* braided streams flow around channel bars and islands, and *multithread* braided streams make an anastomosing pattern of individually meandering channels across *flood plain surfaces* many times the size of channel bars and islands (Schumm, 1977; Bridge, 1993).

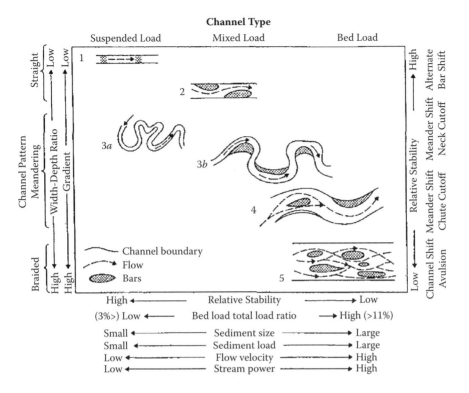

FIGURE 5.10 Classification of channel patterns based on type of sediment load and associated variables. (Source: Figure 4 in Shen, W. H., and Schumm, S. A., *Methods for Assessment of Stream-Related Hazards to Highways and Bridges*, Report FHWA/RD-80/160. [Washington, DC: Federal Highway Administration, Office of Research and Development, 1981]. With permission from Federal Highway Administration.)

Single-Channel Braided Streams

Single-channel braided streams are characteristic of high and variable discharge, steep channel gradient, erodible banks, and a width-to-depth ratio of 50 or more (Fredose, 1978; Fukuoka, 1989; Bridge, 1993). A single-channel braided pattern is generally thought to result from channel aggradation. Experience, however, suggests that aggradation does not necessarily imply braiding, nor is the absence of braiding sufficient proof that channels are not aggrading (Leopold and Wolman, 1957). The braided Kosi River over the years has, for example, scoured then aggraded its bed between Belka Hills in India and Hanuman Nagar in Nepal (Gole et al., 1974).

A single-channel braid pattern establishes by the processes of channel aggradation and channel avulsion. In sand-bedded *flume channels*, the braid pattern evolves by the mid-channel deposition of bedload as longitudinal bars at high flows, gradual growth of the bars by vertical and lateral accretion processes, dissection of the bars into smaller units at intermediate flows, and deflection of the flow that causes bank erosion and channel widening (Figure 5.11). Spatial continuity of this activity in channels of erodible bed and bank materials establishes a field of dissected longitudinal

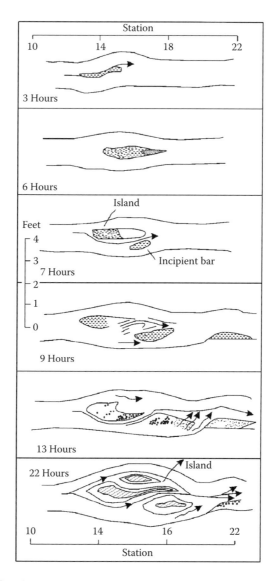

FIGURE 5.11 Development of braid in flume channels. (Source: Figure 4 in Leopold, L. B., and Wolman, M. G., *River Channel Patterns; Braided, Meandering and Straight*, U.S. Geological Survey Professional Paper 282B [U.S. Geological Survey, 1957]. With permission from U.S. Geological Survey.)

bars, some of which evolve as *islands* by the process of local channel accretion. Whether longitudinal bars are permanent bed forms (Leopold and Wolman, 1957) or change with time (Smith, 1974), though, is debated. Data on width, depth, and bed slope of divided and undivided braided channel segments suggest that channel division is an essential compensatory mechanism for efficient transport of discharge and sediment load that otherwise cannot be moved as efficiently through single, undivided stream reaches (Leopold and Wolman, 1957).

Field data on *gravel-bedded streams* suggest that flow convergence and divergence around longitudinal bars, and geometry of confluence and diffluence zones are fundamental yet inadequately understood processes of braid bar development (Ashmore, 1993; Bridge, 1993; Bristow and Best, 1993; Ferguson, 1993; Leddy et al., 1993). The accelerating flow converging in the lee of gravel bars scours the bed, and the decelerating diffluent flow upstream of bars promotes channel aggradation, which process segments the channel reach respectively into sediment transport and deposition zones. The significance of the two zones to braid form development, however, is far from understood.

Multithread Braided Streams

Multithread braided streams appear in an anastomosing pattern of both short straight and long sinuous channels interspersed by flood plain surfaces of sizable extent. The anastomosed (or anabranched) pattern is common to streams of vast alluvial plains, alluvial fans, river deltas, and subsiding basins.

Multithread braided streams develop by processes of channel dispersion and channel avulsion. Channel dispersion is common to streams that suddenly lose *competence* for the transport of discharge and sediment load across foothill-type alluvial fans and subsiding basins, evolving a system of short and straight flow channels. Short, straight channels are also common to gravel-bedded streams, which possibly evolve by avulsion caused by rapid sedimentation in diffluence zones upstream of gravel bars. Rapid siltation of active channels on vast alluvial plains and deltaic surfaces of gentle gradient (Chapter 6) likewise initiates channel avulsion, developing a typical braid pattern comprised of long segments of meandering channels interspersed by flood plain surfaces of sizable extent. The cyclic siltation of active channels hundreds to thousands of years apart has similarly evolved a vast system of old, young, and youngest anastomosing channels in a part of southeastern Australia (Schumm et al., 1996).

MEANDERING CHANNELS

Alluvial streams have a natural tendency to evolve successive loops of sufficient geometric symmetry almost proportional to the channel size. These curves are called meanders. Meanders develop independent of surface slope, terrain roughness, vegetation, and other conditions.

Geometry of Meanders

Wavelength, channel width, radius of channel curvature, and amplitude are interdependent attributes of meandering streams (Figure 5.12). The meander wavelength

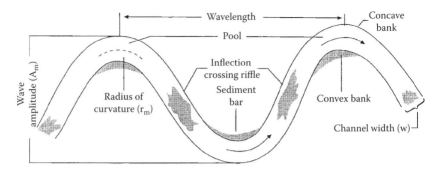

FIGURE 5.12   Geometry of meanders.

and meander amplitude, in particular, are related to other channel shape aspects as (Leopold and Wolman, 1960)

$$\lambda = c \ w^m$$

$$\lambda = f \ r_m^{\ f}$$

$$A_m = d \ w^n$$

in which λ is the meander wavelength, A_m is the meander amplitude, w is the channel width, and r_m is the radius of channel curvature. The letters c, f, and d are constants and m, f, and n are exponents in the given empirical relationships. The exponents are close to unity, suggesting that the relationships are almost linear.

Observations suggest that meander wavelength, and meander amplitude are related to channel width. The meander wavelength is between seven and ten times the mean channel width, and the amplitude of meander loops is determined more by bank stability than other channel conditions (Leopold and Wolman, 1960). However, the proportion between the radius of channel curvature and channel width, called *curvature ratio*, is diagnostic of the meandering process (Bagnold, 1960; Leopold and Wolman, 1960; Hickin, 1974; Hooke, 1987). Bagnold demonstrated that the flow of water in closed pipes of a uniform cross section becomes unstable and recedes from the main boundary with decreasing pipe curvature, affecting velocity distribution over the entire cross section. Flow receding from the inside region of greatest curvature reduces the bed shear, boundary resistance, and flow velocity at the inside bend, causing a compensatory velocity increase at the opposite convex bend. At a critical curvature ratio of 2, the effective channel width is reduced and the effective radius is so increased that the flow becomes unstable and breaks down into eddies, reaching a minimum value for the resistance to flow. Hence, a critical curvature ratio exists for streams at which the resistance to flow is minimum. The general physical principles relating resistance to flow with the curvature ratio, however, are far from understood.

Cause of Meandering

The tendency of streams to meander is related to the *curvature ratio* of 2, at which a deficit of pressure at the inside and an excess of pressure at the outside channel bend institute a helical or transverse flow circulation in the direction of pressure gradient (Bagnold, 1960). The magnitude of *transverse flow* is commonly 10 to 20% of the downstream flow velocity (ASCE, 1998; Jia, 1999).

Field data suggest that the line of maximum downstream flow velocity in meandering streams is closest to the inside bend and crosses over to the next downstream inside bend at the point of inflection on the channel bed. The deviation of maximum velocity from the centerline develops a gradient of *centrifugal force* of water near channel bends (Bridge and Jarvis, 1982; Richards, 1982; Thorne et al., 1985; Reid and Frostick, 1994), initiating a *helical circulation system* at the inside and outside channel banks (Figure 5.13). The highest velocity at the inside bank erodes the bank sediments. These sediments are carried across the point of inflection and laid to rest at the outside bend immediately downstream as *point bar deposits* of the *lateral accretion process*. Hence, the meandering pattern is related to helical-flow-induced bank erosion and lateral accretion of sediments within channels (Jia, 1999).

Leopold and Langbein (1966) observed that the water surface slope or *energy gradient* is uniform and steeper for meandering and stepped for straight-channel segments of Baldwin Creek in Wyoming (Figure 5.14). Hence, a meandering channel pattern closely approximates a uniform rate of energy loss per unit of channel distance and large total energy loss in the curved path.

STRAIGHT CHANNELS

Straight channels are uncommon in nature, rarely exceeding ten times the channel width (Leopold and Langbein, 1966). They are also characteristic of a large width-to-depth ratio, a channel gradient almost equal to the valley slope, and an undulatory bed alternating between a riffle and pool sequence spaced five to seven times the local channel width (Figure 5.15). Riffles are comprised of coarser and pools of finer bed materials along channel banks. The riffle and pool sequence is related to stress-induced *transverse flow* in wide-shallow channels, which generates a weak component of secondary velocity 1 to 2% of the total stream velocity in the system (Rhodes and Knight, 1994; ASCE, 1998). The transverse flow results in back-to-back convergent and divergent eddy flow circulation in the system, respectively evolving a riffle and pool sequence (Figure 5.16). The spacing of riffles is almost similar to the wavelength in meandering streams. Hence, straight and meandering channels possibly evolve by similar mechanisms.

SUMMARY

Channel flow is initiated when the fluid stress of water exceeds the resistance to flow from channel roughness elements. The fluid stress varies with the type of channel flow and size and shape of channel-forming sediments. The resistance to flow is directly proportional to the fluid stress at the bed of streams and inversely proportional to the flow velocity. The dynamic activity of frictionally dominated flow is central to stream erosion and deposition of sediments and, thereby, to the evolution of fluvial landscape.

The character of fluid motion bounded by solid surfaces is given as laminar and turbulent. The type of fluid flow depends on a number of interrelated parameters

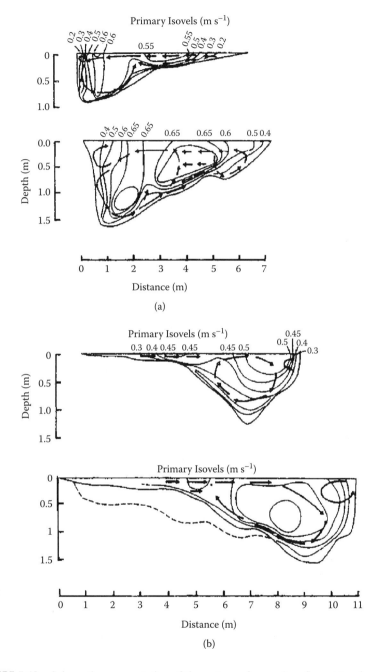

FIGURE 5.13 Schematic representation of the pattern of secondary flow at two bends of a meandering stream at discharges of (a) 1.7 and (b) 4.0 m^3s^{-1}. (Source: Figures 1 and Figure 2 in Thorne, C. R., Zevenbergen, L. W., Pitlick, J. C., Rais, S., Bradley, J. B., and Julien, P. Y., "Direct Measurements of Secondary Currents in Meandering Sand-Bed Rivers," *Nature* 315 [1985]: 746–47. With permission from Macmillan Magazines Ltd.)

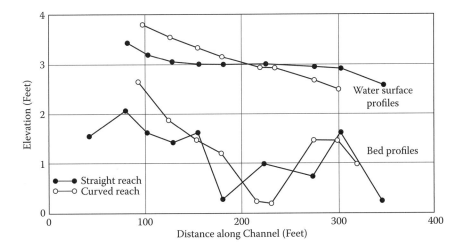

FIGURE 5.14 Water surface and bed profile of meandering and straight-channel segments of Baldwin Creek in Wyoming. (Source: Figure on p. 69 in Leopold, L. B., and Langbein, W. B., "River Meanders," *Sci. Am.* 214 [1966]: 60–70. With permission from Scientific American © 1966.)

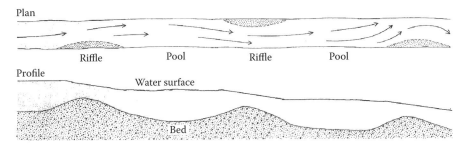

FIGURE 5.15 Uneven bed of a straight-channel segment consists of pools and riffles situated alternatively on each side of the stream at intervals roughly five to seven times the local stream width. Consequently, the stream at low flows follows a coarse wandering from one side of the channel to the other in a manner similar to that of meandering channels. (Source: Figure on p. 69 in Leopold, L. B., and Langbein, W. B., "River Meanders," *Sci. Am.* 214 [1966]: 60–70. With permission from Scientific American © 1966.)

that are summarized in the widely used dimensionless Reynolds number and Froude number. The Reynolds criterion differentiates flow types by the viscous effect in fluid flow, and the Froude criterion is a measure of the gravity effect on fluid motion. The type of flow, in general, governs the deformation behavior of sediments in alluvial streams.

Streams transport quantity-limited clastic and solute loads of respective drainage basins. The clastic load is commonly classified as bedload, suspended load, and wash load, and the solute load comprises a variety of dissolved elements and compounds in the stream flow. Bedload is the coarser fraction of bed material load that travels by rolling, sliding, and saltation at the bed of streams. Suspended load is held in

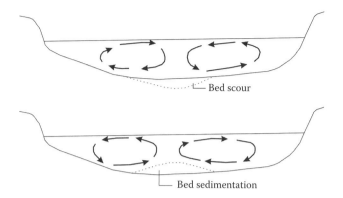

FIGURE 5.16   Possible mechanism for the evolution of pool and riffle sequence in straight-channel segments. (Source: Based on Figure 4.40 in Reid, I., and Frostick, L. E., "Fluvial Sediment Transport and Deposition," in *Sediment transport and depositional processes*, ed. K. Pye [Oxford: Blackwell Scientific Publications, 1994], 89–155.)

suspension by the turbulence in channel flow and is never in contact with the bed for a sufficient time. The suspended load arrives into the system from overland flow during storm events. The concentration of suspended load in the flow is a usable index for otherwise poorly understood channel conditions and the total sediment load moved by streams. Wash load is comprised of colloidal clay particles that always remain in suspension irrespective of the channel condition. Chemical composition of the dissolved load varies with basin lithology. Its concentration in the stream flow progressively increases from arid to the humid climatic regimes.

Flow velocity is the expression of fluid stress on the streambed and the resistance to flow. It varies through the water depth and across channels. The mean velocity is conveniently determined by the Chézy and Manning equations. The Chézy equation relates the mean flow velocity to the hydraulic radius and slope of energy gradient. The Manning equation expresses the mean flow velocity by water depth, channel slope, and resistance to flow due to channel roughness elements.

The bed particles are set in motion by the fluid stress just exceeding the grains' resistance to shear. The deforming bed, however, continuously changes the bed configuration and thus influences conditions of roughness elements, channel shape, and water and material flow through streams. Hence, the physical laws governing sediment transport can only be conservatively applied to fluvial systems. Several theoretical and semiempirical models have, therefore, been developed for the estimation of sediment transport rate.

Frictional interaction between the channel flow and bed materials generates a variety of downstream migrating microrelief bed forms. The range of flows producing similar resistance to flow, sediment transport rate, and bed configuration are called regimes of flow. Regimes of flow, given as lower, intermediate, and upper, explain channel conditions for bed accretion.

Interrelated hydraulic variables of channel width, depth, velocity, and suspended load vary systematically with channel discharge. The hydraulic geometry interprets the rate of mutual adjustment among hydraulic variables with at-a-station and downstream discharge. Field data suggest that the hydraulic variables do not universally respond by a like amount of change for the given magnitude of at-a-station discharge. In general, channel depth increases at a somewhat faster rate than channel width, and the rate of velocity increase is somewhat smaller for channels of a higher rate of depth increase. For downstream equal-frequency discharge, however, the channel width increases rapidly, depth does not change by as much of a rate, and flow velocity increases at a moderate rate.

Suspended load affects the channel shape through mutual adjustments in velocity and depth to discharge. The concentration of suspended load increases with the decrease of channel width at constant velocity or the increase of velocity at constant width. The concentration of suspended load in channel flow, though, varies throughout a flood event. It dampens the resistance to flow and increases the flow velocity. Consequently, bed scour in the rising stage of higher concentration of suspended load and channel siltation in the falling stage of lower concentration of suspended load modifies the channel shape during flood events.

Channel slope is also a hydraulic variable in adjustment with channel roughness elements and resistance to flow in alluvial streams. The channel slope progressively declines downstream, evolving a typical concave-up bed gradient. This gradient is adjusted to channel discharge in ways that provides an efficient slope of transport in equilibrium with the increase in channel width, depth, and velocity downstream.

Channel patterns depict the mutual adjustment of channel slope and sediment load to discharge over long segments of alluvial streams. They are commonly single and multithread channels, giving braided, meandering, and straight-channel patterns in plan view. Braided streams comprise single and multithread channels. Single-channel braided streams are characteristic of variable discharge, erodible banks, steep channel gradient, large sediment load, and width-to-depth ratio of 50 or more. A single-channel braid pattern evolves by local accretion of the bedload as longitudinal bars at high flows, dissection of the bars at intermediate flows, deflection of flow toward banks, and bank erosion. This process establishes a network of intertwined short straight channels around and through longitudinal bars. Multithread braided channels represent an anastomosed or anabranched pattern of sinuous channel segments interspersed with flood plain surfaces of sizable extent. This pattern evolves by flow dispersion and channel avulsion. A meandering channel pattern, which is common to alluvial streams over surfaces of gentler gradient, is somehow related to a critical ratio of 2 for the radius of curvature to channel width at which the resistance to flow is minimum and helical circulation is set up in the system. Straight channels are uncommon, seldom exceeding ten times the local channel width. They possess a gentler gradient, and an undulatory bed alternating between a riffle and pool sequence spaced five to seven times the local channel width. This sequence is possibly related to a gentle secondary flow circulation somewhat similar to that in meandering streams.

REFERENCES

ASCE Task Committee on Hydraulics, Bank Mechanics, and Modeling of River Width Adjustment. 1998. River width adjustment. I. Processes and mechanisms. *J. Hydraulic Eng.* 124:881–902.

ASCE Task Force on Bed Forms in Alluvial Channels. 1966. Nomenclature for bed forms in alluvial channels. *J. Hydraulics Div.* 92:51–64.

Ashmore, P. 1993. Anabranch confluence kinetics and sedimentation processes in gravel-braided streams. In *Braided rivers*, ed. J. L. Best and C. S. Bristow, 129–46. Geological Society Special Publication 75. Geological Society.

Ashmore, P. E. 1991. How do gravel-bed rivers braid? *Can. J. Earth Sci.* 28:326–41.

Bagnold, R. A. 1960. *Some aspects of the shape of river meanders.* U.S. Geological Survey Professional Paper 282E. U.S. Geological Survey.

Bogárdi, J. 1978. *Sediment transport in alluvial streams.* Budapest: Akadémiai Kiadó.

Bridge, J. S. 1993. The interaction between channel geometry, water flow, sediment transport and deposition in braided rivers. In *Braided rivers*, ed. J. L. Best and C. S. Bristow, 13–71. Geological Society Special Publication 75. Geological Society.

Bridge, J. S., and Jarvis, J. 1982. The dynamics of a river bend: A study in flow and sedimentary processes. *Sedimentology* 29:499–541.

Brice, J. C. and Blodgett, J. C. 1978. Counter measures for hydraulic problems at bridges. Vol. 1, Analysis and Assessment, Federal Highway Administration, Dept. No. FHWA-RD-78-162, Washington, D.C.

Bristow, C. S., and Best, J. L. 1993. Braided rivers: Perspectives and problems. In *Braided rivers*, ed. J. L. Best and C. S. Bristow, 1–11. Geological Society Special Publication 75. Geological Society.

Chang, H. H. 1985. River morphology and thresholds. *J. Hydraulic Eng.* 111:503–19.

Chitale, S. V. 1970. River channel patterns. *J. Hydraulics Div.* 96, 201–21.

Coleman, S. E., and Melville, B. W. 1994. Bed-form development. *J. Hydraulic Eng.* 120:544–60.

Fahnestock, R. K. 1963. *Morphology and hydrology of a glacial stream—White River, Mount Rainier, Washington.* U.S. Geological Survey Professional Paper 422A. U.S. Geological Survey.

Ferguson, R. I. 1993. Understanding braiding processes in gravel bed rivers: Progress and unsolved problems. In *Braided rivers*, ed. J. L. Best and C. S. Bristow, 73–87. Geological Society Special Publication 75. Geological Society.

Foley, M. G. 1978. Scour and fill in sand-bedded ephemeral streams. *Bull. Geol. Soc. Am.* 89:559–70.

Fredose, J. 1978. Meandering and braiding of rivers. *J. Fluid Mechanics* 84:609–24.

Fukuoka, S. 1989. Finite amplitude development of alternate bars. In *River meandering*, ed. S. Ikeda and G. Parker, 237–65. Water Resources Monographs 12. American Geophysical Union.

Garde, R. J., and Ranga Raju, K. G. 1977. *Mechanics of sediment transportation and alluvial stream problems.* New Delhi: Wiley Eastern Ltd.

Gomez, B. 1991. Bedload transport. *Earth Sci. Rev.* 31:89–132.

Graf, W. H. 1998. *Fluvial hydraulics.* Chichester: John Wiley & Sons.

Hack, J. T. 1957. *Studies on longitudinal stream profiles in Virginia and Maryland.* U.S. Geological Survey Professional Paper 249B. U.S. Geological Survey.

Hickin, E. J. 1974. The development of meanders in natural river channels. *Am. J. Sci.* 274:414–42.

Jia, Y. 1999. Numerical model for channel flow and morphological change studies. *J. Hydraulic Eng.* 125:924–33.

Karim, F. 1999. Bed-form geometry in sand-bed flows. *J. Hydraulic Eng.* 125:1253–61.

Knighton, A. D. 1974. Variation in width-discharge relation and some implications for hydraulic geometry. *Bull. Geol. Soc. Am.* 85:1069–76.

Knighton, A. D. 1975. Variations in at-a-station hydraulic geometry. *Am. J. Sci.* 275:186–218.

Langbein, W. B., and Leopold, L. B. 1964. Quasi-equilibrium states in channel morphology. *Am. J. Sci.* 262:782–94.

Leddy, J. O., Ashworth, P. J., and Best, J. L. 1993. Mechanisms of anabranch avulsion within gravel-bed braided rivers: Observations from a scaled physical model. In *Braided rivers*, ed. J. L. Best and C. S. Bristow, 119–27. Geological Society Special Publication 75. Geological Society.

Leopold, L. B., and Langbein, W. B. 1966. River meanders. *Sci. Am.* 214:60–70.

Leopold, L. B., and Maddock, T., Jr. 1953. *The hydraulic geometry of streams and some physiographic implications*. U.S. Geological Survey Professional Paper 252. U.S. Geological Survey.

Leopold, L. B., and Wolman, M. G. 1957. *River channel patterns; braided, meandering and straight*. U.S. Geological Survey Professional Paper 282B. U.S. Geological Survey.

Leopold, L. B., and Wolman, M. G. 1960. River meanders. *Bull. Geol. Soc. Am.* 71:769–94.

Leopold, L. B., Wolman, M. G., and Miller, J. P. 1964. *Fluvial processes in geomorphology*. San Francisco: W. H. Freeman and Co.

McLean, S. R., Wolfe, S. R., and Nelson, J. M. 1999. Predicting boundary shear stress and sediment transport over bed forms. *J. Hydraulic Eng.* 125:725–36.

Millar, R. G., and Quick, M. C. 1998. Stable width and depth of gravel-bed rivers with cohesive banks. *J. Hydraulic Eng.* 124:1005–13.

Morisawa, M. 1968. *Streams*. New York: McGraw-Hill Book Company.

Powell, D. M., Reid, I., and Laronne, J. B. 1999. Hydraulic interpretation of cross-stream variations in bed-load transport. *J. Hydraulic Eng.* 125:1243–52.

Reid, I., and Frostick, L. E. 1994. Fluvial sediment transport and deposition. In *Sediment transport and depositional processes*, ed. K. Pye, 89–155. Oxford: Blackwell Scientific Publications.

Rhodes, D. G., and Knight, D. W. 1994. Distribution of shear force on the boundary of smooth rectangular duct. *J. Hydraulic Eng.* 120:787–807.

Richards, K., Chandra, S., and Friend, P. 1993. Avulsive channel systems: Characteristics and examples. In *Braided rivers*, ed. J. L. Best and C. S. Bristow, 195–203. Geological Society Special Publication 75. Geological Society.

Richards, K. S. 1982. *Rivers: Form and process in alluvial channels*. London: Methuen.

Schick, A. P., Lekach, J., and Hassan, M. A. 1987. Vertical exchange of coarse bedload in desert streams. In *Desert sediments: Ancient and modern*, ed. L. E. Frostick and I. Reid, 7–16. Oxford: Blackwell Scientific Publications.

Schumm, S. A. 1963. Sinuosity of alluvial rivers of Great Plains. *Bull. Geol. Soc. Am.* 74:1089–99.

Schumm, S. A. 1977. *The Fluvial System*. New York: John Wiley & Sons.

Schumm, S. A. 1981. Geomorphic thresholds and complex response of drainage systems. In *Fluvial geomorphology*, ed, M. Morisawa, 299–310. London: George Allen & Unwin.

Schumm, S. A., Erskine, W. D., and Tilleard. 1996. Morphology, hydrology and evolution of anastomosing Ovens and King rivers, Victoria, Australia. *Bull. Geol. Soc. Am.* 108:1212–24.

Schumm, S. A., and Khan, H. R. 1971. Experimental study of channel patterns. *Nature* 233:407–9.

Shen, W. H., and Schumm, S. A. 1981. *Methods for assessment of stream-related hazards to highways and bridges*. Report FHWA/RD-80/160. Washington, DC: Federal Highway Administration, Office of Research and Development.

Simons, D. B., and Richardson, E. V. 1966. *Resistance to flow in alluvial channels*. U.S. Geological Survey Professional Paper 422J. U.S. Geological Survey.

Smith, N. D. 1974. Sedimentology and bar formation in the Upper Kicking Horse River, a braided outwash stream. *J. Geol.* 82:205–23.

Sundborg, Å. 1956. The dynamics of flowing water. *Geogr. Annlr.* XXXVIII:133–64.

Sundborg, Å. 1967. Some aspects on fluvial sediments and fluvial morphology. I. General views, and graphic methods. *Geogr. Annlr.* 49-A:333–43.

Swamee, P. K., and Ojha, C. S. P. 1994. Criteria for evaluating flow classes in alluvial channels. *J. Hydraulic Eng.* 120:652–57.

Thorne, C. R., Zevenbergen, L. W., Pitlick, J. C., Rais, S., Bradley, J. B., and Julien, P. Y. 1985. Direct measurements of secondary currents in meandering sand-bed rivers. *Nature* 315:746–47.

Wolman, M. G. 1955. *The natural channel of Brandywine Creek, Pennsylvania*. U.S. Geological Survey Professional Paper 271. U.S. Geological Survey.

6 Fluvial Processes and Depositional Landforms

Weathering, surface erosion, and mass movement activities within drainage basins generate sufficient sediments of diverse origin and size composition for transport and deposition by streams as *alluvium*. This alluvium is deposited as flood plains on land, alluvial fans commonly on land, and river deltas partly on land and partly in adjacent water bodies of sizable extent. *Flood plains* are the hydrologic response of sediment transport processes within and beyond channel margins. *Alluvial fans* are a feature of local deposition on surfaces over which streams suddenly lose competence for the transport of sediment load from upland basin sources. *River deltas* develop in the transitional environment as subaerial and subaqueous deposits of alluvium at or near the mouth of streams draining into the sea or large freshwater lakes.

FLOOD PLAINS

Streams deposit the alluvium within channels and beyond channel banks during normal and overbank flow conditions, developing a low-relief linear topographic surface to the level of *mean annual flood* that statistically reoccurs once every 1 to 2 years. This surface of *alluvial fill* is called flood plain, the composition and thickness of which varies with discharge and sediment load characteristics of streams, frequency and magnitude of overbank flows, lateral distance from channel banks, channel pattern, and pattern of channel migration.

FLOOD PLAIN SEDIMENTS

The alluvial fill of most flood plains comprises the bed material of active and inactive channels, sediments laid down at channel margins, sediments deposited overbank, and sediments of mass movement activity adjacent to the valley-side slopes (Vanoni, 1971). These sediments are diagnostic of lateral and vertical accretion processes and processes of slope failure (Figure 6.1), providing for four major sedimentary environments and eight subenvironments of fluvial deposition (Table 6.1). *Lateral accretion* is the deposition process of coarser sediments within channels, and *vertical accretion* refers to the sedimentation of finer fractions in overbank flows at and beyond channel margins. The deposits of *slope failure*, called colluvium, are external to channel processes of flood plain evolution. They only contribute to the total depth of valley fills constrained by valley walls. The subenvironments of lateral and vertical accretion domains, like channel, channel margin, and overbank sedimentation, evolve units or *facies* that are distinct in grain size composition (Figure 6.2).

139

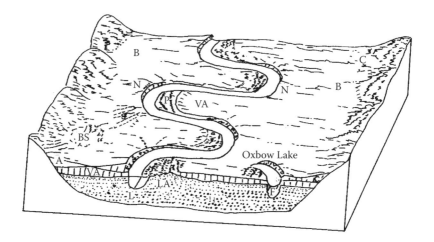

FIGURE 6.1 Typical association of valley sediment deposits: VA, vertical accretion; N, natural levee; B, backland; BS, backswamp; LA, lateral accretion; P, point bar; S, splay; A, alluvial fan; T, transitory deposit; F, channel fill; C, colluvium. (Source: Figure 2-Q.1 in Vanoni, V. A., "Sediment Transportation Mechanics: Q. Genetic Classification of Valley Sediment Deposits," *J. Hydraulics Div.*, No. HYI, 43-53, 1971, p. 44. With permission from American Society of Civil Engineers.)

Channel Deposits

Channel deposits are classified as transitory coarser bedload materials, heavier and denser lag fractions, and finer channel fill sediments, which move or had moved at the bed of streams. Transitory bedload sediments consist of downchannel migrating small-sized bed forms in equilibrium with the flow velocity and sediment transport rate (Chapter 5), and longitudinal and transverse bars are typically large-sized bed forms of *braided streams.* *Longitudinal bars* established in the period of rising flood as permanent submerged bedload deposits of sand or gravel composition, or both, which, however, become dissected in the falling flood stage into smaller-sized units. *Transverse bars* are semipermanent bed forms of sand composition. They are deposited in the period of low to intermediate channel flow but are destroyed in flood events.

Lag deposits are heavier clastic fractions and denser mineral placers within channel deeps. They move infrequently in exceptional floods.

Channel fill is generally identified with the bed material of abandoned courses of *meandering streams.* Segments of active meanders are abandoned by chute cutoff, neck cutoff, and channel avulsion (Figure 6.3). *Chute cutoff* develops when a surging flood opens up a straight channel segment across *point bars* on the outside bend of meander loops. A small angular difference between the old and new channels, however, initially allows movement of water and sediment load through the old channel until it is completely plugged and abandoned in time. *Neck cutoff* evolves from the joining of the channel of an exaggerated meander loop across a narrow meander neck. This process leaves a large angular difference between the old and new channels that does not permit much movement of the bedload through the segment of abandoned course. *Avulsion* is the process of those streams, which rapidly silt large segments of

TABLE 6.1
Classification of Valley Fill

Place of Deposition	Name	Characteristics
Channel	Transitory deposits	Primarily bedload temporarily at rest; part may be preserved in channel, more durable channel fills, or lateral accretions
	Lag deposits	Segregation of larger and heavier particles; more persistent than transitory channel deposits; includes heavy mineral placers
	Channel fills	Accumulations in abandoned or aggrading channel segments; ranging from relatively coarse bedload to fine-grained oxbow deposits
Channel	Lateral accretion deposits	Point and marginal bars, which may be preserved by channel margin shifting and added to overbank flood plain by vertical accretion deposits at top
Overbank flood plain	Vertical accretion deposits	Fine-grained sediment deposited from suspended load of overbank floodwater; includes natural levee and backland (backswamp) deposits
	Splays	Local accumulation of bedload materials spread from channels onto adjacent flood plains
Valley margin	Colluvium	Deposits derived chiefly from unconcentrated slope wash and soil creep on adjacent valley sides
	Mass movement deposits	Earthflow, debris avalanche, and landslide deposits commonly intermixed with marginal colluvium; mudflows usually follow channels but also spill overbank

Source: Table 2-Q.1 in Vanoni, V. A., "Sediment Transportation Mechanics: Q. Genetic Classification of Valley Sediment Deposits," *J. Hydraulics Div.*, No. HYI, 43-53, 1971, p. 44. With permission from American Society of Civil Engineers.

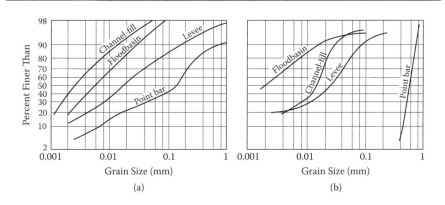

FIGURE 6.2 Cumulative size frequency curves of representative deposits from alluvial environments. (a) Sacramento River, California (after Lorens and Thronson, 1955, pl. 7). (b) Mississippi River (after Fisk, 1947, pl. 68, 69; Frazier and Osanik, 1961, Figure 3). (Source: Figure 23 in Allen, J. R. L., "A Review of the Origin and Characteristics of Alluvial Sediments," *Sedimentology* 5 [1965]: 89–191. With permission from Blackwell Science.)

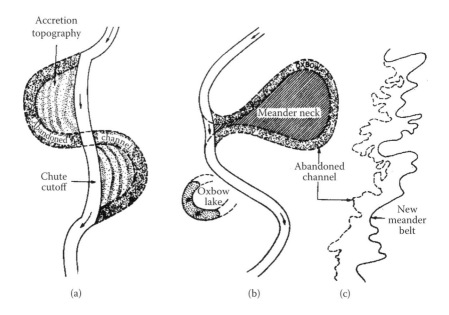

FIGURE 6.3   Models of channel shifting. (a) Chute cutoff. (b) Neck cutoff. (c) Development of new meander belt following avulsion. Abandoned channels are shown by dotted lines.

the active channel to the level of adjacent flood plain surface (Smith, 1983; Richards et al., 1993). Consequently, the aggraded channel segment is altogether abandoned by the opening of a new channel, called *anabranch*, upstream of the affected reach. Such abandoned channel courses represent the alluvial fill only of the former stream regime. Gravel-bedded braided streams also avulse, possibly by the effect of the adjustment between sediment transport rate and flow structure around braid bars and channel islands (Bristow and Best, 1993; Thorne et al., 1993; Warburton et al., 1993). Channels can also silt from other causes that may be internal and external to the fluvial system. Thus, enlarged meander loops of *flattened gradient* rapidly silt (Schumm et al., 1996), and *neotectonic movements* that enhance the sediment supply rate of drainage systems silt channels at an accelerated rate (Ramasamy et al., 1991).

Channel Margin Deposits

Sediments eroded from the inside bank of meandering streams are carried across the point of channel inflection even at a low flow velocity of 0.08 m s^{-1} (Wolman and Leopold, 1957), and laid down by a like amount immediately downstream on the outside bank of meandering streams as moderately sorted and stratified mud through to pebble-sized point bar deposits of the lateral accretion process. Such a regular pattern of channel erosion and deposition maintains uniformity of channel width through time (Figure 6.4).

Point bar deposits appear in a series of small-sized curved ridges and intervening depressions aligned to the bank in conformity with channel sinuosity and pattern of channel migration. The thickness of point bar deposits, in general, varies directly with the size of meandering streams. The point bars of the small-sized Watts Branch

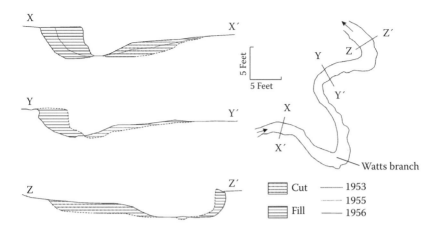

FIGURE 6.4 Progressive erosion and deposition on sections of Watts Branch. Such a process of erosion and deposition maintains a channel at constant width. (Source: Figure 4 in Wolman, M. G., and Leopold, L. B., *River Flood Plains: Some Observations on Their Formation*, U.S. Geological Survey Professional Paper 282-C [U.S. Geological Survey, 1957]. With permission from United States Geological Survey.)

in Maryland are 1 m thick (Wolman and Leopold, 1957), and those of the large-sized Mississippi River in Louisiana are 12 to 18 m thick at places (Fisk, 1947).

Overbank Flood Plain Deposits

Turbulent flow in flood spasms places a large volume of fine-grained sediments in suspension for overbank flood plain sedimentation adjacent to and beyond channel margins. The overbank sedimentation evolves *vertical accretion* deposits of natural levees adjacent to channel margins, and splay and backland deposits outward from stream banks. Observations suggest that the *concentration* of suspended particles in channel overflow decreases in an inverse logarithmic fashion with distance from stream banks (Sundborg, 1965; Kesel et al., 1974). Hence, the depth of sedimentation from overbank suspended particles and also the sediment size of the deposit decrease logarithmically outward from stream banks (Figure 6.5). Studies also suggest that the overbank sedimentation rarely accounts for more than 10% of the flood plain height (Schmudde, 1968).

Natural levees are wedge-shaped channel bank deposits of coarsest fractions in the overflow escaping channels. The size of natural levees, in general, depends on the frequency of overbank flow, concentration of suspended particles in overflow, and postdepositional compaction and subsidence of sediments. The natural levees of the lower Mississippi River are more than 1.6 km wide and 4.5 to 7.5 m high (Fisk, 1947), and those of Brahmaputra River, India, are comparable in size (Coleman, 1969). Natural levees hold along stable channel banks but are destroyed when related streams migrate laterally, causing bank erosion. Few natural levees, which accidentally breach in flood events, offer spillways for carrying the flood flow and its sediment load for deposition in the form of lobe-shaped deposits called *splays*. Splays present an irregular surface outward from the levees.

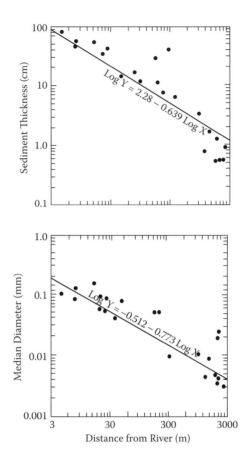

FIGURE 6.5 Variation of sediment thickness (top) and median grain size (bottom) as a function of lateral distance from channel margin. (Source: Figures 2 and 3 in Kesel, R. H., Dunne, K. C., McDonald, R. C., and Allison, K. R., "Lateral Erosion and Overbank Deposition on the Mississippi River in Louisiana Caused by 1973 Flooding," *Geology*, September, 461–64. With permission from Geological Society of America.)

The overbank flow routed through breached natural levees reaches the limit of valley walls but returns along the surface gradient to the contributing channel. *Backswamps* evolve in this lateral extent of the surface with slow draining conditions. Such sufficiently large-sized surface depressions, however, accommodate fairly thick backswamp deposits, which in Yazoo County, Mississippi, are 3 to 30 m thick and rich in organic matter. *Backlands* represent thick clay deposits in backwater stilling basins of near-stagnant water conditions.

Valley Margin Deposits

Valley-side slopes release a variety of mass movement deposits, discussed in Chapter 4. They variously intermix vertically and horizontally with flood plain sediments and add to the total volume of valley fill.

FLOOD PLAIN PROCESSES

Flood plains are *valley fills* of lateral and vertical accretion, in which the relative contribution of two processes to the overall development of flood plains varies with flood frequency, flood velocity, rate of channel migration, and channel pattern of individual drainage systems. Studies on the dynamics of flood plains from the United States and Canada suggest that they evolve from dominant and co-dominant processes of lateral and vertical accretion of sediments, and valley margin deposits that are external to channel processes only contribute to the total volume of the alluvial fill of flood plain surfaces.

LATERAL ACCRETION PROCESS

The flood plain of Watts Branch, a small-sized *meandering stream* in Maryland, is comprised mostly of *point bar deposits* of the lateral accretion process. The relative insignificance of the vertical accretion process to the alluvial fill of this flood plain is related to low concentration of suspended sediments in the channel overflow, return of much of the overbank sediments to the channel in receding floods, and a regular pattern of channel migration (Wolman and Leopold, 1957).

Flood plains of *gravel-bedded braided streams* also evolve from lateral accretion of sediments. Experimental and field studies on such streams suggest that the sediment transport rate adjusts rapidly to variations in discharge and channel roughness elements. This form of adjustment distinctly segments the channel into sediment transport zones in the area of confluent flow downchannel of braid bars and channel islands, and of sediment storage zones in the area of diffluent flow upchannel of longitudinal bars and channel islands. Lateral accretion of bed sediments in this storage zone evolves flood plain surfaces to the level of water height in channels (Ashmore, 1993; Bridge, 1993; Bristow and Best, 1993; Ferguson, 1993; Passmore et al., 1993; Thorne et al., 1993; Warburton et al., 1993).

VERTICAL ACCRETION PROCESS

Vertical accretion is the dominant morphodynamic control of the flood plains of braided streams developed in noncohesive bed and bank materials. The flood plain of *sand-bedded braided streams* in semiarid environments evolves through the mediation of vegetation that colonizes braid bars in periods of above-average rainfall, and helps trap channel-moving sediments in moderate flows. These sediments attach to one or the other bank, evolving a flood plain surface of sizable extent within wide shallow channels (Schumm and Lichty, 1963; Thorne et al., 1993). The crest and lee of channel bends, mouths of tributary junction, and abandoned channel courses are additional sites for the vertical accretion of sediments in moderate channel flow conditions (Figure 6.6). Flood plains of sand-bedded braided streams, however, are destroyed in flood spasms.

The *backwater effect* upstream of reservoirs and in the zone of stream confluence rapidly silts channels, developing flood plains to the level of the high water mark.

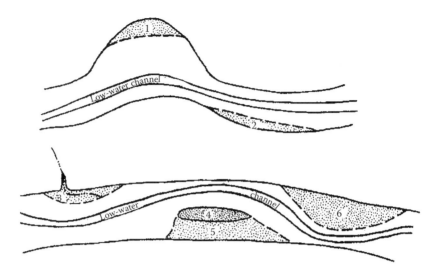

FIGURE 6.6   Channel areas favorable for flood plain formation: 1, crest of valley meanders; 2, lee of valley bends; 3, mouth of tributaries; 4, island; 5, lee of island (abandoned channel); 6, surface above low-water channel. (Source: Figure 50 in Schumm, S. A., and Lichty, R. W., *Channel Widening and Flood-Plain Construction along Cimarron River in Southwestern Kansas*, U.S. Geological Survey Professional Paper 352-D [U.S. Geological Survey, 1963]. With permission from United States Geological Survey.)

A flood plain of the North Saskatchewan River at its confluence with the Alexandra River in the Canadian Cordillera clearly illustrates the role of the backwater effect on vertical accretion of sediments in fluvial systems (Smith, 1981).

LATERAL AND VERTICAL ACCRETION PROCESSES

Many flood plains comprise lateral and vertical accretion sediments through to the depth of valley fills, suggesting a change in the hydrologic regime of related stream or channel processes, or both. The alluvial fill of the Mississippi River in Louisiana comprises a substratum of sand and gravel as channel deposits and silt and clay as overbank sediments. This flood plain also carries extensive back-swamp deposits at the surface. The *stratigraphy* of the flood plain, therefore, suggests a change in channel pattern from braided to meandering and channel avulsion in recent times (Kesel et al., 1974). The alluvial fills of a few British streams similarly suggest a change in channel pattern due to the *climate change* (Passmore et al., 1993). The sequence of lateral and vertical accretion sediments in alluvial fills of Duck River, Tennessee (Brakenridge, 1984), and that of Gasconade River, Missouri (Ritter and Blakley, 1986), however, attest to channel conditions altered by *sediment loading* of drainage systems from the effects of cultivation, deforestation, construction, and mining activities within catchments. This sediment rapidly silts channels and increases the frequency of discharge exceeding the bankful stage.

VALLEY MARGIN PROCESSES

Flood plains constrained by valley walls carry deposits of *slope failure* at and within the depth of valley fill. The valley fill of Beaverdam Run in Pennsylvania carries *colluvium*, which alone is some 20% of the total volume of flood plain sediments (Lattman, 1960). This colluvium is identified with locally slumped bedrock along valley-side slopes.

RIVER TERRACES

Terraces are *abandoned flood plains*. Flood plains are abandoned when streams downcut into the parent channel sufficiently rapidly to a depth for which the *mean annual flood* cannot top the flood plain surface. Streams trench the bed when the *fluvial system* becomes unstable by the effects of climate change, tectonic instability, and eustatic rise and fall of the sea level (Chapter 1). Signatures of the environmental change, which normally can be traced to the beginning of the Quaternary period, are preserved in the morphology and stratigraphy of alluvial fills.

NOMENCLATURE

Terraces exist at several levels along active channels. They are generally classified by the height above related streams and the sequence of trenching and aggradation of alluvial fills along valleys. Terrace remnants at matching levels on either side of streams are paired, and at nonmatching levels on opposite sides of the valley are nonpaired (Figure 6.7). *Paired terraces* represent alluvial fills of continuous lateral planation, *nonpaired terraces* suggest progressive trenching of existing alluvial fills, and terraces of *mixed type* are composite in origin (Thornbury, 1969).

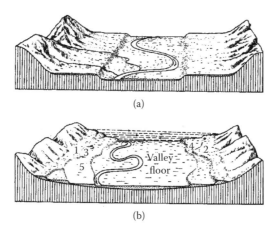

(a)

(b)

FIGURE 6.7 Diagrams showing differences between paired terraces (a) and nonpaired terraces (b). (Source: Figure 6.9 in Thornbury, W. D., *Principles of Geomorphology*, 2nd ed. [New York: John Wiley & Sons, 1969]. With permission.)

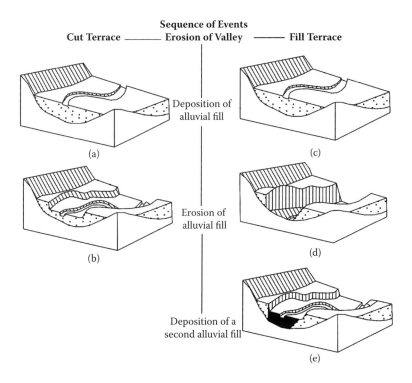

FIGURE 6.8   Block diagrams illustrating stages in the development of a cut terrace (a and b) and fill terrace (c, d, and e). (Source: Figure 1 in Leopold, L. B., and Miller, J. P., *A Postglacial Chronology for Some Alluvial Valleys in Wyoming*, U.S. Geological Survey Water-Supply Paper 1261 [U.S. Geological Survey, 1954]. With permission from United States Geological Survey.)

Terraces evolve from a cut-and-fill sequence of alluvial fills within valleys (Leopold and Miller, 1954). *Cut terraces* describe a trenched topographic surface of preexisting channel fills (Figure 6.8(a),(b)), and *fill terraces* refer to the surface of subsequent channel alluviation not exceeding the channel depth (Figure 6.8(c),(d)). Depending on the magnitude and sequence of alluviation and trenching, any number of terraces can develop at different heights along streams (Figure 6.8(e)). The cut-and-fill terrace sequences, however, are difficult to identify as channel trenching and channel migration (Figure 6.9(a)), and overlapping alluvial fills (Figure 6.9(b),(c)), obliterate the stratigraphic continuity of alluvial fills within drainage systems.

ENVIRONMENTAL CONTROLS

Climate change, tectonic instability of the landmass, and Quaternary sea level oscillations affect aggradation and trenching of alluvial fills in many different ways, evolving several terraces along streams. Human activities also alter the environment of watersheds by affecting the channel flow and sediment yield of drainage basins in long-term perspectives.

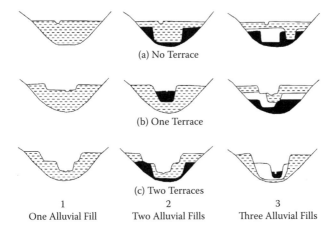

(a) No Terrace

(b) One Terrace

(c) Two Terraces

1	2	3
One Alluvial Fill	Two Alluvial Fills	Three Alluvial Fills

FIGURE 6.9 Examples of valley cross sections showing some possible stratigraphic relations in valley alluvium. (Source: Figure 2 in Leopold, L. B., and Miller, J. P., *A Postglacial Chronology for Some Alluvial Valleys in Wyoming*, U.S. Geological Survey Water-Supply Paper 1261 [U.S. Geological Survey, 1954]. With permission from United States Geological Survey.)

Climatic Controls

The erosive potential of climate varies with spatial variations in climate-dependent *biomass density* (Langbein and Schumm, 1958). Data on the sediment yield of a large number of drainage basins similar in size, topography, and geology across the United States suggest remarkable covariance of the sediment yield with indices of climate and associated density of vegetal cover (Figure 6.10). The per unit basin yield of channel flow and sediment load is lowest for vegetation-sparse arid climates of

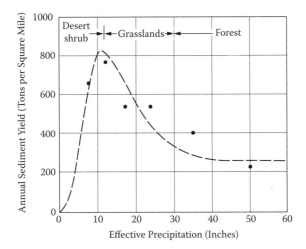

FIGURE 6.10 Climatic variation of sediment yield as determined from records at sediment stations. (Source: Figure 2 in Langbein, W. B., and Schumm, S. A., "Yield of Sediment in Relation to Mean Annual Precipitation," *Trans. Am. Geophys. Union* 39 [1958]: 1076–84. With permission from Institute of American Geophysical Union.)

10 in. (25 cm) or less effective mean annual rainfall. The basin yield of detritus load likewise is low in the forest environment of humid climates of more than 30 in. (75 cm) effective mean annual rainfall, where higher flow volume relative to the sediment yield trenches channels and preexisting valley fills. However, vegetation density favorably combines with the semiarid climate of 10 to 14 in. (25 to 35 cm) effective mean annual rainfall for optimum surface erosion and, thereby, of sediment yield for channel aggradation. Hence, sequences of channel trenching and aggradation follow from the change of climate between humid and semiarid phases of sustained length.

The morphology and association of terrace remains with other geomorphic features of terrain and datable organic matter within the terrace alluvium are useful indices of the climate control on the evolution and chronology of Quaternary terrace sequences. Thus, several terrace sequences in the Alpine region of Western Europe (Zeuner, 1968) and in the Canadian Cordillera (Stalker, 1968) are held to be glacial-interglacial equivalents of the Pleistocene climate. ^{14}C-dated terrace remnants in parts of the Prairies of Canada (Sharma, 1973), Great Plains of the United States (Brice, 1964) and New South Wales, Australia (Erskine, 1994) similarly attest to episodic changes in the Holocene climate.

Tectonic Controls

Tectonic instability affects the established order of equilibrium between process rates and components of the geomorphic landscape. Tectonic revival rejuvenates the relief, and streams adjust to the new energy level by trenching into the bed at higher gradient. Thus, the valley fill subsequent to tectonic uplift is laid down at a gradient steeper than the valley fills of earlier periods of channel aggradation. In basins of tectonic instability, a number of terraces evolve with profiles converging at the level of channel beds (Figure 6.11). In the case of tectonic subsidence, older valley fills are either buried or diverge from the stream thalweg.

The morphology of Quaternary terraces of the middle and lower Rhine presents evidence of tectonic instability along the valley (Zeuner, 1964). The terraces of the middle valley indicate *differential uplift* of the basin and of the lower valley suggest *subsidence* of the North Sea basin. Two major and three minor Pleistocene terraces of the Ganga-Brahmaputra system in the deltaic zone of Bangladesh rest on unstable rock pavements. They are explained as a feature of *complex tectonic activity* of block

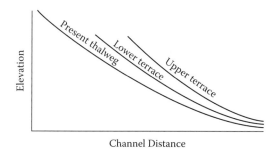

FIGURE 6.11   Idealized illustration of rejuvenation-controlled terrace profiles in relation to stream thalweg.

faulting, tilting of fault blocks, movement along subsurface faults, and subsidence of the Bengal basin (Morgan and McIntire, 1959). This tectonic activity was initiated in the Eocene (36 Ma to 58 Ma), and is still continuing.

Eustatic Controls

Pleistocene sea level oscillations or *eustacy* periodically affected the pattern of channel aggradation and trenching of alluvial fills in the estuarine segment of streams. Alluvial fills accumulated in these stream segments at higher interglacial sea stands and trenched in the period of lower sea stands of the glacial periods, evolving a sequence of *thalassostatic terraces* worldwide. Present-day marine erosion, however, has nearly destroyed thalassostatic terraces, but a few sequences have survived in Europe and South Africa, and along the east coast of North America and the west coast of South America (Zeuner, 1964, Chapter IX).

GEOMORPHIC SIGNIFICANCE

Morphology and stratigraphy of fluvial terraces carry the signal of the environmental change. The longitudinal profile of terrace remnants is diagnostic of the climate change and tectonic history of drainage basins. Terrace levels parallel to each other, and to the channel gradient, suggest rapid adjustment of the fluvial system to changes in the *base level* of streams. This type of adjustment, as in Weed Creek near Edmonton, Canada (Figure 6.12), is typical of *multicyclic streams* (Culling, 1957).

Volcanic ash, paleosols, and remains of organic matter within terrace fills, and artifacts at terrace surfaces enable the dating of terraces, reconstruction of the environment, and regional correlation of terrace sequences. The Holocene terraces east of the Cordillera of North America carry *pyroclastic material* of a volcanic eruption in the alluvium, which for a 6 to 9 m terrace of the North Saskatchewan River in Edmonton, Canada, is 6600 calendar years old (Westgate et al., 1969). The presence of the volcanic ash of this period in other terraces of the region suggests synchronous alluviation and correlation in the age of terrace fills.

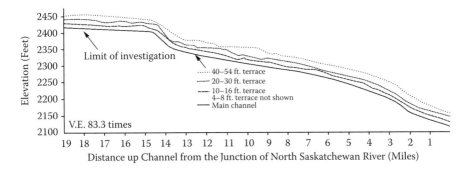

FIGURE 6.12 Longitudinal profile of the main channel and terrace remnants of Weed Creek. (Source: Figure 3 in Sharma, V. K., "Post-Glacial Terraces of the Weed and Willow Creeks, Central Alberta, Canada," *Quaternaria* XVII [1973]: 417–28. With permission from Instituto Italiano di Paleontologia.)

A temporary cessation of flood plain building activity evolves soils in the valley fill, which isolate from the *pedogenic regime* by burial beneath subsequent alluvium. These buried soils are called *paleosols*. The chemical composition of the matrix of paleosols is diagnostic of the then mega-scale climatic regime. The *remnantal magnetism* in ferromagnesian grains of the paleosols is an additional source of data for interpreting the chronology of geologic events (Zhou et al., 1990).

The mid-latitude valley fills are generally rich in diversified faunal remains, which indicate a certain *paleo-ecological habitat*. The faunal remains in the alluvial fills of Pleistocene terraces at Cochrane, Canada, provide a suitable proxy for the paleo-climatic environment (Stalker, 1968). Here, the absence of faunal remains in the upper group of terraces suggests a severely cold climate for the valley fills. The lower group of terraces, which carry the remains of Mexican ass and Western bison in the alluvium, however, suggest a warm climate at the end of the Great Ice Age 11,000 years BP. Remains of vole molar in the alluvium of a 3 m high Holocene terrace of Weed Creek near Edmonton, Canada, suggest a humid climate for the valley fill, and the presence of 1075 ± 80 carbon-14 year old charcoal in a 7 m high terrace of the system suggests a burnt forest horizon in the drainage basin (Sharma, 1973).

Potsherds, implements, and other imprints of human habitation of the past are important *archaeological references* for dating the environment and establishing the chronology of alluvial fills. The hearths and other organic remains at and within alluvial fills provide ^{14}C dates on the specimen (Brice, 1966) and pottery can be thermoluminescence (TL) dated (Chapter 1), providing for the period of habitation and age of terrace sequences. The artifacts in Hunter Valley, NSW, Australia, confine only to the Holocene rather than Pleistocene terraces of the system, in which valley fill of the Holocene terraces is 11,400 ± 100 carbon-14 years old (Erskine, 1994).

ALLUVIAL FANS

Highland drainage systems generate sediment-gravity and fluid-gravity sediments, a part of which is retained at the foot or some downchannel distance on adjacent lowland surfaces of gentle gradient in the form of alluvial fan deposits (Figure 6.13). The subaerial fan sedimentation extending into adjacent water bodies of sizable extent is called *fan delta* (Nemec and Steel, 1988; Parker et al., 1998).

CLASSIFICATION

Alluvial fans widely occur across arid through to periglacial environments. They are called dry and wet by the *climatic regime* (Schumm, 1977), sediment-gravity and fluid-gravity by the *process domain* (French et al., 1993; Blair and McPherson, 1994; Parker et al., 1998), distributary and avulsive by the *sediment distribution system*, and active and relict by the *process activity* of the present and past fluvial domains, respectively.

Dry and wet fans are associated respectively with arid and humid climates. This differentiation, however, is viewed with caution, as alluvial fans are built over several thousands of years, during which the climate may have fluctuated between dry and humid phases of sustained length (Dorn, 1994). In general, *dry fans* are built

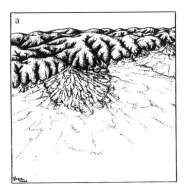
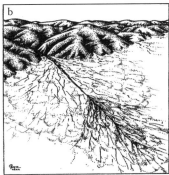

FIGURE 6.13 Two typical types of alluvial fans. (a) Area of deposition at fan head. (b) Fan head trench with deposition at fan toe. (Source: Figure 2 in Bull, W. B., "Alluvial Fan, Cone," in *The Encyclopedia of Geomorphology*, ed. R. W. Fairbridge [New York: Reinhold Book Corporation, 1968], 7–10. With permission from Kluwer Academic Publishers.)

predominantly of sediment-gravity mudflow and debris flow deposits, and *wet fans* comprise fluid-gravity finer fluvial sediments.

Alluvial fans of the *sediment-gravity process* in the arid environment of the Middle East and the United States are comprised of active overriding lobes of mass movement deposits (Whipple and Dunne, 1992; Parker et al., 1998). Alluvial fans of similar material composition in the humid environment of the northwestern Himalayas, however, are the relic of a fluvial landscape (Figure 6.14). By comparison, the alluvial fans of the *fluid-gravity process* are built of poorly sorted and stratified stream deposits that had traveled only a short distance from the source area before being laid to rest.

Alluvial fans also evolve from surface distribution of sediments by a distributary network, channel avulsion, and paraglacial sedimentation. These fan-building activities are discussed in a subsequent section of this chapter.

FIGURE 6.14 A foothill-type relict alluvial fan comprising huge granitoidal gneiss boulders near Dharmsala (32° 16' N), Himachal Pradesh, India.

Alluvial fans are an active and relict form of fluvial sedimentation. Active fans present a network of discontinuous and in-filled channels, and relict fans describe a dissected surface of tributary drainage and exposed bedrock within abandoned channel segments (French et al., 1993).

MORPHOLOGIC ZONES

Classic foothill alluvial fans comprise process-differentiated proximal, mid-fan, and distal zones of sedimentation. The *proximal zone* extends along the fan apex for up to one-half of the radial fan length, and carries a *trench* that possibly is the expression of climate change or regional tectonic instability of the landmass, or both. The *mid-fan segment* is the zone of distributary development or channel avulsion, whose activities ensure widespread deposition and evolution of a fan-shaped zone of sedimentation. The *distal zone* carries several inactive channels as reminiscent of past flow paths, and a feeder channel conveying the upchannel flow and sediments for building this part of the fan surface. Alluvial fans from adjacent highland drainage systems coalesce in this zone, evolving *compound fans*.

Morphologic Relationships

The size of the alluvial fan and the fan gradient depend upon the flow volume and sediment yield of contributing highland drainage basins. All other conditions being equal, fan size varies with the basin area and fan gradient depends on the average discharge and sediment load characteristics of drainage systems. The fan area is a power function of highland basin area as (Table 6.2)

$$A_f = c\, A_d^n$$

where A_f is the fan area, A_d is the upstream drainage basin area, c is a constant, and n is the exponent. This relationship suggests that fan size does not increase in the same proportion as the basin area increase (Figure 6.15). Hence, other basin variables of relief, lithology, tectonic history, and climate of the past and present affect the given bivariate relationship in many different ways (Bull, 1964; Hooke, 1968, 1972; Allen and Allen, 1990; Parker et al., 1998).

TABLE 6.2
Fan Area (A_f) and Drainage Basin Area (A_d) for Selected Fans

Relation ($A_f = cA_d^n$)	Source	Region/Country
$A_f = 2.1A_d^{0.91}$	Bull (1964)	Fresno County, United States
$A_f = 0.5A_d^{0.8}$	Denny (1965)	Death Vally, United States
$A_f = 1.781A_d^{0.946}$	Beaumont (1972)	Elburez Mountains, Iran
$A_f = 1.92A_d^{0.89}$	Williams (1982)	Sapta Kosi Gorge, Nepal

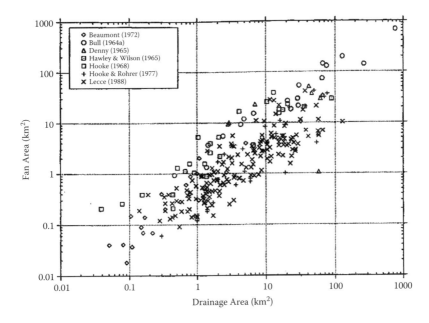

FIGURE 6.15   Log-log plot of drainage basin area versus fan area based on a compilation of data from published sources. (Source: Figure 14.22 in Blair, T. C., and McPherson, J. C., "Alluvial Fan Processes and Landforms," in *Geomorphology of Desert Environments*, ed. A. D. Abrahams and A. J. Parsons [London: Chapman & Hall, 1994], 354–402. With permission from Kluwer Academic Publishers.)

Morphology

Alluvial fans are either concave up or segmented along the radial profile (Figure 6.16). The *concave-up* alluvial fans are commonly between 3 and 20° steep, in which the average slope varies inversely with the area of contributing drainage basin (Figure 6.17) and directly with the sediment size. Other observations, in general, suggest that alluvial fans of sheet flow process and mass movement deposits are steeper than the fans of fluid-gravity or channel-bound sediments (Wells and Harvey, 1987; Schumm et al., 1987; Jackson et al., 1987; Parker et al., 1998). The alluvial fans of *segmented* longitudinal profile generally depict steepening of the channel gradient due to tectonic causes and shifting loci of sedimentation (Bull, 1964). The segmented alluvial fans of the Sapta Kosi drainage system, Nepal, which are situated over tectonically unstable tilted basement rocks, suggest the effect of subsurface geology on the manner of subaerial deposition by the stream (Williams, 1982).

ALLUVIAL FAN PROCESSES

Sediment delivery systems, sedimentation processes, and surface properties supporting fan sedimentation are diverse. *Sediment delivery systems*, in general, are climate dependent. However, a change in sediment–water ratio during transport can locally modify the nature of deposits (Wells and Harvey, 1987). Such deposits, therefore, imitate a process domain unrelated to the climatic regime. Hence, alluvial fans are

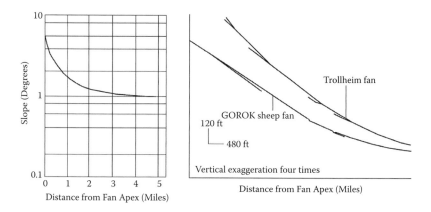

FIGURE 6.16 Types of longitudinal profile. Concave-up profile of Santa Catalina fan, Arizona (left), and segmented profile of Trollheim fan, California (right). (Source: Figure 5 in Blissenbach, E., "Geology of Alluvial Fans in Semi-arid Regions," *Bull. Geol. Soc. Am.* 65 [1954]: with permission from the Geological Society of America, 175–90. Figure 5 in Bull, W. B., *Geomorphology of Segmented Alluvial Fans in Western Fresno County, California,* U.S. Geological Survey Professional Paper 352-E [U.S. Geological Survey, 1964]. With permission from U.S. Geological Survey.)

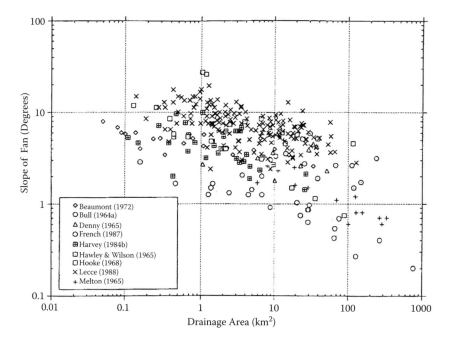

FIGURE 6.17 Log-log plot of average fan slope versus drainage basin area based on a compilation of data from published sources. (Source: Figure 14.23 in Blair, T. C., and McPherson, J. C., "Alluvial Fan Processes and Landforms," in *Geomorphology of Desert Environments,* ed. A. D. Abrahams and A. J. Parsons [London: Chapman & Hall, 1994], 354–402. With permission from Kluwer Academic Publishers.)

explained as a feature of readily observable distributary, avulsion, and paraglacial sedimentation processes. It is also understood that surface properties as slope steepness and roughness elements variously affect the pattern of subaerial deposition. Thus, alluvial fans in areas of tectonic subsidence are generally smaller in size and built of thicker sediments than those in other areas of similar hydrologic conditions.

Distributary Channel Process

Alluvial fans of the arid environment are built predominantly of *mudflow* and *debris flow* deposits, which plug the active channel and form levees (Bull, 1964, 1968). The *levees* accidentally breach in floods, opening distributary channels through which fan sedimentation shifts to surfaces topographically lower than that at the point of bifurcation. A study on a small-sized alluvial fan of an extinct drainage system in the northern plains of India suggests that the form of fan-shaped surface of sedimentation owes to the progressively increasing angle of divergence of distributary channels down the fan surface (Mukerji, 1976).

Channel Avulsion Process

Rapidly aggrading meandering stream segments build up the channel bed to the level of adjacent *flood plain surface* (Chapter 5). Hence, channels upstream of silted segments avulse across the flood plain by opening *anabranches* in the direction of a lower topographic surface (Allen, 1965, 1978; Gole and Chitale, 1966; Wells and Dorr, 1987). The mega-sized Kosi fan in the state of Bihar, India, has similarly evolved by rapid siltation and flattening of the Kosi bed (Figure 6.18(a)), evolving more than twelve anabranches in a westward direction (Figure 6.18(b)). The Kosi fan is described as the product of excessive sediment load of the system derived from highland sources in Nepal, rapid channel siltation, and westward migration of the channel by 120 km in the past 200 years (Gole and Chitale, 1966).

Alluvial fans of channel avulsion are also common to gravel-bedded braided stream*s* draining the Siwalik Hills of weak lithologic composition in northwestern India. A major alluvial fan on either side of the Jamuna River in Haryana and Uttar Pradesh has evolved by channel avulsion (Figure 6.19), suggesting that braided streams transporting a large volume of highland detritus load in floods rapidly aggrade channels, and avulse across the flood plain surface. The effect of *neotectonic movements* of the region on fan sedimentation, though, is not known at the present time.

Paraglacial Sedimentation Process

Paraglacial sedimentation is the process of snow- and ice-fed alpine streams. In this environment, glacial debris and mass movement deposits periodically block channels, and impound the discharge and sediment load of contributing drainage systems (Chapter 1). In time, the dammed channels burst and release a short-lived peak flood and abundant sediments for the building of alluvial fans (Ryder, 1971; Church and Ryder, 1972). Observations suggest that the rate of fan sedimentation, as in drainage basins of the Canadian Cordillera (Desloges and Gardner, 1981) and the Bhagirathi Valley in the northwestern Himalayas, India (Owen and Sharma, 1998), depends directly on the *frequency* of rockfall and avalanche events.

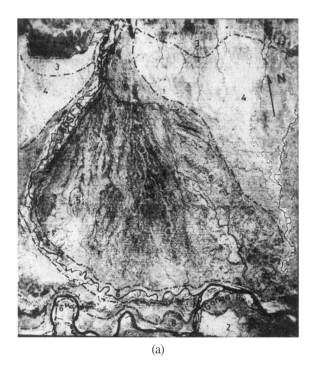

(a)

FIGURE 6.18   Kosi mega-fan (a) and shifting channel courses on the Kosi fan (b). (Source: Landsat imagery courtesy Dr. A. K. Roy. Figure 1 in Parker, G., Paola, C., Whipple, K. X., and Mohrig, D., "Alluvial Fans Formed by Channelized Fluvial and Sheet Flow. I. Theory," *J. Hydraulic Eng.* 124 [1998]: 985–95. With permission from American Society of Civil Engineers.)   *Continued*

RIVER DELTAS

River deltas are a feature of a transitional terrestrial–aquatic environment, in which fluvial sediments are laid partly on land and partly in the seawater (Moore and Asquith, 1971). Therefore, the morphology and sedimentary attributes of river deltas are largely governed by discharge and sediment load of streams draining into the sea, and the manner of redistribution of subaqueous sediments by waves, tides, and currents (Wright, 1978). Many of the terms used in this section are discussed in Chapter 11.

MORPHOLOGY

River deltas represent process-differentiated upper, intermediate, and subaqueous zones of sedimentation (Figure 6.20). The *upper zone* of fluvial sedimentation begins a little upstream of first distributary development or channel avulsion, and terminates downchannel at the limit of tidal influence. This zone is typical of a fluvial landscape of inactive channels, and features of a point bar, natural levee, crevasse splay, and backswamp deposit. The *intermediate zone* lies between the upper limit of tidal influence and the shoreline. It supports coastal dunes, mudflats, and tidal marshes

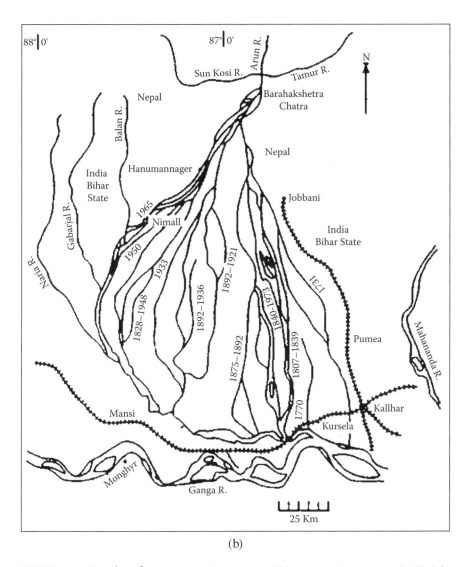

(b)

FIGURE 6.18 (Continued) Kosi mega-fan (a) and shifting channel courses on the Kosi fan (b). (Source: Landsat imagery courtesy Dr. A. K. Roy. Figure 1 in Parker, G., Paola, C., Whipple, K. X., and Mohrig, D., "Alluvial Fans Formed by Channelized Fluvial and Sheet Flow. I. Theory," *J. Hydraulic Eng.* 124 [1998]: 985–95. With permission from American Society of Civil Engineers.)

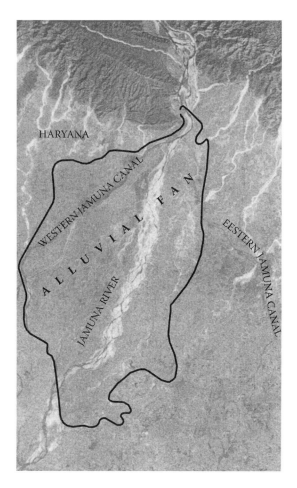

FIGURE 6.19   Alluvial fan of avulsion process in a gravel-bedded braided stream at the foot of Siwalik Hills, northwest India. (Source: IRS LISS-I imagery.)

as features of interacting fluvial and marine environments, and a limited number of fluvial forms that appear in the upper delta zone. The *subaqueous zone* is the foundation of deltas. It is the area of a high rate of riverine sedimentation and progradation of the delta into the sea. This zone comprises a delta front and a prodelta zone of sedimentation. The *delta front* is identified with the zone of subaqueous bars and sedimentary jetties of coarsest sediments in the stream flow. These subaqueous forms subsequently grow into spits and develop coastal lagoons by the effect of wave refraction and longshore currents in shallow waters offshore. The *prodelta* extends farthest over the continental shelf as a subaqueous zone of fine-grained sediments.

The size and shape of river deltas are unique to each drainage system. In general, the *delta size* varies directly with the average discharge and sediment load of streams, the manner of sediment distribution on land and in the sea, and the width and structural behavior of the continental shelf. Morphologic attributes of the *upper delta zone*, in general, depend on surface slope and extent of channel division and

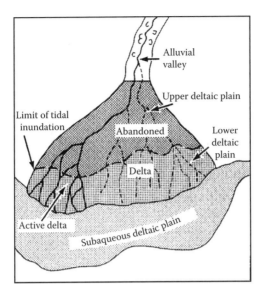

FIGURE 6.20 Generalized diagram showing the upper, intermediate, and subaqueous morphologic zones of a delta system. (Source: Figure 3 in Wright, L. D., "River Deltas," in *Coastal Sedimentary Environments*, ed. R. A. Davis, Jr. [New York: Springer Verlag, 1978], 5–68. With permission from Springer-Verlag.)

avulsion, of the *intermediate delta zone* on the strength of tidal currents, and of the *distal delta zone* on tectonic stability and width of the continental shelf. The *delta shape* is an equilibrium form of interacting fluvial, tidal, and wave activities (Figure 6.21), evolving a great variety of shapes, a few of which are illustrated in Figure 6.22. The *elongate delta* of the Mississippi in the Gulf of Mexico, the *lobate delta* of the Mahanadi, and the *cuspate delta* of the Krishna River in the Bay of

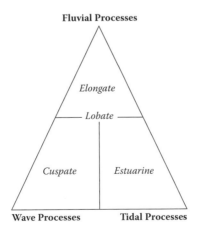

FIGURE 6.21 Classification of river deltas by dominant processes that control the delta shape. The three end members are fluvial processes, tidal processes, and wave processes.

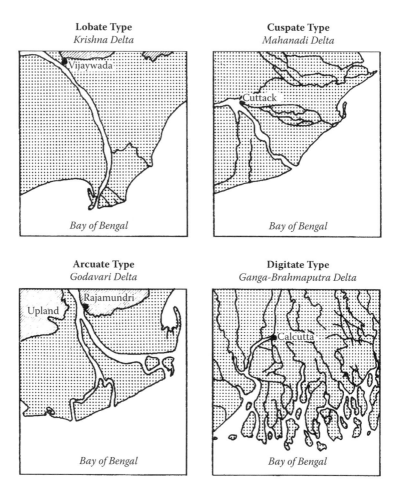

FIGURE 6.22 Morphology of selected river deltas in relation to the relative importance of fluvial, tidal, and wave processes in the Bay of Bengal. (Source: Figure 3 in Vaidyanadhan, R., *Quaternary Deltas of India*, Geological Society India Memoir 22 [Geological Society, 1991]. With permission.)

Bengal are products of stronger fluvial activity. The *arcuate delta* of the Godavari is wave dominated, and the *digitate delta* of the Ganga-Brahmapurta system, also in the Bay of Bengal, is tide dominated.

Delta Processes

The *upper delta zone* is the surface of subaerial sedimentation. It evolves from the development either of a distributary channel network or of channel avulsion. The opening of *distributary channels* is a compensatory mechanism for the transport of discharge and sediment load of drainage systems, which otherwise cannot be efficiently conveyed through a single channel on the flat surface of this zone. These channels also silt in time, a process that increases the frequency of overbank flows

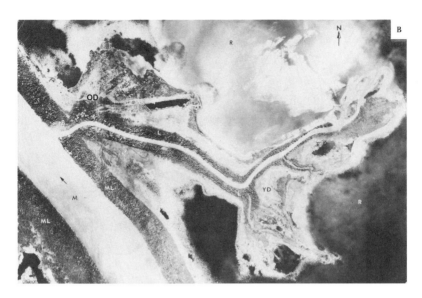

FIGURE 6.23 Delta of the Rombebai Lake, New Guinea, showing Memberamo River (M) and its well-developed levees. A breach (B) in the east levee of the river has evolved a small delta in the lake, which comprises splay deposits of old delta surface (OD) and finer sediment of active delta surface (YD). Plumes of sediment in the lake (R) suggest the direction of prograding delta front. (Source: Plate B in Verstappen, H. Th., "Geomorphology in Delta Studies," *Int. Inst. Aerial Survey Earth Sci.*, Sr. B, No. 24, 1964. With permission.)

and develops natural levees along channels. These levees accidentally breach in flood events, and similarly open up a network of minor distributary channels and shifting loci of sedimentation at a lower elevation on the deltaic surface (Figure 6.23). Tectonic instability of the coastal tract due to the growth of *morphostructures* also initiates channel division (Rao and Vaidyanadhan, 1979). This process is the suggested cause of distributary development for the Holocene delta of the Godavari River in the Bay of Bengal, India (Figure 6.24). By comparison, the delta of the Mississippi River in the Gulf of Mexico has developed from repeated *avulsion* of the river and subsidence of the continental shelf. The avulsion has evolved five major and two minor deltaic lobes of sedimentation, which together resemble a bird foot shape (Figure 6.25).

The *intermediate zone* is the surface of terrestrial and marine sediments moved landward in the *flood current* and seaward in the *ebb current*. Hence, disruption in the strength of the two tidal currents affects the growth pattern of this zone of sedimentation. The weakening of the ebb current from recent hydrographic changes in the fluvial system has thus set the Bhagirathi-Hoogli delta in the Bay of Bengal, India, to decay (Bhattacharya, 1973).

Morphologic and sedimentary attributes of the *subaqueous zone* depend on the depth of the sedimentation basin and the water density characteristics of effluent discharge and those of the basin of sedimentation (Silvester and de la Cruz, 1970; Wright and Coleman, 1974; Wright, 1978; Elliott, 1978). Small-sized shallow water bodies restrict lateral expansion of the effluent stream flow. They, therefore,

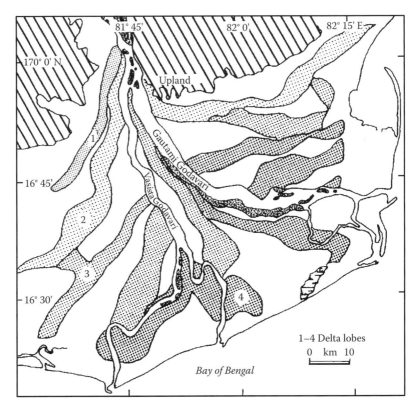

FIGURE 6.24 Stages in the growth of lobes in Godavari delta. (Source: Figure 1 in Rao, M. S., and Vaidyanadhan, R., "Morphology and Evolution of Godavari Delta, India," *Z. Geomorph*. 23 [1979]: 243–55. With permission.)

evolve axial jet flow and semicircular deltaic lobes of sedimentation. Large-sized deep water bodies, however, develop plane jet flow and dissipate the effluent discharge laterally. Hence, elongated deltaic lobes of sedimentation are typical of such water bodies.

Water density characteristics of the effluent discharge and those of the basin of sedimentation also affect the nature of subaqueous sedimentation, generating homopycnal, hyperpycnal, and hypopycnal flows within the receiving basin. *Homopycnal* (equal density) flow is typical of stream-fed large freshwater lakes, wherein the coarser stream load settles at or near the mouth of streams and the finer charge is carried farther into the water body for sedimentation. This process evolves top-, fore-, and bottom-set beds of subaqueous sedimentation. *Hyperpycnal* or dense flow sticks to the bottom of water bodies and, therefore, prohibits the mixing of sediments in transport. This type of flow occurs seaward of the continental shelf, evolving deltas at the mouth of submarine canyons. *Hypopycnal* flow is characteristic of less dense stream flow moving over more dense seawater. Hence, fluvial sediments are carried farthest into shallow coastal waters for sedimentation by the plane jet flow.

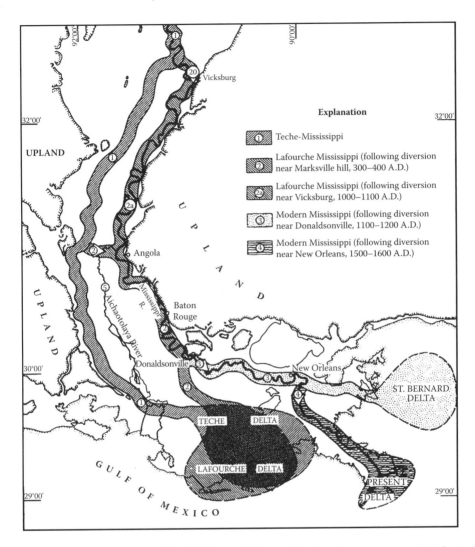

FIGURE 6.25 Subdeltas of the Mississippi River, which together represent a bird foot shape. (Source: Figure 7.6 in in Thornbury, W. D., *Principles of Geomorphology*, 2nd ed. [New York: John Wiley & Sons, 1969]. With permission.)

SUMMARY

Flood plains, alluvial fans, and river deltas are major forms of fluvial sedimentation. Flood plains are low-relief linear topographic surfaces of an alluvial fill within channels, at channel margins, and beyond to the level of mean annual flood in streams. Flood plains evolve by lateral accretion of coarser bed material load of streams at low to moderate channel flow conditions, and by vertical accretion of finer suspended particles in overflow channel conditions. Most flood plains evolve by the activity of co-dominant lateral and vertical accretion processes. Mass movement deposits, though unrelated to the fluvial activity, are also a part of flood plain surfaces constrained by valley walls.

Flood plains are abandoned, and termed fluvial terraces, when the long-established relationship between the mean annual flood and flood plain height above channels is disturbed by channel trenching due to the effects of climate change, tectonic instability of the landmass, and sea level oscillations. The signature of the environmental change, though difficult to isolate by the cause, is preserved in the morphology and stratigraphy of terrace surfaces. Streams trench the channel and preexisting valley fills in the humid phase of climate and aggrade channels in the phase of semiarid climate. Regional tectonic instability also affects channel processes of trenching and alluviation. Eustacy or Quaternary sea level oscillations have affected aggradation and trenching of the estuarine segment of streams, evolving thalassostatic terraces. Volcanic ash, paleosols, and organic remains within valley fills, and artifacts at terrace surfaces, are sensitive indicators of the environmental change.

Alluvial fans are concave-up or segmented surfaces of sedimentation at some downchannel distance of contributing highland basins. Fans are built of mass movement deposits, channel-bound finer fractions, and sheet flow sediments. The fans are wet and dry by the climatic regime, sediment gravity, and fluid gravity by the process domain, and active and relict in relation to the activity of present or past processes. It is easy, however, to classify alluvial fans as surfaces of distributary network, channel avulsion, and paraglacial sedimentation. The fan area and highland drainage area are positively related to each other, and fan gradient and sediment size bear an inverse relationship with the basin area. These relationships are somewhat modified by basin lithology, topography, relief, surface properties of sediment accumulation area, and tectonic and climatic history of contributing drainage basins.

River deltas are a geomorphic feature of fluvial deposition partly on land and partly in adjacent water bodies of sizable extent. The delta surface represents process-differentiated upper, intermediate, and distal zones of sedimentation of fluvial, tidal, and wave environments, respectively. The shape and size of river deltas depend on the area of contributing drainage, extent of distributary development, manner of channel avulsion, size and depth of coastal waters, tectonic aspects of the continental shelf, wave, tide, and fluvial influences, and density characteristics of stream water and those of the basin of sedimentation.

REFERENCES

Allen, J. R. L. 1965. A review of the origin and characteristics of alluvial sediments. *Sedimentology* 5:89–191.

Allen, J. R. L. 1978. Studies in fluviatile sedimentation: An exploratory quantitative model for the architecture of avulsion-controlled alluvial suites. *Sedimentary Geol.* 21:129–47.

Allen, P. A., and Allen, J. R. 1990. *Basin analysis: Principles and applications.* Oxford: Blackwell Scientific Publications.

Ashmore, P. 1993. Anabranch confluence kinetics and sedimentation processes in gravel-braided streams. In *Braided rivers*, ed. J. L. Best and C. S. Bristow, 129–46. Geological Society Special Publication 75. Geological Society.

Beaumont, P. 1972. Alluvial fans along the foothills of Elburez Mountains, Iran. *Palaeoecol. Palaeogeogr. Palaeoclimat.* 12:251–73.

Bhattacharya, S. K. 1973. Deltaic activity of Bhagirathi-Hooghly River system. *J. Wtrwy. Harb. Coast. Eng.* 99:69–87.

Blair, T. C., and McPherson, J. C. 1994. Alluvial fan processes and landforms. In *Geomorphology of desert environments*, ed. A. D. Abrahams and A. J. Parsons, 354–402. London: Chapman & Hall.

Blissenbach, E. 1954. Geology of alluvial fans in semi-arid regions. *Bull. Geol. Soc. Am.* 65:175–90.

Brakenridge, G. R. 1984. Alluvial stratigraphy and radiocarbon dating along the Duck River, Tennessee: Implications regarding flood-plain origin. *Bull. Geol. Soc. Am.* 95:9–25.

Brice, J. C. 1964. *Channel patterns and terraces of the Loup Rivers in Nebraska.* U.S. Geological Survey Professional Paper 422-D. U.S. Geological Survey.

Brice, J. C. 1966. *Erosion and deposition in the loess-mantled Great Plains, Medicine Creek drainage basin, Nebraska.* U.S. Geological Survey Professional paper 352-4. U.S. Geological Survey.

Bridge, J. C. 1993. The interaction between channel geometry, water flow, sediment transport and deposition in braided rivers. In *Braided rivers*, ed. J. L. Best and C. S. Bristow, 13–17. Geological Society Special Publication 75. Geological Society.

Bristow, C. S., and Best, J. L. 1993. Braided rivers: Perspectives and problems. In *Braided rivers*, ed. J. L. Best and C. S. Bristow, 1–11. Geological Society Special Publication 75. Geological Society.

Bull, W. B. 1964. *Geomorphology of segmented alluvial fans in western Fresno County, California.* U.S. Geological Survey Professional Paper 352-E. U.S. Geological Survey.

Bull, W. B. 1968. Alluvial fan, cone. In *The encyclopedia of geomorphology*, ed. R. W. Fairbridge, 7–10. New York: Reinhold Book Corporation.

Church, M., and Ryder, J. 1972. Paraglacial sedimentation: A consideration of fluvial processes conditioned by glaciation. *Bull. Geol. Soc. Am.* 83:3059–72.

Coleman, J. M. 1969. Brahmaputra River: Channel processes and sedimentation. *Sediment. Geol.* 3:129–239.

Culling, W. E. H. 1957. Equilibrium states in multicyclic streams and the analysis of river-terrace profiles. *J. Geol.* 65:451–67.

Denny, C. S. 1965. *Alluvial fans in the Death Valley region of California and Nevada.* U.S. Geological Survey Professional Paper 466. U.S. Geological Survey.

Desloges, J., and Gardner, J. 1981. Recent chronology of an alpine alluvial fan in southwestern Alberta. *Albertan Geogr.* 17:1–18.

Dorn, R. I. 1994. The role of climatic change in alluvial fan development. In *Geomorphology of desert environments*, ed. A. D. Abrahams and A. J. Parsons, 593–615. London: Chapman & Hall.

Elliott, T. 1978. Deltas. In *Sedimentary environments and facies*, ed. H. G. Reading, 97–142. New York: Elsevier.

Erskine, W. D. 1994. Late Quaternary alluvial history of Nowlands Creek, Hunter Valley, NSW. *Austr. Geogr.* 25:50–60.

Ferguson, R. I. 1993. Understanding braiding processes in gravel-bed rivers: Progress and unsolved problems. In *Braided rivers*, ed. J. L. Best and C. S. Bristow, 73–87. Geological Society Special Publication 75. Geological Society.

Fisk, H. N. 1947. *Fine grained alluvial deposits and their effects on Mississippi River activity.* Vol. 1. Mississippi River Commission, U. S. Corps of Engineers.

French, R. H., Fuller, E., and Waters, S. 1993. Alluvial fan: Proposed new process-oriented definitions for arid southwest. *J. Water Resources Planning Management* 119:588–98.

Gole, C. V., and Chitale, S. V. 1966. Inland delta building activity of Kosi River. *J. Hydraul. Div.* 92:111–26.

Hooke, R. LeB. 1968. Steady-state relationships on arid-region alluvial fans in closed basins. *Am. J. Sci.* 266:609–29.

Hooke, R. LeB. 1972. Geomorphic evidence for late-Wisconsin and Holocene tectonic deformation, Death Valley, California. *Bull. Geol. Soc. Am.* 83:2073–98.

Jackson, L. E., Kostaschuk, R. A., and MacDonald, G. M. 1987. Identification of debris flow hazard on alluvial fans in the Canadian Rocky Mountains. *Rev. Eng. Geol.* 7:115–24.

Kesel, R. H., Dunne, K. C., McDonald, R. C., and Allison, K. R. 1974. Lateral erosion and overbank deposition on the Mississippi River in Louisiana caused by 1973 flooding. *Geology*, September, 461–64.

Langbein, W. B., and Schumm, S. A. 1958. Yield of sediment in relation to mean annual precipitation. *Trans. Am. Geophys. Union* 39:1076–84.

Lattman, L. H. 1960. Cross section of a flood plain in a moist region of moderate relief. *J. Sediment. Petrol.* 30:275–82.

Leopold, L. B., and Miller, J. P. 1954. *A postglacial chronology for some alluvial valleys in Wyoming.* U.S. Geological Survey Water-Supply Paper 1261. U.S. Geological Survey.

Moore, G. T., and Asquith, D. O. 1971. Delta: Term and concept. *Bull. Geol. Soc. Am.* 82:2563–68.

Morgan, J. P., and McIntire, W. G. 1959. Quaternary geology of the Bengal Basin, East Pakistan and India. *Bull. Geol. Soc. Am.* 70:319–42.

Mukerji, A. B. 1976. Terminal fans of inland streams in the Sutlej-Yamuna Plain, India. *Z. Geomorph.* 20:190–204.

Nemec, W., and Steel, R. J. 1988. *Fan deltas: Sedimentary and tectonic settings.* Glasgow: Blackie.

Owen, L. A., and Sharma, M. C. 1988. Rates and magnitudes of paraglacial fan formation in the Garhwal Himalaya: Implication for landscape evolution. *Geomorphology* 26:171–84.

Parker, G., Paola, C., Whipple, K. X., and Mohrig, D. 1998. Alluvial fans formed by channelized fluvial and sheet flow. I. Theory. *J. Hydraulic Eng.* 124:985–95.

Passmore, D. G., Macklin, M. G., Brewer, P. A., Lewin, J., Rumsby, B. T., and Newson, M. D. 1993. Variability of late Holocene braiding in Britain. In *Braided rivers*, ed. J. L. Best and C. S. Bristow, 205–229. Geological Society Special Publication 75. Geological Society.

Ramasamy, S. M., Bakliwal, P. C., and Verma, R. P. 1991. Remote sensing and river migration in western India. *Int. J. Remote Sensing* 12:2597–609.

Rao, M. S., and Vaidyanadhan, R. 1979. Morphology and evolution of Godavari delta, India. *Z. Geomorph.* 23:243–55.

Richards, K., Chandra, S., and Friend, P. 1993. Avulsive channel systems: Characteristics and examples. In *Braided rivers*, ed. J. L. Best and C. S. Bristow, 195–203. Geological Society Special Publication 75. Geological Society.

Ritter, D. F., and Blakley, D. S. 1986. Localised catastrophic disruption of the Gasconade River flood plain during the 1982 flood, southeast Missouri. *Geology* 14:472–76.

Ryder, J. 1971. The stratigraphy and morphology of paraglacial alluvial fans in south-central British Columbia. *Can. J. Earth Sci.* 8:279–98.

Schmudde, T. H. 1978. Flood plain. In *The encyclopedia of geomorphology*, ed. R. W. Fairbridge, 359–62. New York: Reinhold Book Corp.

Schumm, S. A. 1977. *The fluvial system.* New York: John Wiley & Sons.

Schumm, S. A., Erskine, W. D., and Tilleard, J. W. 1996. Morphology, hydrology, and evolution of the anastomosing Ovens and King rivers, Victoria, Australia. *Bull. Geol. Soc. Am.* 108:1212–24.

Schumm, S. A., and Lichty, R. W. 1963. *Channel widening and flood-plain construction along Cimarron River in southwestern Kansas.* U.S. Geological Survey Professional Paper 352-D. U.S. Geological Survey.

Schumm, S. A., Mosley, M. P., and Weaver, W. E., 1987. *Experimental Fluvial Geomorphology.* New York, John Wiley & Sons.

Sharma, V. K. 1973. Post-glacial terraces of the Weed and Willow creeks, central Alberta, Canada. *Quaternaria* XVII:417–28.

Silvester, R., and de la Cruz, D. de R. 1970. Pattern forming processes in deltas. *J. Wtrwy. Harb. Coastal Eng.* 96:201–17.

Smith, D. G. 1981. Aggradation of the Alexandra-North Saskatchewan River, Banff Park, Alberta. In *Fluvial geomorphology*, ed. M. Morisawa, 201–19. London: George Allen & Unwin.

Smith, D. G. 1983. *Anastomosed fluvial deposits: Modern examples from western Canada*, 155–68. International Association of Sedimentologists Special Publication 6. Oxford: Blackwell Scientific Publications.

Stalker, A. MacS. 1968. Geology of the terraces at Cochrane, Alberta. *Can. J. Earth Sci.* 5:1455–66.

Sundborg, Å. 1956. Morphological activity of flowing water, chapter II. *Georg. Annlr.* XXXVIII:165–316.

Thornbury, W. D. 1969. *Principles of geomorphology*. 2nd ed. New York: John Wiley & Sons.

Thorne, C. R., Russell, P. G., and Alam, M. K. 1993. Planform pattern and channel evolution of the Brahmaputra River, Bangladesh. In *Braided rivers*, ed. J. L. Best and C. S. Bristow, 257–76. Geological Society Special Publication 75. Geological Society.

Vaidyanadhan, R. 1991. *Quaternary deltas of India*. Geological Society India Memoir 22. Geological Society.

Vanoni, V. A. 1971. Sediment transportation mechanics: Q. Genetic classification of valley sediment deposits. *J. Hydraulics Div.*, No. HYI, 43-53.

Verstappen, H. Th. 1964. Geomorphology in delta studies. *Int. Inst. Aerial Survey Earth Sci.*, Sr. B, No. 24.

Warburton, J., Davies, T. R. H., and Mandl, M. G. 1993. A meso-scale field investigation of channel change and floodplain characteristics in an upland braided gravel-bed river, New Zealand. In *Braided rivers*, ed. J. L. Best and C. S. Bristow, 241–55. Geological Society Special Publication 75. Geological Society.

Wells, N. A., and Dorr, J. A., Jr. 1987. Shifting of the Kosi River, northern India. *Geology* 15:204–7.

Wells, S. G., and Harvey, A. M. 1987. Sedimentological and geomorphic variations in storm-generated alluvial fans, Howgill Fells, northwest England. *Bull. Geol. Soc. Am.* 98:182–98.

Westgate, J. A., Smith, D. G. W., and Nichols, H. 1969. *Late Quaternary pyroclastic layers in Edmonton area, Alberta*. Geology Department Contract 464, 179–86, University of Alberta.

Whipple, K. X., and Dunne, T. 1992. The influence of debris-flow rheology on fan morphology, Owens Valley, California. *Bull. Geol. Soc. Am.* 104:887–900.

Williams, Van S. 1982. Tectonic tilting of mountain-front alluvial fans near the Sapta Kosi gorge, eastern Nepal. In *Himalaya landforms and processes*, ed. V. K. Verma and P. S. Saklani, 115–32. New Delhi: Today and Tomorrows.

Wolman, M. G., and Leopold, L. B. 1957. *River flood plains: Some observations on their formation*. U.S. Geological Survey Professional Paper 282-C. U.S. Geological Survey.

Wright, L. D. 1978. River deltas. In *Coastal sedimentary environments*, ed. R. A. Davis, Jr., 5–68. New York: Springer Verlag.

Wright, L. D., and Coleman, J. M. 1974. Mississippi River mouth processes: Effluent dynamics and morphologic development. *J. Geol.* 82:751–88.

Zeuner, F. E. 1964. *The Pleistocene period*. London: Hutchinson.

Zhou, L. P., Oldfield, F., White, A. G., Robinson, S. G., and Wang, J. T. 1990. Partly pedogenic origin of magnetic variations in Chinese loess. *Nature* 346:737–39.

7 Glacial Processes and Landforms

In parts of North America, Europe, and Asia, massive glaciers periodically appeared in colder glacial and thawed in warmer interglacial phases of the Pleistocene climate. This ice of continental dimension reached a massive thickness of 3 km or more at places by slow accretion over thousands of years in glacial phases (Gates, 1976), but thawed rapidly in less than a decade in intervenient interglacial phases of the Pleistocene epoch (Lehman, 1997). These recurrent events eventually culminated in 10,000 years BP, bringing the 2-million-year-old *Great Ice Age* to a sudden end. The *multiple glaciation* has modeled the landscape beneath sliding glaciers and the yield of meltwater discharge beneath and beyond the vast expanse of glacier ice.

GLACIERS

Glaciers are a large mass of perennial ice. The ice comprises several centimetres across interlocked ice crystals of variable size and orientation. This ice begins as 0.02 to 0.08 density hexagonal-shaped snowflakes. The snowflakes progressively increase in density by compaction, melting, air expulsion and recrystallization processes, evolving 0.85 to 0.9 gm cm^{-3} density ice through intermediate stages of conversion to granular snow and firn. Massive glaciers that flow outward from their centres of ice accumulation are called *ice sheets*. Glaciers occupy some 11% of the earth's land surface, but hold roughly 75% of its fresh water.

THERMAL PROPERTIES OF GLACIERS

The thermal gradient of the ice distinguishes temperate or warm-based and polar or cold-based glaciers as two fundamental types with several transitions in between. The *thermal gradient*, in general, is a function of the air temperature, thickness of the ice mass, heat conductivity of the ice, and geothermal heat escape through the ice. In *temperate glaciers*, the relationship between the gradient of the ice temperature and geothermal heat, except in the winter season, is such that the geothermal heat cannot escape the glacier ice and is consumed in melting the ice. Such glaciers, therefore, are at 0° C or close to an appropriate temperature for melting due to the pressure of ice. Hence, temperate glaciers are at *pressure melting temperature* throughout, and *meltwater* exists at the glacier-bedrock interface. Meltwater at the base of glacier ice enhances the *sliding rate* of temperate glaciers and provides for the *regelation-controlled* basal ice transfer in glacier morphodynamic systems. In winter however, the glacier ice to the depth of seasonal temperature fluctuation becomes temporarily colder. This phenomenon affects thermal gradient in the surface ice somewhat, but not the morphologic activity of temperate glaciers.

By comparison, the relationship between gradients of ice temperature and geothermal heat flux in *polar glaciers* is such that the geothermic heat escapes the ice. Hence, meltwater is absent in the ice and the ice is frozen to the depth of bedrock. Therefore, polar glaciers require a relatively greater *shear stress* to induce movement at the bed of ice. Theoretically, the cold-based glaciers deform by the *creep of ice*, which process generates *frictional heat* almost equal in amount to that of the geothermal heat flow. The frictional heat also raises the temperature of basal ice to the *pressure melting point*, and generates *meltwater* in the system. The meltwater activity is a suggested process of *regelation-controlled* basal ice movement, evolving the *end moraines* of high-latitude glaciers.

The high-latitude glaciers of Spitsbergen, Baffin Island, and Greenland comprise an inner thicker ice at pressure melting temperature and an outer thinner ice frozen to the bed. The glaciers of such thermal behavior are called *polythermal*. They are intermediate between the temperate and polar glaciers in thermal property of the ice.

GLACIER DEFORMATION

Glaciers deform by gliding or creep of individual ice crystals of varying shape and orientation along crystal boundaries and basal sliding of the ice. The dynamics of glacier flow is known and understood in theory, but theoretical presuppositions on glacial flow find support in field and test conditions. The motion of glacier ice and associated meltwater activity generates a variety of glaciogenic sediments for transport and evolution of the glaciated landscape.

CREEP

Glacier ice deforms internally by the stress of its own weight, producing a continuous slow motion of the glacier ice called creep. The relationship between internal deformation of the ice and the stress producing it is called the flow law of ice. The *flow law*, which refers to the physical properties of materials, predicts that the creep rate depends on the thickness and surface slope of the glacier ice as

$$\tau_b = \rho \, g \, h \, \sin \theta$$

in which τ_b is the basal shear stress of the ice, ρ is the density of ice, h is the thickness of ice, and θ is the surface slope of glacier ice. The value of τ_b can be determined from ice thickness (h) and surface slope (θ); measurements on many glaciers suggest that the basal shear stress varies between 0.5 and 1.5 bars. Thus, the ice is regarded as a perfectly plastic substance with a yield stress of 1.0 bar (Paterson, 1969). The rate law suggests that the shear stress is zero at the surface and increases linearly with depth of the ice mass, such that the stress reaches *yield stress* only at the base of the glacier ice. Thus, all the deformation in glacier ice occurs in its lowest layers. Furthermore, the creep rate increases with the thickness of ice mass, and glacier ice deforms in the direction of maximum ice surface slope. Theoretically, the creep rate is proportional to the fourth power of ice thickness and to the third power of ice surface slope (Paterson, 1969).

Experimental data on the deformation of ice suggest that the strain rate varies with temperature-dependent ice hardness as (Paterson, 1969)

$$\dot{\varepsilon} = B \, \sigma^n$$

in which $\dot{\varepsilon}$ is the strain rate, B is a constant for temperature-dependent ice hardness, σ is the normal stress, and n is the exponent for creep with a mean value of about 3. In test conditions, the strain rate of ice at $-22°$ C is only one-tenth of that at $0°$C. Hence, the ice of *polar glaciers* withstands a higher stress for deformation than the ice of *temperate glaciers* at pressure melting point.

Valley glaciers deform over the irregular bed by two states of creep movement called *extending flow* and *compressing flow* (Nye, 1952). By the flow law, the ice descending a steep bed is thin and that ascending a gentle subglacial bed is thick. This variation of ice thickness affects per unit discharge rate and velocity of the ice flow. The velocity is highest at the surface and decreases linearly with depth, such that it is downward in extending and upward through the ice in compressing flow. The flow, which is tensile in upper layers of extending and compressive at all depths in compressing flow, evolves stress trajectories or *slip lines* tangential to the bed and at 45° to the surface ice (Figure 7.1). They express as *crevasses* in extending flow and as thrust or shear planes in compressing flow near the snout of valley glaciers and at the edge of temperate and subpolar glaciers. In general, the flow is extending in the *accumulation zone* and compressing in the *ablation zone* of glaciers.

BASAL SLIDING

Basal sliding is the act of a glacier sliding over its bed. It is essentially related to the presence of an ideal friction-reducing film of *meltwater* at the glacier-bedrock interface (Weertman, 1964; Stupavski and Gravenor, 1974; Raymond, 1978). Besides, *saturated basal till* is also viewed as a suitable friction-reducing medium for the sliding of glaciers over their bed (Boulton, 1979). *Temperate glaciers* at pressure melting temperature thus readily deform by sliding at the glacier-bedrock interface, wherein basal sliding is some 90% and creep 10% of the total ice movement (Stupavski and Gravenor, 1974). Parts of *polar glaciers* at pressure melting temperature similarly deform by sliding at the bed (Waller and Hart, 1999) and by sliding and streaming of the ice within ice sheets (Fitzsimons, 1996).

Weertman (1964) developed the theory of basal sliding to perfection. Basal sliding comprises pressure melting and enhanced plastic flow mechanisms, the theoretical postulates of which have since been verified in test and field conditions (Kamb and LaChapelle, 1964; Raymond, 1978; Echelmayer and Wang, 1987). *Pressure melting* is the melting and refreezing of basal ice against small-sized bedrock obstacles, and *enhanced plastic flow* is the deformation of ice at an enhanced creep rate across large-sized subglacial obstacles.

Pressure Melting

Weertman theorized that pressure of the sliding ice against the stoss face of bedrock obstacles less than 1 m across lowers the melting temperature of the isothermal ice.

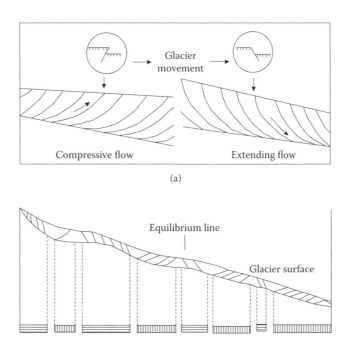

(a)

(b)

FIGURE 7.1 Compressing and extending flow and associated slip lines (a), and distribution of compressing and extending flow in an idealized glacier (b). (Source: Figure 9 in Nye, J. F., "The Mechanics of Glacier Flow," *J. Glaciol.* 2 [1952]: 82–93. With permission from International Glaciological Society.)

Therefore, the basal ice melts in this zone. The subglacial meltwater moves over and across obstacles and freezes in the lower pressure zone leeward of obstacles as bubble-free *regelation ice* by releasing the latent heat of fusion. This heat is conducted upglacier through subglacial obstacles, where it is consumed in melting the ice (Figure 7.2(a)). Pressure melting and related sliding velocity, thus, depend on the thermal gradient across obstacles at the interface of glacier ice. Theoretically, the sliding velocity is inversely proportional to the length of subglacial obstacles. Pressure melting provides for heat transfer within glaciers and enables the basal ice to flow by melting and refreezing processes (Paterson, 1969).

Enhanced Plastic Flow

Bedrock obstacles more than 1 m across create additional longitudinal stress in glaciers, causing the basal ice to deform at an enhanced creep rate (Figure 7.2(b)). The longitudinal stress is *compressive* on the stoss and *tensile* on the leeward face of obstacles, producing a strain rate and sliding velocity proportional to the obstacle length.

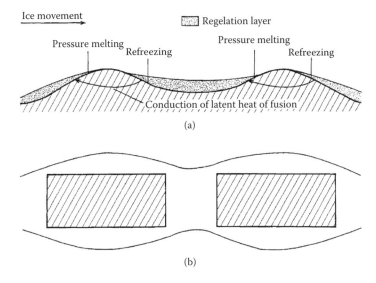

FIGURE 7.2 Illustration of the mechanisms of basal sliding (a), and enhanced plastic flow (b). The ice in the basal sliding process overrides small-sized obstacles, producing a regelation layer. In the enhanced plastic flow, the ice circumvents large-sized bedrock obstacles.

EROSION IN A GLACIAL ENVIRONMENT

Glacier is a mass of ice, but meltwater nonetheless is an integral component of the glacial system. The deformation of ice over its bed, and meltwater flow at and beneath the ice and beyond the margins of glacier ice, variously sculptures the landscape of glacial erosion. Glacial erosion is a subglacial mechanical process associated with the deformation of ice over its bed, and meltwater erosion is a subglacial and subaerial activity of mechanical and chemical domains.

GLACIAL EROSION

Glacial erosion happens by the sliding of glacier ice at its bed. Boulton and Clark (1990) observed that subglacial erosion occurs throughout the expanse of temperate glaciers, reaches an optimum magnitude at the *equilibrium line* separating accumulation and ablation zones of the glacier ice, and decreases thereafter to the limit of ablation zone (Figure 7.3). The ice flow in the ablation zone is *compressive* at all depths and decelerating. Therefore, subglacial sediments are laid to rest in a wide range of depositional forms near the margins of glacier ice. Glacial erosion is affected by abrasion and plucking of the glacial bed.

ABRASION

Abrasion occurs by the sliding of clean ice over its bed and by the sliding of rock fragments firmly held at the sole of glacier ice. The sliding of clean ice evolves a

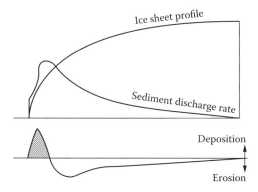

FIGURE 7.3   Subglacial erosion and deposition are functions of sliding velocity and the rate of sediment discharge. The theoretical rate of subglacial erosion progressively increases outward to the limit of glacier ice, reaching its highest value near the equilibrium line. The sediment discharge rate thereafter increases suddenly and deposition occurs. (Source: Figure 3a in Boulton, G. S., and Clark, C. D., "A Highly Mobile Laurentide Ice Sheet Revealed by Satellite Images of Glacial Lineations." *Nature* 346 [1990]: 813–17. With permission from Macmillan Magazines Ltd.)

polished subglacial bed, and the dragging of basal rock fragments against the bed develops polished, pitted, striated, and grooved pavements. Theoretically, the rate of abrasion by *clean ice* is proportional to the normal and shear stress at the bed of glacier ice and sliding velocity of glaciers as (Drewry, 1986)

$$W \propto \tau \, F_N \, U_s^{1/3}$$

in which W is the rate of abrasion, τ is the shear stress, F_N is the normal stress, and U_s is the sliding velocity of clean ice. The *normal stress* at the subglacial bed varies directly with the density and thickness of glacier ice and inversely with the hydrostatic pressure of meltwater at the glacier-bedrock interface, called the effective pressure of ice. The *shear stress*, which drives glacier movement, acts parallel to the bed and is a function of ice thickness and ice surface slope. In test conditions, abrasion rate increases up to a certain limit of the *effective ice pressure* and tapers off sharply once this threshold is reached (Figure 7.4).

Sharp-edged clastic fragments firmly held at the sole of *temperate glaciers* plow the bed, evolving small-sized striae and grooves parallel to the direction of ice flow. The abrasive wear due to the frictional drag of asperities harder than the bedrock is given as (Drewry, 1986)

$$A_b = (2 \cot \theta_a / \pi) \, (W^* / \sigma_y) \, U_p$$

in which A_b is the rate of abrasion, θ_a is the half angle of a single asperity tip, W^* is the volume of clast load in contact with the subglacial bed, σ_y is the yield stress of bedrock, and U_p is the relative velocity of rock fragments. The abrasive wear, thus, varies directly with the volume of subglacial clastic load and the distance basal fragments drag in the direction of flow, and inversely with the yield stress of subglacial

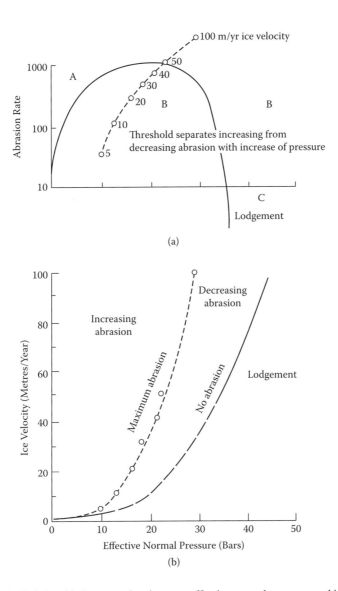

FIGURE 7.4   Relationship between abrasion rate, effective normal pressure, and ice velocity in determining the threshold between increasing and decreasing abrasion rate, and between abrasion and lodgment, for 50 m y⁻¹ ice velocity (partly after Boulton). (a) Subglacial abrasion initially increases with the rate of ice discharge, but tapers-off sharply once a certain threshold of the velocity of overriding ice is reached. This stage is followed by the lodgment of glacial till onto its bed. (b) Subglacial abrasion reaches a maximum with the increasing effective normal pressure of the ice over its bed, but decreases thereafter leading to the lodgment of till over the glacial bed. (Source: Figure 2a in King, C. A. M., "Thresholds In Glacial Geomorphology," in *Thresholds in Geomorphology*, ed. D. R. Coates and J. D. Viteks [London: George Allen & Unwin, 1980], 297–321. With permission from George Allen & Unwin.)

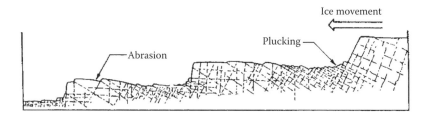

FIGURE 7.5 Development of a glacial stairway by selective quarrying. Bodies of closely jointed rock are readily plucked, while abrasion predominates the sparsely jointed rock surface. (Source: After Figure 11 in Matthes, F. E., *Geological History of the Yosemite Valley*, U.S. Geological Survey Professional Paper 160 [U.S. Geological Survey, 1930], 54–103.)

rock. A higher concentration of basal fragments in the ice, though, does not increase the abrasion rate as the ice becomes loaded and loses on the sliding rate. Hallet (1979, 1981) predicts that the abrasion rate is highest only for the glacial load between 10 and 30% of the mass of glacier ice. The abrasive wear generates *rock flour*, which in test conditions is even known to polish the subglacial bed at the base of *striations* (Rea, 1996). Efficient removal of the rock flour by the glacier melt, though, is an essential precondition for abrasion to happen and to sustain the given rate of subglacial abrasion.

PLUCKING

Plucking is the most potent process of subglacial erosion. It excels in areas where rock surfaces stand weakened by the effect of offloading, dry permafrost, frost wedging, hydration shattering, and rock rotting. Glaciers advancing over such pavements pluck loosened rock blocks from within high-density joint segments and abrade the remaining intact rock surface in low-density joint systems. The ice, while plucking the rock bed, also releases *subglacial till* (Hart, 1996). This till provides additional finer clastic fractions in the system for subglacial *abrasion* of rock pavements (Evans et al., 1998). Spatial variations in fissure-governed activity of plucking and abrasion evolve glacial stairways (Figure 7.5), such as in the granite terrain of Yosemite Valley, California.

MELTWATER EROSION

A large volume of meltwater widely exists in the environment of temperate and parts of polar glaciers, much of which flows subglacially before escaping the ice. The meltwater of temperate ice sheets drains through a network of bedrock channels (Booth and Hallet, 1993; Jones and Arnold, 1999). The meltwater of the 4 km thick Antarctic ice sheet, however, is held in isolated subglacial lakes, which flush periodically by transferring water to the next lake downslope (Clarke, 2006; Wingham et al., 2006). The volume of meltwater flow, which progressively increases in the direction of glacier terminus (Figure 7.6), also accompanies a simultaneous increase in the sediment load of temperate glaciers (see Figure 7.3). Hence, the rate of abrasive wear

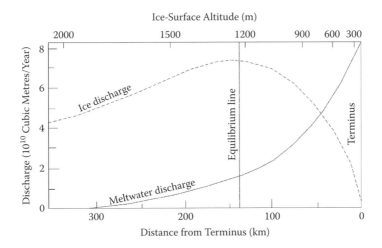

FIGURE 7.6 Spatially averaged total flux of ice and water across the entire Puget lobe as a function of longitudinal position. Equilibrium line position and mass flux data are from Booth (1986). (Source: Figure 12 in Booth, D. B., and Hallet, B., "Channel Networks Carved by Subglacial Water: Observations and Reconstruction in the Eastern Puget Lowland of Washington," *Bull. Geol. Soc. Am.* 105 [1993]: 671–83. With permission from Geological Society of America.)

of subglacial pavements and channel erosion tends to be highest near the *ablation zone* of temperate glaciers. Besides, the *suspended particles* in turbulent subglacial meltwater flow groove pavements and the walls of bedrock channels. In addition, meltwater discharge exceeding 8 m s^{-1} in velocity develops airless water bubbles in the channel flow (Drewry, 1986). These bubbles burst against bedrock channels and generate a strong shock wave pressure to cause *cavitation erosion* of the walls and bed of subglacial channels. Cavitation erosion is also said to be the process of small-sized *potholes* called bowls in the glaciated terrain of variable lithologic composition.

Meltwater escaping continental glaciers erodes outlet channels. In certain situations, the glacier melt is blocked behind *ice dams* near the margins of ice sheets. These ice dams periodically burst, releasing short-lived *glacier floods* that severely erode the surface and preexisting glaciogenic sediments beyond the margins of glacier ice.

Meltwater carries diffused carbon dioxide from the atmosphere. Hence, it is chemically reactive to some extent. Pavements that carry *calcite* as the rock-forming mineral, and are exposed by the retreat of glaciers, show solutional furrows (Chapter 10) and a surface coating of the carbonate precipitate, signifying *regelation slip* in temperate glaciers (Hallet, 1976). The *precipitates* of silica and ferromanganese minerals on rock pavements additionally suggest that the meltwater discharge is rich in the solute load of diverse chemical composition (Drewry, 1986).

EROSIONAL FEATURES OF GLACIAL ENVIRONMENT

Sliding of alpine and continental glaciers erodes the subglacial bed. Subglacial erosion occurs by processes of abrasion and plucking, and nivation beneath the margins

TABLE 7.1
Erosional Features of Glacial Environment

Locus	Dominant Processes	Characteristic Landforms
Glacial erosion at or inside glacier margins	Abrasive wear	Polished surfaces, chatter marks, striations, grooves
	Abrasion and plucking	Crescentic gouges, roches moutonnées, glaciated valleys, fjords
	Nivation	Cirques
Meltwater erosion	Subglacial and subaerial meltwater erosion	Plastically molded surfaces, meltwater channels

of lingering snow covers on mountain slopes, and by meltwater erosion at the glacier bed (Table 7.1). Roches moutonnées, cirques, glaciated valleys, and fjords are major erosional landforms of *alpine glaciation*, and rock basins of *continental glaciation*. Meltwater erosion evolves meltwater channels beneath and beyond the margins of glacier ice.

FEATURES OF ABRASIVE WEAR

Subglacial abrasion produces polished, pitted, striated, and grooved bedrock surfaces. The sliding glaciers *polish* rock pavements, and the vibratory motion of clastic fragments loosely held in the basal ice pits the subglacial bed, evolving *chatter marks*. Chatter marks can also evolve beneath stop and jerky or slip-stick movement of the ice over its bed. *Striations* are up to 1 m long discontinuous scratches parallel to the ice flow on suitable rock surfaces. They form from the dragging of finer basal sediments in the moving ice. Being parallel to the direction of ice flow, striae are an excellent indicator of the local direction of ice flow. Palimpsest of striae similarly indicates local changes in the direction of ice flow. Striae of a larger dimension are probably caused by abrasion and plucking of the subglacial bed (Wintges, 1985). *Grooves* are still larger striae, which possibly develop from the streaming of ice within ice sheets and from the plowing action of large boulders firmly held in the basal position of continental ice. The Mackenzie Valley in northwestern Canada supports the largest grooves known: they are up to 30 m deep, 100 m wide, and 12 km long in the direction of ice movement. Such gigantic grooves are the product possibly of the plowing action of large ice-cemented boulders firmly held in the basal ice.

FEATURES OF ABRASION AND PLUCKING

Subglacial abrasion and plucking develop small-sized crescentic gouges, medium-sized roches moutonnées, and large-sized linear troughs called glaciated valleys and fjords. Crescentic gouges are half-moon-shaped shallow gouges on rock pavements, and roches moutonnées are scores of meters long and tens of meters high asymmetric bedrock eminences aligned to the direction of ice flow. Glacial troughs are

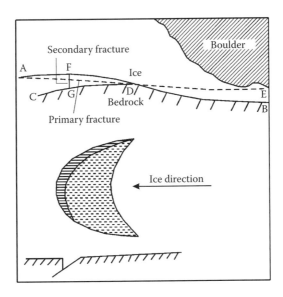

FIGURE 7.7 Theoretical origin of fractures producing crescentic gouge (above), and a crescentic gouge in plan view and cross section (below). AE, original profile of rock bed; AFDB, deformed profile of rock bed; DGC, concoid fracture; FG, vertical fracture. (Source: Figures 2, 3, and 9 in Gilbert, G. K., "Crescentic Gouges on Glaciated Surfaces," *Bull. Geol. Soc. Am.* XVII [1906]: 303–13. With permission from Geological Society of America.)

exclusive to alpine glaciers. *Crescentic gouges* probably develop from oblique transfer of the contact force of large-sized basal boulders in the ice sliding over its bed, causing localized compression and elastic deformation of the bed ahead of advancing glaciers (Gilbert, 1906). The compression opens up primary and nearly vertical secondary fractures, and a wedge of loosened rock between them (FGD) that, when plucked by the advancing ice, evolves into a crescentic gouge (Figure 7.7). Crescentic gouges also suggest a *stick-slip movement* of the ice that sets up greater shear force against bedrock surfaces (Rastas and Seppälä, 1981), and a frequent contact of basal boulders in rapidly moving ice against the subglacial bed (Wintges, 1985).

Roches moutonnées are asymmetric bedrock hills of a long and polished stoss surface and a steep shattered leeward face (Figure 7.8). The stoss and leeward knoll faces are the product of abrasion and plucking processes, respectively. It is generally believed that the ice riding the stoss face pressure melts and generates meltwater in the system, which flows over and across obstacles and freezes in rock joints leeward of the knoll. Repeated *freeze-thaw activity* here loosens joints and prepares the rock

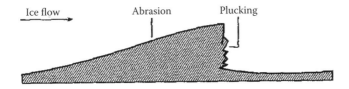

FIGURE 7.8 Morphology of a roches moutonnée.

for plucking by the advancing ice. At times, the presence of subglacial till and crescentic gouges on the stoss face suggests a stick-slip movement of the ice and loss of physical contact of the descending ice with obstacles. Subglacial modification of smaller-sized resistant rock outcrops that present a streamlined form of smooth short stoss and a long tapering leeward face of till material are called *crag and tail*. Crag and tail may be an intermediate stage to *erosional drumlins*.

Glaciated valleys are deeply excavated broad troughs of alpine glaciers (Figure 7.9). They are parabolic or loosely U shaped in transverse and undulatory in longitudinal profile. The *parabolic shape* results primarily from the higher stress

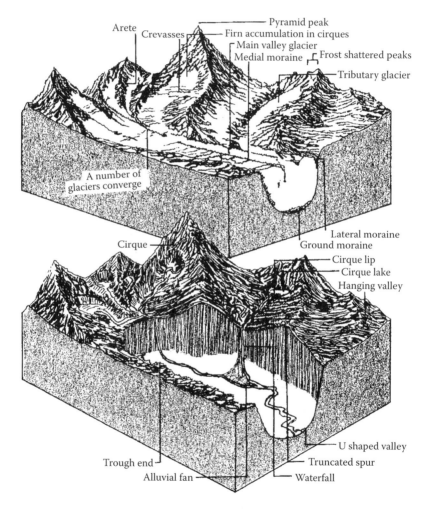

FIGURE 7.9 Block diagram showing various features of alpine glaciation. Glacial cirques and U-shaped valleys are illustrated in the lower part of the diagram. (Source: Redrawn from figure on p. 110 in Bisacre, M., Carlisle, R., Robertson, D., and Ruck, J., eds., the *Illustrated Encyclopedia of the Earth's Resources* [London: Marshall Cavendish Ltd., 1975–84]. With permission from Marshall Cavendish Books Ltd.)

and faster flow of thickest ice in the middle of valley glaciers than outward to the valley walls. The *undulatory thalweg* of basins and rises is inherited from preglacial troughs through which valley glaciers develop and expand. This thalweg configuration is understood in the activity of extending and compressing flow in the ice. The *extending flow* erodes the subglacial bed, evolving a concave-up form in conformity with *slip lines* in the ice. The ice, however, loses capacity to deepen this segment of the subglacial bed further on reaching equilibrium with the *geometry of ice flow*. Hence, the irregular thalweg persists as a permanent feature of glacial troughs. The Finger Lakes, New York, and the lakes in Okanagan Valley, British Columbia, Canada, are water-filled glacial troughs.

The Pleistocene highland glaciers, which had descended into the adjacent sea of a stand 100 m or so lower than the present (Chapter 11), excavated wide and deep troughs called *fjords* over and beyond the continental shelf to the limit of a threshold bar in the open sea. Fjords, covered with hundreds of meters deep seawater, are common along uplifted coasts of Alaska, Canada, Greenland, Norway, Iceland, Spitzbergen, New Zealand, and Chile. However, they are deepest in Antarctica, where fjord troughs lie buried beneath 2 km or so thick glacier ice (King, 1980). Fjords, like glaciated valleys, are irregular in profile, which suggests an equilibrium form of *negative feedback* between the subglacial bed and creep deformation of the ice in troughs. Fjords also randomly expand and contract, which suggests strong divergent and convergent flow of rapidly moving thick ice in active troughs (Shoemaker, 1986). These linear troughs terminate at a threshold bar ahead of deep basins, suggesting that the floating ice thinned considerably, dissipated by rotational movement, and failed to excavate the seabed further. Embleton and King (1968) predict that 1308 m deep Sognefjord in Norway is excavated by 1600 to 1800 m thick ice.

FEATURES OF NIVATION

Nival processes evolve steep-sided and deep amphitheatrical hollows called *cirques* at or above the head of many glaciated valleys (Figure 7.10). These hollows comprise a shattered wall, polished deep basin, and rounded lip at the outlet of ice (Figure 7.11). Cirques, which cut back to back on opposing slopes evolve a sharp crested ridge called *arête*, and those that cut toward each other leave a high pyramidal peak called *horn* between them. The basin of relict cirques sometimes supports a lake called *tarn*.

Cirques develop in topographic depressions and geologic structures that can hold sufficiently thick snow on steep mountain slopes. These depressions evolve into a mature cirque of an ideal height-to-length ratio between 2 and 3 to 1 by nival processes of frost shattering, hydration shattering, snow creep, basal sliding of the ice, and meltwater erosion around and beneath lingering snow covers of sizable extent (Chapter 8). Cirques probably evolve by a combination of nival activities, which provide arguments for the evolution of cirques by bergschrund, meltwater, and rotational slip hypotheses (Embleton and King, 1968). The *bergschrund hypothesis* envisages that ice melt and rainwater draining through a bergschrund penetrate deep into the ice, freeze against the backwall, and dislodge the debris by freeze-thaw activity, evolving the shattered backwall. These clastic fragments drag in the direction of ice

FIGURE 7.10   Two cirque glaciers in the Canadian Rocky Mountains, Banff area, Alberta. Bow River is in the foreground.

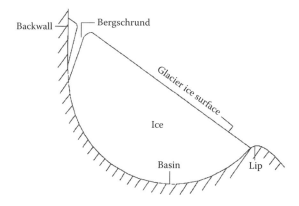

FIGURE 7.11   Elements of a typical cirque glacier.

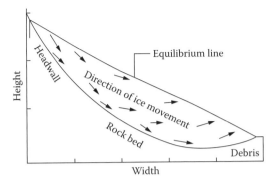

FIGURE 7.12   Schematic pattern of the ice flow as revealed by measurements in Vesl-Skautbreen cirque glacier in Norway. (Source: After J. G. McCall, Norwegian Cirque Glaciers (ed.) W. V. Lewis, *R. Geog. Soc. Res. Ser.* 4, 39–62, 1960.)

flow, causing abrasive wear of the cirque basin. However, bergschrund is not universal to cirque glaciers, and temperature at the depth of this deep crevasse in the ice is not known to be compatible enough for the freeze and thaw activity. The *meltwater hypothesis* argues that cirque basins enlarge by meltwater sapping. This hypothesis applies to cirques that comprise a thick *temperate ice* beneath the *polar ice*. In this type of thermal stratification of the ice, meltwater from the bottom layer of isothermal ice at *pressure melting point* rises by hydrostatic pressure along the cirque wall and freezes at the boundary of cold ice. Repeated freeze-thaw with its attendant effects and meltwater sapping in this zone, therefore, enlarges the cirque basin. The *rotational slip hypothesis* is based on measured flow velocities in a small Norwegian cirque glacier, which suggests that the ice flow is downward in the *accumulation zone*, parallel at the *névé*, and upward in the *ablation zone* (Figure 7.12). This is rotational movement of the ice that entrains debris from upslope segments and abrades and deepens the cirque basin. The rotational slip hypothesis widely appeals to the evolution of cirques.

FEATURES OF MELTWATER EROSION

Subglacial meltwater discharge dominates the environment of temperate and parts of polycrystalline polar glaciers. The subglacial meltwater and its charge of sediment load evolve polished and striated pavements and bedrock channels that additionally present evidence of *cavitation erosion*. The meltwater escaping the ice evolves erosinal features in the manner of any other fluvial system.

Subglacially abraded hard rock pavements present evidence of having been molded into small-sized complex morphologic forms of channels oriented in the direction of ice flow, grooves of rounded edges, shallow troughs with horns pointing downglacier called sichelwannen, curved and winding channels arising through the merging of sichelwannen, glacial striae moving in and out of potholes, and somewhat larger-sized potholes. These forms of possible fluvial scouring by high-velocity flow in englacial and subglacial tunnels are collectively called *plastically molded*

FIGURE 7.13 Supraglacial drift forming a lateral moraine in one of the glacial valleys near Banff, Alberta, Canada.

surfaces, or p-forms (Embleton and King, 1968). Several forms of plastically molded surfaces occur in the glaciated landscape of Norway and Sweden. In North America, bowls in different rock types may be associated with initial cavitation erosion. Such bowls of smooth undercut walls in a part of New Hampshire show evidence of having been enlarged by the swirling action of running water. Elsewhere, fairly large and deep hydraulic potholes at Taylor's Falls, Minnesota, and near Devil's Lake, Wisconsin, have probably developed from the high-energy impact of englacial or subglacial streams at an angle to the rock bed.

GLACIAL TRANSPORT

The clastic load of *alpine glaciers* is derived largely from the mass movement activity on valley-side slopes, and that of *continental glaciers* is based essentially in the subglacial bed. The glacial load at the surface and beneath the ice, however, frequently changes position by moving through *shear and thrust planes* in the ice, becoming *supraglacial* at the surface, *englacial* within the ice, and *subglacial* at the sole of the glacier ice. Therefore, the glacial transport system is the most complex of terrestrial sediment transport systems (Derbyshire, 1999).

Alpine glaciers abound in large-sized supraglacial debris of loose rock fragments throughout the lateral and terminal margins of the ice (Figure 7.13). By comparison, the supraglacial debris is compact and finer textured in temperate and parts of polar glaciers. These sediments had been *dragged* in the direction of ice flow as the subglacial load of glaciers, moved in the regelation slip, possibly thrown up the shear and thrust planes near the terminus of *receding glaciers* to become supraglacial prior to deposition, or transferred through the ice in subglacial supercooled meltwater flow in polycrystalline glaciers to become a part of the material of ice marginal moraines. Glacial transport processes, in general, are central to the understanding of the mechanisms of glacier flow and the morphogenesis of depositional landforms.

GLACIAL DEPOSITION

Advancing and retreating glaciers, stagnant glaciers, and glaciers wasting by ablation and sublimation shed their clastic load in many different ways on land and in water bodies. These sediments, which were handled by the ice, are called *glacial drift*. The glacial drift is nonstratified and stratified in nature. *Nonstratified drift* is released from the supraglacial and subglacial ice, and is identified with *till*. The till, in general, is a mixture of sand through to clay fractions with a modest contribution of pebble- to boulder-sized stones, but in certain situations the till could be excessively bouldery and is called stony till. Meltwater activity at and beneath the glacier ice and beyond the glacier margins deposits the glaciogenic sediments in stratified layers called *stratified drift*. The stratified drift is clay to gravel in size composition and, depending upon the environment of sedimentation, is poorly to moderately size sorted and crudely to moderately stratified in nature.

NONSTRATIFIED DRIFT

Nonstratified drift is released from the glacier ice in many different ways as subglacial and supraglacial till (Lawson, 1981). *Subglacial till* is released from the sole of temperate and subpolar glaciers, and *supraglacial till* settles from the ice thinning by ablation and sublimation. The subglacial and supraglacial tills also differ in primary processes of till release and secondary processes of the modification of glacial till prior to deposition at the given site (Table 7.2). The *primary processes* of till

TABLE 7.2
Till Processes and General Deposit Characteristics

Type	Process Primary	Secondary
Meltout	Released from the interstices of slow-melting debris-rich ice as a compact deposit	Carried up shear and thrust planes in the ice to become supraglacial. These fine-grained moist deposits flow down the local slope of the glacier ice as flow till
Sublimation	Sublimation till settles with its original texture onto the bed of polar glaciers wasting by sublimation	Free-fall from the bottom of icebergs and ice shelves onto the sea bottom as crudely-stratified deposit
Lodgement	Released as a compact deposit from the sole of active ice when the frictional resistance at the glacier bed exceeds the drag of ice	Excessively moist fine-textured lodgement till is lodged onto the glacier bed and squeezed into cavities of the basal ice
Supraglacial	Thick mass of angular debris derived from side slopes of valley glaciers and nunatak, and let-down by the ice wasting by ablation	Gravitational slumping of supraglacial debris of low moisture content against the local ice slope and collapse into adjacent depressions on the surface ice as ablation till

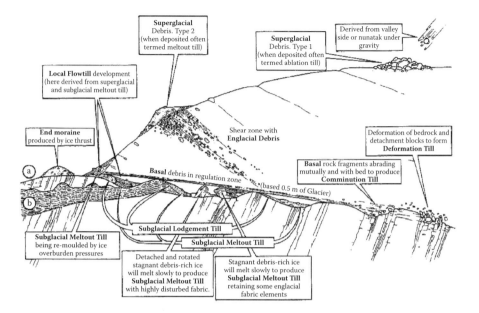

FIGURE 7.14   Generalized relationships of ice and debris in temperate glaciers—vertical scale greatly exaggerated. Two cases of end moraine formation shown. (a) Glacier piles up ridge as it slides over bedrock. (b) Glacier rests on thick saturated till and produces "squeeze-up" moraines. (Source: Figure on p. 436 in Goudie, A. S., Simmons, I. G., Atkinson, B. W., and Gregory, K. J., ed., *The Encyclopaedic Dictionary of Physical Geography*, 2nd ed. [Oxford: Blackwell, 1994]. With permission from Blackwell Publishers.)

release and the *secondary processes* of the modification of till prior to deposition are illustrated in Figure 7.14 from a general point of view.

Till is subglacial and supraglacial relative to the glacier ice. The subglacial and supraglacial tills, the principal deposits of several moraines, are released from the glacier ice in many ways. The till that settles *in situ* in the glacial environment is primary and that transported and deformed in the process of deposition is secondary in origin (Table 7.2).

Primary Processes

The subglacial till is meltout, sublimation and lodgement by the *primary process* of release. The *meltout till* is laid-down from the interstices of a slow melting debris-rich stagnant ice. The meltout till in ice-cored end moraines of the Matanuska Glacier in Alaska is released form *in situ* melting of the upper and lower surfaces of the debris-rich basal ice as compact and, at times, slightly stratified deposit depending upon the moisture content in the till (Lawson, 1981). Sublimation is the process of the wastage of polar glaciers in hyperarid environment. This process releases *sublimation till* onto the glacier bed. The till, therefore, retains its original texture. The till released directly from the base of a debris-rich ice is called lodgement till. The *lodgement till* is released when frictional resistance of the debris load at the glacier bed exceeds drag of the moving ice. The frictional resistance varies in direct proportion to the normal weight of the overlying ice. It is variously given as the clast weight times the effective pressure of the ice at its bed, and the clast weight times the component of the ice flow at the bed. The sediments from the sole of glacier ice lodge onto the rock bed when the frictional resistance at the bed exceeds the normal weight of the overriding ice. This sedimentation increases the *bed roughness*, which additionally accelerates the process of lodgement of the basal ice particles. The debris of *supraglacial* origin is derived from side slopes of valley glaciers and nunatak. It represents a thick mass of non-compact and angular rock fragments at the surface of the glacier ice. The supraglacial debris is 'let-down' onto the floor of glaciers wasting by ablation.

Secondary Processes

The tills of primary process modify in their textural and structural properties by the *secondary processes* of sediment transport to the site of deposition within the glacial systems. The meltout till, carried up *shear* and *thrust planes* in the dynamic ice, becomes *supraglacial* prior to deposition. The *supraglacial meltout till* appears in thick deposits along the margins of valley glaciers, continental glaciers and subpolar glaciers. The fine-grained saturated supraglacial debris, however, flows down the local slope of the surface ice as *flow till*. The flow till and the debris of local mass wasting at the terminus of Matanuska Glacier appear in a combination of various sedimentological units of distinct sedimentary features, clast dispersion and near absence of pebble fabric (Lawson, 1981). An excessively moist fine-textured *lodgement till* collapses into the cavities of sliding ice, and the supraglacial till of coarser fragmentary debris of low moisture content slumps into local depressions on the surface of alpine glaciers.

STRATIFIED DRIFT

Sediments of glacial origin are reworked, and deposited, by the meltwater activity. The sediments laid down in contact with the ice, called *ice-contact deposits*, are only a small part of the glaciogenic load of glaciers. The remaining large proportion of the meltwater load is carried beyond the glacier margins as *proglacial drift*. It is *glaciofluvial* when laid down on land adjacent to the glacier ice, *glaciolacustrine* when deposited in freshwater lakes, and *glaciomarine* when laid down in the seawater.

LANDFORMS OF DRIFT DEPOSITION

Landforms of glaciogenic sediments evolve in response to the activity of glacier ice that releases and deforms its entrained load in many different ways, and the manner of sedimentation from the glacier melt at and within the ice. Hence, the environment of drift deposition offers a suitable framework for genetic classification of the depositional landforms of drift composition (Table 7.3). The glacier ice is active when advancing or retreating over its bed and stagnant when a thin retreating ice breaks away from the main ice lobe and disintegrates *in situ*. The active and stagnant ice evolves a variety of depositional landforms. The melting of glacier ice and attendant supraglacial, englacial, and subglacial meltwater flow in contact with the ice evolves ice-contact landforms of stratified drift composition. The meltwater flow directed

TABLE 7.3

Classification of Glacial Drift and Associated Landforms

Locus	Depositional Environment	Drift Type	Dominant Process	Major Landforms
At and inside glacier margin	Active ice	Nonstratified (till)	Subglacial	Ground moraines, Drumlins
			Supraglacial	Moraines of dumping, pushing/thrusting, and squeezing
	Stagnant ice	Till and stratified drift	Icedisintegration	Hummocky moraines
			Ice contact	Crevasse fillings, Eskers, Kames, Kame terraces
Outside glacier	Proglacial	Stratified	Glaciofluvial	Outwash plains, valley trains
			Glaciolacustrine	Lacustrine plains
			Glaciomarine	Beaches and deltas, presently unrecognizable

away from the glacier margins is called proglacial. The proglacial environment of deposition evolves glaciofluvial landforms on land, glaciolacustrine landforms in freshwater glacial lakes, and glaciomarine landforms in the seawater.

ACTIVE ICE ENVIRONMENT AND LANDFORMS

Glaciers are active when advancing or retreating. The till is released from the base of such glaciers and deformed at the ice-bedrock interface by the moving ice. The deformation of *basal till* in this manner evolves large-sized ground moraine and drumlin forms. The subglacial till, however, also moves up the ice through *shear and thrust planes* near the snout of broad valley glaciers and termini of temperate and subpolar glaciers, in regelation slip in glaciers, and in supercooled basal meltwater flow through the network of ice veins of polythermal glaciers to become *supraglacial* prior to deposition as a system of large-sized morainic ridges.

GROUND MORAINES

In parts of Europe and North America, active continental glaciers had nearly uniformly shed the subglacial load at the bedrock without much deformation. This form of till deposition has evolved an extensive flat to rolling and, at times, fluted ground moraine feature. The *flutes* probably suggest lateral deformation of moist subglacial till in localized low-pressure zones ahead of large boulders held in the basal ice (Figure 7.15). The fluted moraines comprise broad-based till ridges and intervening flutes in the direction of ice flow.

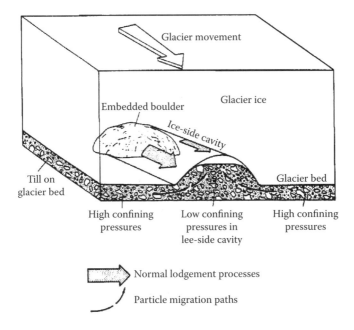

FIGURE 7.15 An explanation of fluted moraine formation. (Source: Figure 12.2 in Sugden, D. E., and John, B. S., *Glaciers and Landscape* [London: Edward Arnold, 1976]. With permission from Edward Arnold Ltd.)

DRUMLINS

Drumlins occur at the inside margin of *end moraines* as depositional and erosional streamlined hills of low height. The hills occur in fields of nearly similar sized and shaped subrounded, oval, and elongated forms that approach an *aerodynamic shape* of a length-to-width ratio of between 2.5 and 4 to 1. The *depositional drumlins* are composed of basal till, and *erosional drumlins* are shaped out of the glacial drift and bedrock surfaces. Drumlins are a conspicuous feature of continental glaciation in Europe, Canada, and the United States, and of *alpine glaciation* in Ireland, Scotland, and Switzerland. Most extensive belts of drumlins occur in the state of New York and in Northern Ireland.

Drumlins of glaciogeneic sediments evolve from the deformation of basal till. The deformation can be initiated by the roughness of the subglacial bed, rheologic behavior and strength variation of the subglacial till, effective normal pressure of the ice on the subglacial bed, and the velocity of glacier flow. The *classic hypotheses* on the evolution of depositional drumlins presume that small-sized bedrock obstacles and patches of frozen till, which invariably increase the *bed roughness*, promote proto-drumlin sites for the accumulation and growth of subglacial till. This till is shaped into a streamlined form by the overriding ice. Erosional drumlins, however, evolve from a readvance of the ice over preexisting subglacial till deposits. More recent hypotheses on the evolution of drumlins theorize why and how, or mechanisms of subglacial till deformation beneath sliding glaciers.

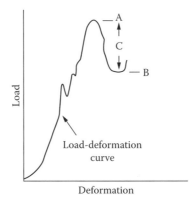

FIGURE 7.16   The load-deformation curve for glacial till. (Source: Figure 1c in Smalley, I. J., and Unwin, D. J., "The Formation and Shape of Drumlins and Their Distribution and Orientation in Drumlin Fields," *J. Glaciol.* 7 [1968]: 377–90. With permission from International Glaciological Society.)

The basal till is a typical *dilatant material* that expands to a certain extent beneath the weight of ice but collapses into rigid heaps at a critical limit of the stress of overlying mass of glacier ice (Smalley and Unwin, 1968). The subglacial till evolves drumlins at a stress magnitude at which it is neither expanding nor rigid but in a state intermediate between the two, indicated as C in Figure 7.16. This stress magnitude is satisfied in a narrow zone of the inner and outer ice, where the subglacial till collapses under the stress of overriding ice into drumlin-forming nuclei (Figure 7.17). These proto-drumlin sites grow by trapping the subglacial till and transforming it into a drumlin form beneath the advancing ice. Hence, drumlin fields occur just inside the margins of end moraines.

Boulton (1979) proposed that the strength of subglacial till increases toward the glacier source, but the drag force of moving glacier is not sufficient here to cause deformation of the subglacial bed. Thus, Boulton's theory is remarkably similar to the dilatancy theory in key respects. Further, composition of the subglacial till is an important effect on the deformation behavior of till beneath sliding glaciers, such

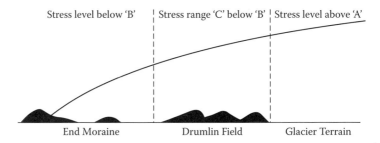

FIGURE 7.17   Cross section at the edge of an ice sheet with critical stress regions indicated. (Source: Figure 2 in Smalley, I. J., and Unwin, D. J., "The Formation and Shape of Drumlins and Their Distribution and Orientation in Drumlin Fields," *J. Glaciol.* 7 [1968]: 377–90. With permission from International Glaciological Society.)

that initially finer fractions of moist basal till deform and collapse into nuclei of high-strength till beneath the weight of active ice. These proto-drumlin sites attract transverse flow of coarser fractions in the till for accretion and evolution of drumlins beneath the moving ice.

Variation in the *moisture content* of subglacial till is another physical attribute of significance to the strength and deformation behavior of the material beneath moving ice. Menzies (1979) and Benn and Evans (1996) hold that the basal till in areas of *permeable bed* and *regelation slip* is relatively dry and high in shear strength. These areas of high-strength till provide nuclei for the accumulation of basal till and evolution of drumlin sites beneath sliding glaciers. Whether these conditions satisfy fully within drumlin fields behind end moraines, however, is not known for certain.

Effective pressure of the ice at its bed varies with the thickness of glacier ice, clastic load held in glaciers, and pore pressure in the subglacial till. Given these assumptions, Stanford and Mickelson (1985) predict that the subglacial till beneath thick and loaded ice laterally flows toward areas of clean and relatively thin ice for accumulation (Figure 7.18) and evolution of drumlins at the glacier-bedrock interface. The hypothesis, therefore, invokes the stress force for explaining drumlins of depositional nature.

1. Prior to Lateral Flow of Sediment

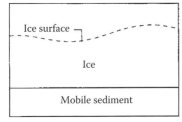

2. Initiation of Lateral Sediment Flow

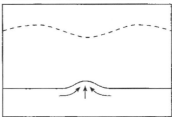

3. Termination of Lateral Sediment Flow and Formation of Proto-drumlins

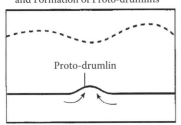

4. Definition of Drumlin by Erosion and Remolding of the Sediment Accumulation into a Streamlined Shape

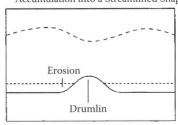

FIGURE 7.18 Diagrammatical cross sections perpendicular to the ice flow, depicting the sequence of events during drumlin formation. These depict great vertical exaggeration. Changes in ice thickness could be due to crevassing. Similar results could be produced by differences in the density of overlying column but with equal ice thickness. (Source: Figure 10 in Stanford, S. D., and Mickelson D. M., "Till Fabric and Deformation Structures in Drumlins Near Waukesha, Wisconsin, U.S.A.," *J. Glaciol.* 31 [1985]: 220–28. With permission from International Glaciological Society.)

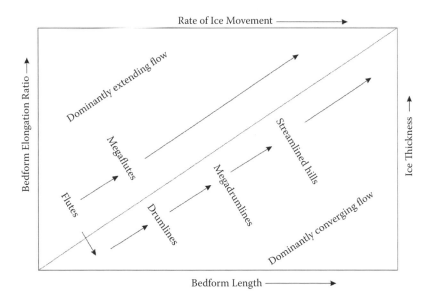

FIGURE 7.19 Continuum of subglacial bed forms in relation to ice thickness and rate of ice movement. (Source: Figure 4 in Derbyshire, E., "Glacial Geomorphology," in *Fourth International Conference on Geomorphology*, Bologna, Italy, August 28–September 3, 1997, pp. 89–97 [published 1999]. With permission.)

Deformation of the ice at its bed depends on the ice thickness and sliding velocity of the glacier ice (Rose, 1987; Boulton, 1996; Derbyshire, 1999). Theoretically, drumlins progress from smaller to larger subglacial depositional form through to erosional bedrock hills beneath increasing thickness of predominantly convergent ice flow and rate of ice deformation (Figure 7.19).

MORAINIC RIDGES OF SUPRAGLACIAL DRIFT

The basal till that becomes supraglacial prior to deposition leaves a series of morainic ridges called *end moraines* parallel to the front of alpine, continental, subpolar, and high-latitude polycrystalline glaciers. The end moraines possibly evolve by dumping, thrusting/pushing, and squeezing of the supraglacial drift (Price, 1973). Subsequent studies, however, suggest that the ice marginal moraines can also evolve from sediments entrained in regelation slip and sediments trapped and transported in supercooled subglacial meltwater flow in temperate and polycrystalline glaciers. The end moraines of the squeezing mechanism possibly evolve in a stagnant ice environment.

Dumping Process

Extensive end moraines that exist near the margins of high-latitude glaciers in Canada and Greenland have possibly evolved from the migration of basal till through shear planes in the ice, *regelation slip*, and migration of basal sediments in supercooled *meltwater flow* at depth in the polycrystalline ice. *Shear planes* evolve from the creep of ice in cold-based glaciers frozen to the bed (Goldthwaite, 1951) and also from the slippage of warm-based glaciers against a mass of dead ice outward to its edge (Bishop, 1957; Evans, 1989). However, observation of a large number of shear planes in the ice frozen to its bed, and migration of 15 mm or more across basal debris constituting the material of end moraines through 1 mm or less wide shear plane openings in the ice, remains speculative and requires alternative explanations. Weertman (1961) theorized that the glacial transport in polycrystalline glaciers is regelation controlled, predicting that the creep deformation of thick ice in the interior of glaciers generates sufficient *frictional heat* to melt the basal ice. The meltwater moving in the direction of thermal and pressure gradients freezes near the edge of thinner outer ice as *regelation debris* one on top of the other, evolving end moraines independent of shear planes in the ice. The *glaciohydraulic supercooling model* of subglacial sediment transport also explains the ice terminal moraines. This theory is based on the premise that rapid increase in the hydraulic potential of subglacial meltwater converts the mechanical energy of flow into *sensible heat*, causing super-cooling of the meltwater in the system (Alley et al., 1998). The sediment-laden super-cooled meltwater from overdeepened subglacial basins flows upward along crystal boundaries and freezes onto the adjacent ice as intraglacial debris-rich basal ice. Studies by Lawson et al. (1998), Knight and Knight (1999, 2004, 2005), Tweed et al. (2005), Cook et al. (2006), and others suggest that supercooling is a significant process of glacier dynamics and subglacial sediment entrainment in the direction of hydraulic and thermal gradients of temperate and polycrystalline glaciers. Knight and Knight (1994) demonstrated experimentally that 0.3 to 0.7 mm sized highly sorted subglacial sediments in supercooled meltwater flow could indeed be entrained upward into the vein network of polycrystalline ice. This distinctive grain size is also a part of the sediment cover of ice marginal moraines in Russell Glacier, Greenland (Adam and Knight, 2003).

Thrusting/Pushing Process

Continental glaciers advancing against topographic obstacles deform the bedrock, and subaquatically advancing glaciers deform the basal till into *ice-thrust ridges* that resemble the end moraines of the dumping process. Broad-based and subparallel ice-thrust ridges of western Canada (Kupsch, 1962) and Poland (Kozarski, 1994) are similarly eminences of the deformed bedrock transverse to the direction of ice flow. The ice-thrust ridges of Canada comprise gentle faults in the downglacier direction, folds in the direction of faults, subglacial till locked in infolded strata, and a veneer of ablation till at the surface of deformed bedrock (Figure 7.20). These geologic attributes suggest a *gentle thrust* of thin continental ice against the bedrock already weakened by the *dry permafrost* (Kupsch, 1962).

All glaciers do not terminate on land; a few also end up in proglacial lakes or the sea. Those terminating in water bodies evolve end moraines from the push and thrust of subaquatic saturated till ahead of the steep ice front. The moraines of subaquatic origin in the *proglacial lake environment* appear in a pair of ridges. They are called De Geer moraines in Sweden (Strömberg, 1965) and annual moraines in northeastren Canada (Andrews, 1963). The morainic ridges of *marine environments* are found below the highest tide level. Hence, they are referred to as moraines of the tidewater glaciers.

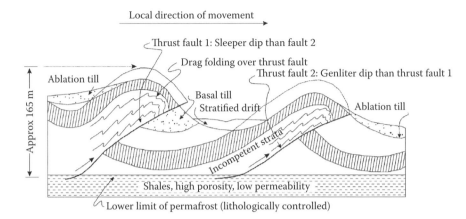

FIGURE 7.20 Conceptual section perpendicular to the strike of ice-thrust ridges, showing the topography and its relation to the structure, stratigraphy, and permafrost. (Source: Figure 4 in Kupsch, W. D., "Ice-Thrust Ridges in Western Canada," *J. Geol.* 70 [1962]: 582–94. With permission from University of Chicago Press.)

The paired morainic ridges in up to 200 m deep *proglacial lakes* probably evolve from the thaw of moraine-based ground on which glaciers rest, squeezing of the water-soaked till ahead of the steep ice front, and deformation of this till into 15 m or so high ridges. The paired moraines possibly result from ablation and advance of the ice together to give two periods in a year when the ice front remains apparently stable for a short period (Embleton and King, 1968). Moraines of the *tidewater glaciers* evolve from the thrust of fjord ice against glaciogeneic and marine sediments. The thrust of fjord ice throws up marine sediments from the fjord floor, folds the supraglacial debris, and squeezes the subglacial till into the ice front (Hambrey et al., 1999). Morainic ridges of the type are common to the tidewater glaciers north of the Arctic Circle.

STAGNANT ICE ENVIRONMENT AND LANDFORMS

In the penultimate stage of deglaciation, the melting continental ice had become stagnant in many parts and disintegrated *in situ* (Gravenor and Kupsch, 1959). The ice disintegration released basal till, ablation till, and stratified drift at the glacier-bedrock interface, evolving a hummocky topography of randomly oriented closed and linear ridges, knobs, plateaus, and shallow surface depressions of *dead-ice moraines* (Figure 7.21). The feature of such a topographic and glacial drift composition is also called a hummocky moraine.

HUMMOCKY MORAINES

Dead-ice moraines possibly evolve by letdown of the ablation till, squeezing of the basal till, and meltwater sedimentation in crevasses and cracks of the wasting glacier ice. The *letdown hypothesis* visualizes that ablation till accumulates within the ice

FIGURE 7.21 The undulatory kettle and knob topography near Edmonton, Canada. A prominent depression in the middle of the photograph is surrounded by low-relief circular and linear disintegration ridges.

crevasses, which upon the thaw of the ice emerges as upstanding ridges of slumped structures and a content of stratified drift above the basal till (Figure 7.22). The *squeezing hypothesis* postulates squeezing of the moist subglacial till into crevasses and cavities of the overlying dead-ice mass (Figure 7.23) that leaves irregularly shaped ridges of basal till upon thaw of the ice. Knob and plateau-like forms of the dead-ice moraine, however, are nonstratified and stratified in drift composition, implying that dead-ice moraines probably evolve by a combination of letdown, squeezing, and glaciofluvial sedimentation processes.

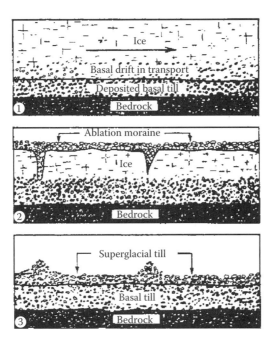

FIGURE 7.22 Disintegration features formed by ablation. (Source: Figure 2 in Gravenor, C. P., and Kupsch, W. O., "Ice-Disintegration Features in Western Canada," *J. Geol.* 67 [1959]: 48–64. With permission from University of Chicago Press.)

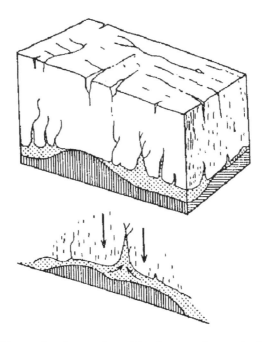

FIGURE 7.23 Disintegration features formed by the squeezing mechanism (originally from Hoppe). (Source: Figure 3 in Gravenor, C. P., and Kupsch, W. O., "Ice-Disintegration Features in Western Canada," *J. Geol.* 67 [1959]: 48–64. With permission from University of Chicago Press.)

ICE-CONTACT ENVIRONMENT AND LANDFORMS

Glaciers generate meltwater, which flows in supraglacial channels, and englacial and subglacial tunnels in the ice. Sedimentation of the glaciofluvial load in surface channels, tunnels of the ice, moulins draining the ice surface flow to englacial routes, and crevasses in the ice evolves ice-contact landforms called eskers, kames, and crevasse fillings. Eskers and kames occur in the landscape of temperate and subpolar glaciers of the present and the past. Crevasse fillings in active crevasses of temperate glaciers (Figure 7.24) would, upon thaw of the ice, appear in a random pattern of short, narrow, and steep ridges of the stratified drift. The glaciofluvial sediments, in general, are moderately sorted, subrounded, stratified, and variable in size composition.

Eskers

Eskers are from a few tens of meters to several kilometers long isolated, and at times intersecting, flat-topped sinuous ridges of glaciofluvial sediments in open-walled supraglacial channels and englacial and subglacial tunnels of *receding* continental, subpolar, and alpine glaciers (Price, 1966). Eskers, in general, are built of moderately sorted and stratified or crudely stratified sand and gravel deposits. Subglacial tunnels receding *subaquatically*, however, evolve oses or beaded eskers (Gorrell and Shaw, 1991). Such eskers comprise alternating beads of coarser and finer sediments at points of inflection in esker ridges. The coarser sediments are deposits of summer

FIGURE 7.24 Ice crevasse deposits in the making near the snout of the Athabasca Glacier, Alberta, Canada.

ablation and finer fractions of winter melting of the ice. Large areas of eskers are found in parts of Ireland, North America, and Scandinavian countries.

The eskers of supraglacial and subglacial origin gradually settle onto the bed of receding glaciers. Such eskers, which incorporate the ice torn from channel and tunnel walls expanding at a higher rate of the ice melt in contact with sediments, grow to a large size from sediments released by enlarging walls (Price, 1966; Gustavson and Boothroyd, 1987). The two types of eskers also grow around ice cores beneath sediments. Melting of the ice fragments and ice cores subsequently results in deformed structures in the stratified sediments.

The eskers of subglacial tunnels are related to melting and refreezing of water-filled subglacial passages along the bed of ice sheets. Shreve (1985) theorized that esker path and esker morphology depend on the subglacial water pressure and the tunnel gradient. The water pressure varies directly with the weight of overlying ice and governs the subglacial tunnel and esker path, such that ascending and descending subglacial channel segments melt and expand at a threshold gradient less than 1.7 times the ice surface slope, and freeze and shrink at more than 1.7 times the ice surface slope. Local expansion and contraction of subglacial tunnels also affect the tunnel cross section and esker morphology (Figure 7.25). Expanding tunnels evolve a sharply arched cross section and freezing tunnels produce a low passage. The expanding tunnel passages release coarse-textured and poorly sorted sediments, evolving sharp-crested esker ridges. The frozen passages, however, are a conduit largely for meltwater and sediment load discharge from the upglacier source. Hence, these tunnel segments of moderate flow velocity develop broad-based eskers of fine-textured and moderately sorted stratified drift.

KAMES

Kames are from a few meters to 30 m or so high mounds of moderately sorted and stratified sand and gravel deposits that were laid down by the supraglacial streams in

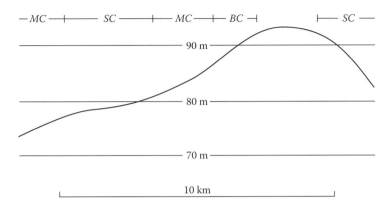

FIGURE 7.25 Longitudinal profile of esker paths in Morrison Ponds area showing correlation of esker type with gradient. As gradient steepens, type changes from sharp crested (SC) to multicrested (MC) to broad crested (BC). Vertical exaggeration, 200 times. (Source: Figure 13 in Shreve, R. L., "Esker Characteristics in Terms of Glacier Physics, Katahdin Esker System, Maine," *Bull. Geol. Soc. Am.* 96 [1985]: 639–46. With permission from Geological Society of America.)

surface pools and crevasses of slow-moving or *stagnant* ice of continental glaciers. Kames generally occur throughout the ground moraine areas of Central Europe and North America. Kames such as those in a part of the highlands of southeastern Scotland, however, have developed in ice-walled troughs above decaying ice cores in terminal moraines of alpine glaciation (Huddart, 1999).

PROGLACIAL ICE ENVIRONMENT AND LANDFORMS

The proglacial meltwater load of alpine and continental glaciers evolves outwash plains and valley trains as major deposits and landforms of glaciofluvial deposition on land. Outwash plains are essentially associated with continental glaciers, and valley trains with alpine or valley glaciers. The proglacial sedimentation in freshwater lakes is known as glaciolacustrine sedimentation. This form of sedimentation evolves a flat surface of glacial varves called lacustrine plain. Glaciomarine sedimentation develops beaches and deltas in the open sea, but these deposits have been reworked by waves to the extent of being unrecognizable. In the present, *ice-rafted* boulders in deep water sediments are the only evidence of glaciation on the adjacent land.

OUTWASH PLAINS AND VALLEY TRAINS

Glacier melt is concentrated in front of ice sheets, and glaciofluvial sedimentation in this area builds a fan-shaped surface of hundreds of meters thick rounded and cross-bedded sand and gravel units called sandur or outwash. The outwash sediments are coarsest closest to the ice front, and become progressively finer with distance down the deposit length. The huge volume of sediments comprising the outwash is derived from *advancing* continental and alpine glaciers that are additionally characteristic of high subglacial erosive potential (Gustavson and Boothroyd, 1987). At times, the turbulent subglacial flow also tears the ice, which then becomes a part of the glaciofluvial load. Subsequent thaw of this ice in sediments produces the morphology of pitted outwash. The Mashpee outwash plain in Massachusetts is one such case in point. Extensive outwash plains occur in many parts of the world. Much of the south coast of Iceland is built of the outwash material. The outwash is also extensive in plains of north Germany and conspicuous in Denmark. The Centerbury Plain in a part of South Island, New Zealand, is another outwash of some 10,000 km² extent. It is the product of sediment-laden meltwaters from glaciers in the Southern Alps.

Outwash plains are convex up in transverse and concave up or segmented along the radial profile. They generally represent aggradational segments of proximal, intermediate, and distal zones of sedimentation. The *proximal zone* is identified with steeply sloping trenched surface of coarsest sediments, the *intermediate zone* represents the surface of braided channels in substantially thick sand deposits, and the *distal zone* is characteristic of finer sediments and coalescing outwash of adjacent ice lobes.

The evolution of contemporary outwash is related to the phenomenon called *glacier floods* that are associated with the burst of some 30 to 60 m deep ice-dammed lakes. *Ice-dammed lakes* had formed when the Pleistocene ice invaded ice-free tributaries to glaciers and blocked the meltwater flow and clastic load of the glacier ice behind ice barriers. Such dams become periodically unstable on reaching a certain depth of the ponded water. They then lift and drain subglacially when the hydrostatic pressure of ponded water exceeds the overburden pressure of the ice. Once drained, the subglacial outlets close by the *creep* of the ice. The ice-dam burst events, called jökulhlaup (pronounced yokel-laup), and katlahlaup if severe, release catastrophic floods and a large volume of glaciogenic sediments that had accumulated in lakes for building the outwash. Glacier floods reoccur once every 5 to 6 years in the ice-dammed Grimsvöten in southeastern Iceland (King, 1980) and once every 1 to 5 years in the ice-dammed Strandline Lake in Alaska (Sturm and Benson, 1985).

Valley trains represent the outwash of meltwater streams confined to valley walls and the outwash of broad alpine valleys leading away from the lobes of advancing glaciers. These trains extend for several kilometers down valley in parts of Iceland, continental Europe, North America, and New Zealand (Figure 7.26). They comprise crudely bedded and angular to subrounded coarse sand and pebble units derived from the parent valley glaciers. Valley trains are also expected to carry a slight volume of sediments derived from the erosion of preexisting morainic debris and mass movement activity on adjacent valley-side slopes.

LACUSTRINE PLAINS

Lacustrine plains are monotonously flat surfaces of *cyclic sedimentation* of the meltwater load of continental glaciers in freshwater glacial lakes, evolving a sequence of texturally and genetically distinct light-colored silt and comparatively thin dark-colored clay layers one on top of the other (Figure 7.27). A pair of silt and clay lamina is called a *varve*.

FIGURE 7.26   A valley train underlying parabolic sand dunes in the Canadian Cordillera near Banff, Alberta, Canada.

FIGURE 7.27   A varve sequence overlying two tills is exposed in an escarpment adjacent to the north Saskatchewan River, Edmonton, Canada. The varves appear in narrow bands of lighter and darker sediments deposited in the proglacial Lake Edmonton

The difference in the thickness and texture of sedimentary units is primarily governed by the density difference in the waters of glacier melt and that of the receiving basin of sedimentation (Ashley, 1971). The summer season meltwater flow that carries a large volume of suspended load is denser than the lake water. Hence, it moves as a *hypopycnal flow* to the distal part of the lake for sedimentation. The silt units evolve by this turbid flow in glacial lakes. The finer clay fractions in the sediment load, however, stay in *suspension*. They gradually settle by the period approaching winter freezing of the lake. This process of sedimentation evolves a uniformly thick sequence of clay units. Varves are annual rhythmities. Hence, they are excellent *timekeepers* of the continental ice retreat from glaciated regions.

SUMMARY

Extensive continental and alpine glaciers had repeatedly built up in the Pleistocene epoch, moved outward from the accumulation centers, and thawed. The continental glaciers reached a massive ice thickness of up to 3 km over thousands of years of accretion in cold glacial phases but thawed remarkably rapidly in the intervening warm interglacial phases of the Pleistocene climate. Deformation of the continental and alpine glaciers and the glacier melt beneath and beyond the ice has evolved a variety of glaciogenic sediments and landforms. The glaciogenic sediments have also variously contributed to the evolution of contemporary landforms of other geomorphic environments.

Glacier ice deforms by the weight of its own mass or creep, and sliding of the ice over its bed, or basal sliding. Warm-based temperate glaciers at pressure melting temperature throughout the depth of isothermal ice largely deform by sliding over the lubricating film of meltwater at the glacier-bedrock interface. The basal sliding comprises the components of subglacial pressure melting called regelation slip and enhanced flow mechanisms. Polar glaciers, which are frozen to the depth of ice, largely deform by the creep of ice. High-latitude polythermal glaciers of a thicker inner ice at pressure melting and an outer thinner ice frozen to the bed deform respectively by sliding in the interior and creep in the exterior of the ice mass.

Temperate glaciers erode the subglacial bed by abrasion, plucking, and crushing processes that are broadly known in outline and theory. Their effects, however, are known on rock pavements exposed by the retreat of glacier ice. The subglacial erosion, in general, depends on the sliding velocity of ice, the type and distribution pattern of clasts in the ice, the effective normal pressure of ice on its bed, and geologic attributes of the subglacial bed. Nival processes beneath and around lingering snow covers on mountain slopes evolve cirques, and movement of thick alpine ice in preexisting valleys develops glacial valleys and fjords.

Meltwater erosion evolves subglacial channels and spillways, and develops scabland topography beyond the margins of glacier ice. Glacially abraded hard rock surfaces also present evidence of having been scoured by the englacial and subglacial meltwater flow, evolving a variety of small-sized forms called plastically molded surfaces, or p-forms.

Temperate glaciers transport a large volume of clastic load throughout the thickness of ice. It is laid down in many different ways from the surface and sole of advancing, retreating, and stagnant glaciers on land and from advancing or retreating glaciers in water bodies as nonstratified and stratified drift. The release and deformation of subglacial till beneath an advancing ice evolves ground moraine and drumlin forms of glacial deposition. Ground moraines comprise a uniformly thick and compact basal till. Drumlins are smooth, streamlined hills of low height at the inside margin of end moraines. They are largely built of the subglacial till, but few drumlins are shaped out of the bedrock, or comprise stratified drift, or nonstratified and stratified drift deposits. Drumlins are explained in terms of bed roughness, rheologic behavior of moist subglacial till, effective normal pressure of the glacier ice, and the thickness and rate of ice deformation. End moraines commonly occur near the termini of temperate and subpolar glaciers as broad-based arcuate ridges of subglacial debris that may have moved up the ice to become supraglacial by diverse mechanisms involving shear and thrust planes in the ice, regelation slip of the basal debris, and migration of debris in supercooled meltwater flow through the vein network of polycrystalline ice. The push and thrust of continental ice against frost-weakened pavements deforms the bed into ice-thrust ridges that appear morphologically similar to the end moraines. The push of continental ice readvancing subaquatically deforms the preexisting till into paired morainic ridges called De Geer or annual moraines. The ice thrust in fjords, however, deforms the glacier bed and overturns the marine sediments over the ice surface. In the final stages of the glacier retreat from the continents, large sections of the ice had thawed considerably and disintegrated, releasing the glacier load as nonstratified and stratified drift by the debated letdown, squeezing, and meltwater processes of drift deposition. These deposits make dead-ice or hummocky moraines of the glaciated landscape.

The surface thaw and basal melting of the glacier ice release meltwater, which moves at, within, and beneath the glacier ice. The clastic load of glacier melt is deposited in contact with the ice and beyond the margins of glacier ice on land, in freshwater lakes, and the sea. These glaciofluvial sediments in retreating glaciers evolve ice-contact crevasse fillings, eskers, and kames. Crevasse fillings evolve within crevasses of the surface ice and the basal cracks of continental ice terminating subaquatically. Eskers are winding ridges of meltwater sedimentation in supraglacial streams, englacial tunnels, and subglacial channels of receding alpine, temperate, and subpolar glaciers. The esker processes are debated. Kames are mounds of stratified sand and gravel deposits of supraglacial streams within pools, crevasses, and open ice-walled troughs of a stagnant glacier ice environment.

The glacier melt carried beyond the margins of continental and alpine glaciers is called proglacial. The proglacial sedimentation from advancing glaciers evolves massive deposits and landforms called outwash and valley train. Contemporary outwash pains are built in periods of the glacial burst of ice-dammed lakes. Valley train is a form of the outwash of alpine glaciers. It is laid down by supraglacial streams along the valley margins of alpine glaciers. Freshwater glacial lakes receiving the meltwater load of temperate glaciers evolve a rhythmic sequence of glacial varves and a flat surface of sedimentation called lacustrine plain. The glacier melt discharging into the sea represents a glaciolacustrine environment of sedimentation. Beaches and deltas, therefore, are the characteristic forms of sedimentation. However, these deposits and landforms have been reworked so extensively by the waves since sedimentation that they now are practically difficult to identify.

REFERENCES

Adam, W. G., and Knight, P. G. 2003. Identification of basal layer debris in ice-marginal moraines, Russell Glacier, West Greenland. *Quat. Sci. Rev.* 22:1407–14.

Alley, R. B., Larson, E. B., Evenson, E. B., Strasser, J. C., and Larson, G. J. 1998. Geohydraulic supercooling: A freeze-on mechanism to create stratified, debris-rich basal ice. II. Theory. *J. Glaciol.* 44:563–69.

Andrews, J. T. 1963. Cross-valley moraines of the Rimrock and Isortoq river valleys, Baffin Island, North West Territories. *Geogr. Bull.* 19:49–77.

Ashley, G. M. 1971. Rhythmic sedimentation in glacial lake Hitchcock, Massachusetts-Connecticut. *Univ. Mass. Geol. Dept. Contrib.* 10:131–38.

Benn, D. I., and Evans, D. J. A. 1996. The interpretation and classification of subglacially-deformed materials. *Quat. Sci. Rev.* 15:23–52.

Bishop, B. C. 1957. *Shear moraines in the Thule area, northwest Greenland.* Research Report 17. U.S. Snow, Ice and Permafrost Research Establishment.

Booth, D. B., and Hallet, B. 1993. Channel networks carved by subglacial water: Observations and reconstruction in the eastern Puget Lowland of Washington. *Bull. Geol. Soc. Am.* 105:671–83.

Boulton, G. S. 1979. Processes of glacial erosion on different substrata. *J. Glaciol.* 23:15–28.

Boulton, G. S. 1996. Theory of glacial erosion, transport and deposition as a consequence of subglacial sediment deformation. *J. Glaciol.* 42:43–62.

Boulton, G. S., and Clark, C. D. 1990. A highly mobile Laurentide ice sheet revealed by satellite images of glacial lineations. *Nature* 346:813–17.

Clarke, G. K. C. 2006. Ice-sheet plumbing in Antarctica. *Nature* 440:1000–1.

Cook, S. J., Waller, R. I., and Knight, P. G. 2006. Glaciohydraulic supercooling: The process and its significance. *Progress Phys. Geogr.* 30:577–88.

Derbyshire, E. 1999. Glacial geomorphology. In *Fourth International Conference on Geomorphology*, Bologna, Italy, August 28–September 3, 1997, pp. 89–97.

Drewry, D. 1986. *Glacial geologic processes.* London: Edward Arnold.

Echelmayer, K., and Wang, Z. 1987. Direct observations of basal sliding and deformation of basal drift at sub-freezing temperatures. *J. Glaciol.* 33:83–98.

Embleton, C., and King, C. A. M. 1968. *Glacial and periglacial geomorphology.* London: Edward Arnold.

Evans, D. J. A. 1989. Apron entrainment at the margins of sub-polar glaciers, north-west Ellesmere Island, Canadian high arctic. *J. Glaciol.* 35:317–24.

Evans, D. J. A., Rea, B. R., and Benn, D. I. 1998. Subglacial deformation and bedrock plucking in areas of hard bedrock. *Glacial Geol. Geomorph.*, rp04.

Fitzsimons, S. J. 1996. Formation of thrust block moraines at the margin of dry-based glaciers, south Victoria Land, Antarctica. *Ann. Glaciol.* 22:68–74.

Gates, W. L. 1976. Modelling of ice-age climate. *Science* 191:1138–44.

Gilbert, G. K. 1906. Crescentic gouges on glaciated surfaces. *Bull. Geol. Soc. Am.* XVII:303–13.

Goldthwaite, R. P. 1951. Development of end moraines in east-central Baffin Island. *J. Geol.* 59:567–77.

Gorrell, G., and Shaw, J. 1991. Deposition in an esker, bead and fan complex, Lanark, Ontario, Canada. *Sediment. Geol.* 77:285–314.

Goudie, A. S., Simmons, I. G., Atkinson, B. W., and Gregory, K. J., ed. 1994. *The encyclopaedic dictionary of physical geography.* 2nd ed. Oxford: Blackwell.

Gravenor, C. P., and Kupsch, W. O. 1959. Ice-disintegration features in western Canada. *J. Geol.* 67:48–64.

Gustavson, T. C., and Boothroyd, J. C. 1987. A depositional model for outwash, sediment sources and hydrologic characteristics, Malaspina Glacier, Alaska: A modern analog of the southeastern margin of the Laurentide ice sheet. *Bull. Geol. Soc. Am.* 99:187–200.

Hallet, B. 1976. Deposits formed by subglacial precipitation of $CaCO_3$. *Bull. Geol. Soc. Am.* 87:1001–15.

Hallet, B. 1979. A theoretical model of glacial abrasion. *J. Glaciol.* 23:39–50.

Hallet, B. 1981. Glacial abrasion and sliding: Their dependence on the debris concentration in basal ice. *Ann. Glaciol.* 2:23–28.

Hambrey, M. J., Bennett, M. R., Glasser, N. F., Huddart, D., and Crawford, K. 1999. Facies and landforms associated with ice deformation in a tidewater glacier, Svalbard. *Glacial Geol. Geomorph.*, rp07.

Hart, J. K. 1996. Subglacial deformation associated with a rigid bed environment, Aberdaron, north Wales. *Glacial Geol. Geomorph.*, rp01.

Huddart, D. 1999. Subglacial trough fills, southern Scotland: Origin and implications for deglacial processes. *Glacial Geol. Geomorph.*, rp04.

Jones, H., and Arnold, N. 1999. Modelling the entrainment and transport of suspended sediment in subglacial hydrologic systems. *Glacial Geol. Geomorph.*, rp09.

Kamb, B., and LaChapelle, E. 1964. Direct observation of the mechanism of glacier sliding over bedrock. *J. Glaciol.* 5:159–72.

King, C. A. M. 1980. Thresholds in glacial geomorphology. In *Thresholds in geomorphology*, ed. D. R. Coates and J. D. Viteks, 297–321. London: George Allen & Unwin.

Knight, P. G., and Knight, D. A. 1999. Experimental observations of subglacial debris entrainment into the vein network of polycrystalline ice. *Glacial Geol. Geomorph.*, rp05.

Knight, P. G., and Knight, D. A. 2004. Field observations and laboratory simulations of basal ice formed by freezing of supercooled subglacial water. In *Antarctica ice-sheet dynamics and climatic change: Modelling the ice composition studies*, Brussels, April 6–7, 2004.

Knight, P. G., and Knight, D. A. 2005. Laboratory observations of debris-bearing ice facies frozen from supercooled water. *J. Glaciol.* 51:337–39.

Kozarski, S. 1994. On the origin of the Chodziez end moraine. In *Cold climate landforms*, ed. D. J. A. Evans, 293–312. Chichester: John Wiley & Sons.

Kupsch, W. D. 1962. Ice-thrust ridges in western Canada. *J. Geol.* 70:582–94.

Lawson, D. E. 1981. Distinguishing characteristics of diamictons at the margins of the Matanuska Glacier, Alaska. *Ann. Glaciol.* 2:78–84.

Lawson, D. E., Strasser, J. C., Evenson, E. B., Alley, R. B., Larsen, G. J., and Arcone, S. A. 1998. Glaciohydraulic supercooling: A freeze-on mechanism to create stratified debris-rich basal ice. I. Field evidence. *J. Glaciol.* 44:547–62.

Lehman, S. 1997. Sudden end of an interglacial. *Nature* 390:117–19.

Matthes, F. E. 1930. *Geological history of the Yosemite Valley*, 54–103. U.S. Geological Survey Professional Paper 160. U.S. Geological Survey.

Menzies, J. 1979. The mechanics of drumlin formation with particular reference to the changes in pore-water content of the till. *J. Glaciol.* 22:373–84.

McCall, J. G. 1960. The flow characteristics of a cirque glacier and their effect on glacial structure and cirque formation. In *Investigations on Norwegian cirque glaciers*, ed. W. V. Lewis, 39–62. Royal Geography Society Research Series IV. Royal Geography Society Research.

Nye, J. F. 1952. The mechanics of glacier flow. *J. Glaciol.* 2:82–93.

Paterson, W. S. B. 1969. *The physics of glaciers*. Oxford: Pergamon Press.

Price, R. J. 1966. Eskers near the Casement Glacier, Alaska. *Geogr. Annlr.* 48A:111–25.

Price, R. J. 1973. *Glacial and fluvioglacial landforms*. Edinburgh: Oliver and Boyd.

Rastas, J., and Seppälä M. 1981. Rock jointing and abrasion forms on roches mounonnées, SW Finland. *Ann. Glaciol.* 2:159–63.

Raymond, C. F. 1978. Mechanics of glacier movement. In *Rockslides and avalanches*, ed. B. Voight, 793–833. Vol. 1. Amsterdam: Elsevier.

Rea, B. R. 1996. A note on the experimental production of a mechanically polished surface within striations. *Glacial Geol. Geomorph.*, tn01.

Rose, J. 1987. Drumlins as part of a glacier bedform continuum. In *Drumlin symposium*, ed. J. Menzies and J. Rose, 103–16. Rotterdam: Balkema.

Shoemaker, E. M. 1986. The formation of fjord thresholds. *J. Glaciol.* 32:65–71.

Shreve, R. L. 1985. Esker characteristics in terms of glacier physics, Katahdin esker system, Maine. *Bull. Geol. Soc. Am.* 96:639–46.

Smalley, I. J., and Unwin, D. J. 1968. The formation and shape of drumlins and their distribution and orientation in drumlin fields. *J. Glaciol.* 7:377–90.

Stanford, S. D., and Mickelson D. M. 1985. Till fabric and deformation structures in drumlins near Waukesha, Wisconsin, U.S.A. *J. Glaciol.* 31:220–28.

Strömberg, B. 1965. Mapping and geochronological investigations in some moraine areas of south-central Sweden. *Geogr. Annlr.* 47A:73–82.

Stupavski, M., and Gravenor, C. P. 1974. Water release from the base of active glaciers. *Bull. Geol. Soc. Am.* 31:272–80.

Sturm, M., and Benson, C. S. 1985. The history of jökulhlaup from Strandline Lake, Alaska, U.S.A. *J. Glaciol.* 31:272–80.

Sugden, D. E., and John, B. S. 1976. *Glaciers and landscape*. London: Edward Arnold.

Tweed, F. S., Roberts, M. J., and Russell, A. J. 2005. Hydrologic monitoring of supercooled meltwater from Icelandic glaciers. *Quat. Sci. Rev.* 24:2308–18.

Waller, R. I., and Hart, J. K. 1999. Mechanics and patterns of motion associated with the basal zone of Russell Glacier, south-west Greenland. *Glacial Geol. Geomorph.*, rp02.

Weertman, J. 1961. Mechanisms for the formation of inner moraines found near the edge of cold ice caps and ice sheets. *J. Glaciol.* 3:965–78.

Weertman, J. 1964. The theory of glacier sliding. *J. Glaciol.* 5:287–303.

Wingham, D. J., Siegert, M. J., Shepherd, A., and Muir, A. S. 2006. Rapid discharge connects Antarctic subglacial lakes. *Nature* 440:1033–36.

Wintges, T. 1985. Studies on crescentic fractures and crescentic gouges with the help of close-range photography. *J. Glaciol.* 31:340–49.

8 Periglacial Processes and Landforms

Lozinski first used the term *periglacial* in 1909 to explain the distribution of *frost-shattered debris* about the cold ice-free margins of *Pleistocene glaciers* (Embleton and King, 1968). As frost activity is not exclusive to the proximity of either former or present-day glaciers, the term *periglacial* nowadays is used for cold region processes and their landforms (Price, 1972; Washburn, 1979; Thorn, 1992). Washburn (1979) adopted the term *geocryology* for processes and landforms of cold environments, but *periglacial* term has come to stay in the literature, and is widely used for the purpose.

IDENTIFICATION

A periglacial environment is characteristic of subfreezing air temperature and a frozen ground to certain depth in the subsoil. This relationship between the air temperature and frozen ground is central to the identification of a periglacial environment. Peltier (1950) defined a *periglacial environment* by a mean annual air temperature between −15 and −1°C and a mean annual precipitation of 125 to 1400 mm, mostly in the subarctic zone of tundra vegetation. Periglacial regions have also been identified with a subfreezing air temperature for at least two consecutive summers and the intervening winter season (Brown and Kupsch, 1974), and with a below freezing subsoil temperature for two consecutive years (Washburn, 1979). Observations, however, suggest that even a slight snow cover effectively insulates the ground beneath from the effect of air temperature fluctuation (Harris, 1974; Embleton, 1980; Harris, 1982; Harris and Brown, 1982; Ødegård et al., 1995; Carey and Ming-ko, 1998). Therefore, periglacial regions are simply identified with perennially and seasonally frozen ground characteristic of a certain magnitude of subfreezing mean annual temperature range. The frozen ground is also continuous, discontinuous, and sporadic with respect to certain amplitudes of seasonal and annual subfreezing mean annual air temperature about the freezing point.

Certain other thermal indices of the periglacial environment have also been proposed. Continentality, air freezing index, and freeze-thaw index are some such derivatives of the thermal potential for freezing that find application in identification and differentiation of the periglacial environment. By the *continentality index*, the interiors of continents are typical periglacial regions of an enormous thickness of permafrost in the subsoil. These areas were either free of Pleistocene glaciers or influenced by glaciers of lesser extent (Péwé, 1969; Demek, 1978). The *air-freezing index* provides a threshold for the freezing potential of *frost-susceptible soils* (Steurer and Crandell, 1995). Derived from the duration and amplitude of winter season air

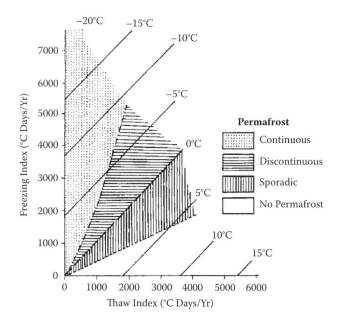

FIGURE 8.1 Freeze-thaw indices for discriminating permafrost types and their thermal controls. (Source: Figure 1 in Harris, S. A., "Identification of Permafrost Zones Using Selected Permafrost Landforms," in *Proceedings of the 4th Canadian Permafrost Conference*, Calgary, Alberta, March 2–6, 1981, pp. 49–58. With permission from the National Research Council of Canada.)

temperature, this index finds application in the design of engineering structures for cold region environments. S. A. Harris (1982) observed that 50 cm or more thick snow effectively insulates the frozen ground, by which it remains slightly warmer at depth than the ambient air temperature regime. *Freezing and thawing indices* (Figure 8.1) overcome this weak relationship between the air temperature and type of permafrost. The *freezing index* is the sum total of daily subfreezing temperatures, and the *thaw index* is the aggregate of daily positive temperatures for the year as a whole. The freeze-thaw indices essentially use distribution of zonal permafrost landforms for mapping limits of permafrost types. These indices have been subsequently used in this chapter for the thermal environment of periglacial landforms.

FROZEN GROUND

A more or less clear ice underlies the frozen ground of adequate moisture in the subsoil. This ice is not distributed uniformly with respect to depth or content. By a conservative estimate, large areas of frozen ground in the northern hemisphere contain more than 50% ice by volume within 2 to 3 m of the subsoil depth (Mackay, 1972). The ice content, thus, sharply diminishes with depth in the frozen ground of *true permafrost*. The frozen ground underlain by rocks, however, is devoid of ground ice for the obvious reason, but the rock body nevertheless remains below the freezing temperature. This form of frozen ground is called *dry permafrost*.

Depth Variation

The depth penetration of permafrost is intricately related to the amplitude of the subfreezing mean annual temperature, rate of geothermal heat escape, time available for the permafrost to build, and hydrologic aspects of terrain. In the periglacial environment of North America, the permafrost depth increases with latitude-dependent amplitude of subfreezing mean annul temperature (Brown, 1960). Here, the actual depth reached by permafrost depends on the rate of subsurface cooling, which, in general, depends on air temperature and heat conductivity of the surface, and on the warming effect of geothermal heat escape at the rate of 1°C per 30 m of the surface depth. Hence, even a slight variation in the geothermal heat flux remarkably affects the depth penetration of frozen ground.

The depth of permafrost penetration also depends on the length of time given for its development. Thus, the frozen ground of *Pleistocene origin* in Russia is thicker than the permafrost of *Holocene origin* in North America (Washburn, 1979).

Water bodies above and adjacent to the frozen ground variously affect the pattern of permafrost penetration in the subsoil (Lachenbruch et al., 1962; Price, 1972). Small shallow lakes, which instantly freeze in winter and rapidly thaw in summer, do not appreciably affect the local thermal regime and behavior of permafrost (Figure 8.2(a), (b)). By comparison, up to 1.5 m deep shallow water bodies freeze slowly in winter, allowing indentation of the frozen ground beneath (Figure 8.2(c)–(e)). Large lakes of a diameter roughly twice as wide as the permafrost depth, however, possess sufficient *heat capacity* to prevent the buildup of permafrost beneath (Figure 8.2(f), (g)). Permafrost is similarly absent beneath shallow coastal waters only a few hundred meters from periglacial regions. The depth pattern of frozen ground also reflects the surface configuration: it rises over hills and sinks beneath valleys (Figure 8.2(h)–(k)).

Thermal Attributes

Frozen ground is a manifestation of subfreezing temperature in the subsoil. The magnitude of subsurface temperature and its depth pattern depend on the amplitude of seasonal temperature rhythm and the mean annual temperature, and the rate of geothermal heat flux through the frozen ground (Figure 8.3). Observations suggest that the seasonal temperature conducting into the ground attains a stable value at a depth of 20 to 30 m in the frozen ground. The depth at which temperature remains constant is called the *level of zero annual amplitude*. The temperature at the level of zero amplitude in the permafrost of recent origin in North America is −15°C (Lachenbruch et al., 1962). Down this depth, the subsurface temperature progressively warms by the effect of geothermal heat escaping the ground and downward cooling of the medium due to the ambient air temperature conducting into the ground. Hence, unfrozen ground or *talik* depicting a state of *heat balance* in the frozen ground appears at a certain depth in the permafrost environment. The term *talik* is applied also to pockets of thawed ground within the permafrost. The seasonal thaw of permafrost at the surface produces a layer of loose mineral soil called *active layer* above the frozen ground. This upper limit of permafrost is called *permafrost table*.

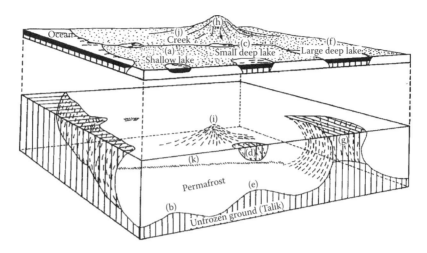

FIGURE 8.2   Frozen ground is also a function of the size of water bodies in periglacial environment. Water bodies of variable depth (a, c, f) freeze at an uneven rate and not at all if water is more than 30 m deep, such as in coastal waters offshore (ocean). The low heat capacity of shallow water bodies (a, c) causes indentation in the permafrost table and rise of permafrost towards the surface (b, e). By comparison, deep water basins of smaller horizontal dimension are normally underlain by unfrozen sediments within a perennially frozen ground. Such unfrozen areas appear as unfrozen chimneys (g) beneath water bodies. The rise of a permafrost table locally updomes the surface, evolving small-sized hills (i). In general, the undulatory permafrost table (b–e) closely follows the ground configuration: it sinks beneath creeks or valleys (k) and rises over hills (j). (Source: Figure 5 in Lachenbruch, A. H., "Permafrost," in *The Encyclopedia of Geomorphology*, ed. R. W. Fairbridge [New York: Reinhold Book Corp., 1968], 833–39. With permission from Kluwer Academic Publishers.)

Depth-temperature profiles in the frozen ground of recent origin at Barrow, Alaska, give temperature variations through the depth of permafrost at three sites (Figure 8.4). The broken lines in this illustration are extrapolated temperature profiles, which suggest an increase in the mean annual surface temperature by 4°C since the middle of the nineteenth century and an overall cooling trend for the last decade or so (Lachenbruch and Marshall, 1969). These observations suggest that permafrost temperature responds to the *climate change* (Mackay, 1975) and is an indicator of the *climatic trend* (Lachenbruch and Marshall, 1969).

The active layer, referred to earlier in this section, freezes in the winter season. This process releases the *latent heat of fusion*, which temporarily maintains a slightly higher temperature in the permafrost at about the permafrost table. The thickness of the active layer varies with the complex response of the frozen ground to diurnal and seasonal temperature fluctuation, thermal conductivity of the surface, soil composition, slope and relief aspects of the terrain, and vegetation density. Therefore, the thickness of the active layer is expected to vary remarkably over short distances within the given permafrost situation (Owens and Harper, 1977). In general, the active layer is thickest in the subarctic. It progressively thins toward the northern latitudes of smaller magnitude of seasonal temperature range about the freezing point, and toward the southern latitudes of increasing thermal efficiency regime (Brown, 1970; Price, 1972).

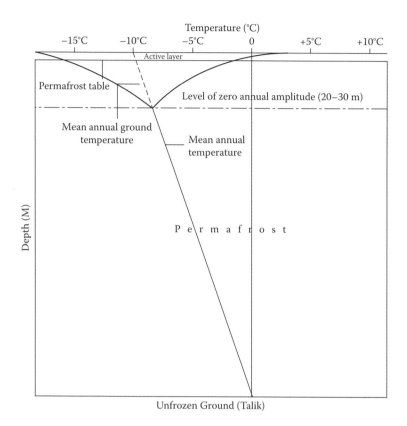

FIGURE 8.3 Idealized geothermal heat gradient in frozen ground, showing active layer, level of zero annual amplitude, and the lower limit of permafrost at which unfrozen ground (*talik*) appears. (Source: Adapted from Figure 1 in Lachenbruch, A. H., Brewer, M. C., Greene, G. W., and Marshall, B. V., "Temperature in Permafrost," in *Temperature—Its Measurement and Control in Science and Industry*, Vol. 3, Pt. I [New York: Reinhold Publishing Corp., 1962], 791–803.)

TYPES

Frozen ground is identified with the ground ice phenomenon. The ground ice evolves by the freezing of water that is either present in sediments or drawn into the medium for freezing. The source and movement of water prior to freezing, thus, offer a suitable framework for the classification of ground ice into types. Mackay (1972) similarly classified the ground ice into types of open-cavity ice, single-vein ice, ice wedge (or wedge ice), tension crack ice, closed-cavity ice, epigenetic ice, aggradational ice, sill ice, pingo ice, and pore ice by the source of water and the manner of its movement for freezing in the environment (Figure 8.5). *Open-cavity ice* evolves from the diffusion and freezing of atmospheric water vapor directly in open cavities or surface cracks in the frozen ground environment. *Single-vein ice* and *ice wedge* essentially differ in size. They evolve from the water drawn from the active layer for freezing into cracks of thermal contraction in the frozen ground. The gravity transfer of surface

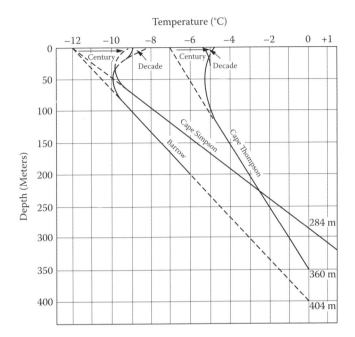

FIGURE 8.4   Temperature change with depth (solid lines) at three locations in arctic Alaska. Broken lines are theoretical extrapolations. Climatic trends can be identified by comparing the actual temperature near the surface with the extrapolated one. These thermal profiles from northern Alaska suggest slightly lower temperatures for the last century and a warming trend for the last decade. (Source: Figure 2 in Lachenbruch, A. H., and Marshall, B. V., "Heat Flow in the Arctic," *Arctic* 22 [1969]: 300–11. With permission from Arctic Institute of North America.)

water for freezing within *tension cracks* of certain periglacial features, however, evolves *tension crack ice*. Besides, the movement and freezing of subsurface water also evolve a variety of ground ice forms. *Closed-cavity ice* results from the freezing of subsurface water vapor into cavities of updoming action, such as of marsh gas. The freezing of subsurface water onto the mineral grains of frost-susceptible soils evolves *segregated ice*, which subsequently develops into epigenetic ice (ice lenses) and aggradational ice (ice layers) during the evolution of certain periglacial landforms. *Sill ice* and *pingo ice* are forms of the intrusive ice, which evolve from the freezing of subsurface water against a pressure potential. *Pore ice* results from processes of cryogenic expulsion of pore water from moist sediments and *in situ* freezing of water in cold region environments.

CLASSIFICATION OF FROZEN GROUND

Frozen ground is continuous, discontinuous, and sporadic by the amplitude of subfreezing mean annual temperature, seasonal and perennial by the climatic rhythm, dry and true by the moisture in subsoil, and active and relict in the context of the present-day climatic regime. In Canada, Alaska, Russia, and China, the distribution of continuous and discontinuous permafrost is closely related to the magnitude of mean

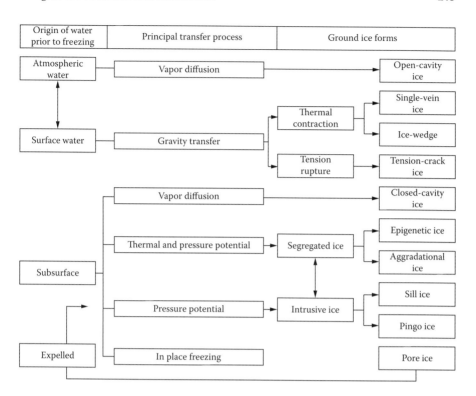

FIGURE 8.5   Classification of ground ice. (Source: Figure 2 in Mackay, J. R., "The World of Underground Ice," *Ann. Assoc. Am. Geogr.* 62 [1972]: 1–22. With permission from Routledge Publishers.)

annual temperature range (Figure 8.6). The *continuous permafrost* exists in areas of −5°C or less mean annual temperature regime. This type of permafrost, however, is normally thicker in Russia than elsewhere at comparable latitudes in North America and China. The continuous permafrost is 1500 m deep in upper reaches of the Markha River (66° N) in eastern Siberia, 609 m deep at Prudhoe Bay (70° N), Alaska, and 548 m deep at Melville Island (75° N), Canada (Price, 1972). The *active layer* in continuous permafrost is generally 15 to 30 cm thick. The southern limit of continuous permafrost is the northern boundary of discontinuous permafrost, at which *pingos* and *ice wedges*, which are major features of ground ice in continuous permafrost, cease to exist (Mackay, 1972). The *discontinuous permafrost* prevails in areas where the mean annual temperature varies between −5 and −1°C. It is shallow and less contiguous than the continuous permafrost, and characteristic of thawed pockets within its vast expanse. The discontinuous permafrost, which is 50 m deep at Norman Wells (65° N) and 10 m deep at Hay River (61° N) in Canada, and only a few meters deep at about 50° N in the region of Amur Darya in Russia, is in a state of delicate balance with the mean annual temperature regime. Thus, the recent thaw of discontinuous permafrost in Manitoba, Canada, is viewed as a consequence of the warming of climate since the sixties (Thie, 1974). The active layer in this type of frozen ground is 2 to 3 m thick. The *sporadic permafrost* is azonal in distribution, devoid of active layer, and occurs in

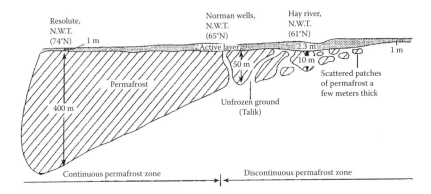

FIGURE 8.6 An idealized cross section of the continuous and discontinuous permafrost zones in Canada. (Source: Figure 4 in Brown, R. J. E., *Permafrost in Canada* [Toronto: University of Toronto Press, 1970]. With permission from University of Toronto Press.)

shallow pockets of frozen subsoil within areas of −1 to 0°C mean annual temperature regime (Embleton and King, 1968; Harris, 1982). This type of permafrost in the Rocky Mountains and parts of eastern United States is a relic of former frozen ground of an extensive nature (Harris and Brown, 1982; Schmidlin, 1988).

Permafrost is seasonal and perennial in nature. *Seasonal permafrost* develops in areas of low-latitude mountain climates, humid climates of mid-latitude zones, and in the active layer that rhythmically freezes in winter and thaws in summer. It is typically shallow in depth. *Perennial permafrost* is identified with continuous and discontinuous types of permafrost, which extends to great depths in porous media and bedrock surfaces.

Permafrost is dry and true by the subsurface moisture content (Shumskii, 1964). *Dry permafrost* evolves in the absence of moisture in subsoil and subsurface rocks, and *true permafrost* requires adequate subsurface moisture for freezing into ground ice. Dry permafrost makes the substratum friable, and true permafrost produces hard subsoil.

Frozen ground is active and relict phenomena. *Active permafrost*, as in North America and the northerly latitudes of interior Russia, is in *equilibrium* with the present-day mean annual temperature. The dynamic association between active permafrost and climate is best illustrated by the recent extension of permafrost in unfrozen sediments of the Arctic of Canada and Alaska (Mackay, 1972; Dean and Morrissey, 1988). The active permafrost in Canada exists in areas of −8 to −1°C mean annual temperature regime (Brown, 1960), but the permafrost toward the southern latitudes of interior Russia has been set to slow degradation by the recent warming of climate (Washburn, 1979). This frozen ground is regarded as a *relic* of the Pleistocene climate.

DISTRIBUTION

Nearly 26% of the land of the earth, comprising a 22.4 million km² area in the northern hemisphere and a 13.1 million km² area in the southern hemisphere, is

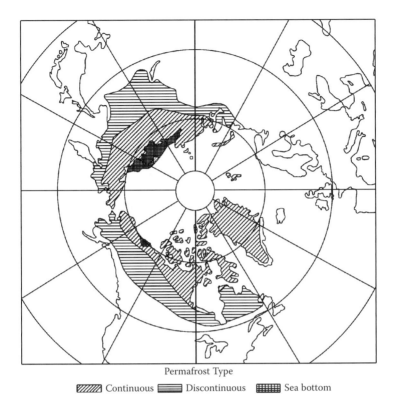

Permafrost Type

///// Continuous ▦ Discontinuous ▦ Sea bottom

FIGURE 8.7 Generalized permafrost map for the northern hemisphere, excluding alpine areas. (Source: Figure 1 in Mackay, J. R., "The World of Underground Ice," *Ann. Assoc. Am. Geogr.* 62 [1972]: 1–22. With permission from Routledge Publishers.)

in permafrost (Black, 1954). Much of the northern circumpolar zone, with 87% area in the territory of Alaska, between 40 and 50% area in Canada, and 47% in the former republics of the Soviet Union, is estimated in permafrost (Figure 8.7). Permafrost distribution and depth pattern of the frozen ground, however, are less well documented for Antarctica, where frozen ground possibly exists at the margins of the ice sheet and beneath ice-free tracts. Borehole data from an ice-free area of Antarctica, however, suggest that the permafrost is 150 m deep at the given site (Price, 1972).

MAJOR FEATURES OF GROUND ICE

Pingos, palsas, ice wedges, and thermokarst are major features of ground ice, which evolve above the level of zero annual amplitude in permafrost. S. A. Harris (1982) observed a general association of ground ice features with continuous, discontinuous, and sporadic types of permafrost, suggesting further that many transgress thermal controls of the three types of frozen ground environment (Figure 8.8). Features of ground ice also exist as relics outside the periglacial environment, but their origin mostly remains suspect.

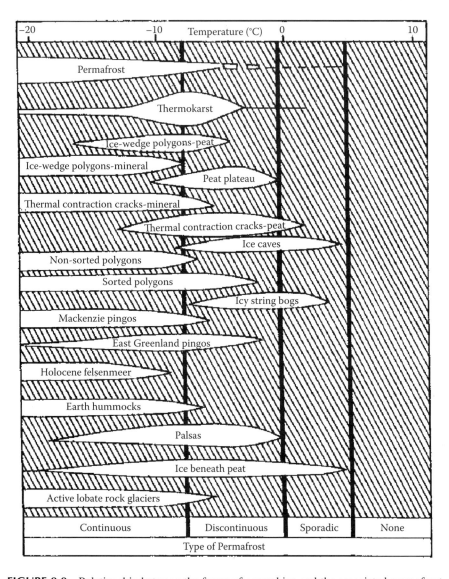

FIGURE 8.8 Relationship between the forms of ground ice and the associated permafrost conditions. (Source: Figure 11 in Harris, S. A., "Identification of Permafrost Zones Using Selected Permafrost Landforms," in *Proceedings of the 4th Canadian Permafrost Conference*, Calgary, Alberta, March 2–6, 1981, pp. 49–58. With permission from the National Research Council of Canada.)

PINGOS

Pingos are up to 70 m high and 600 m across *ice-cored* conical hills of continuous and discontinuous permafrost environments. Those of the *continuous permafrost* are called closed-system or Mackenzie-type pingos, and of the *discontinuous permafrost* are known as open-system or East Greenland–type pingos. The closed- and open-system pingos evolve at limiting mean annual temperatures of − 4 and − 2.5°C that additionally are characteristic, respectively, of 100 and 250 degree-days year^{-1} of thaw index (Harris, 1982). The two types also fundamentally differ in the source of water and manner of its movement for freezing into the *intrusive ice*. The source of water is *internal* for closed- and *external* for open-system pingos (Mackay, 1979). Closed-system pingos evolve from cryogenic expulsion of pore water, and migration of this water to the site of pingo formation under hydrostatic head. By comparison, open-system pingos develop from a distant elevated source of subsurface water moving with a hydraulic head to the freezing site. Pingos not belonging to either of the two types have also been observed from Prince Patric Island (76° N) in the Arctic Ocean, suggesting that processes of pingo formation are inadequately understood at the present time (Price, 1972).

Closed-System Pingos

Closed-system pingos are 3 to 70 m high and 30 to 600 m across isolated ice-cored circular hills, which rise from shallow lake basins, former lake basins, and poorly drained flat lands of the continuous permafrost environment. Smaller pingos are rounded at the top, while larger pingos are characteristic of collapsed summits, tension cracks down the slope, and an outward dipping of coarse sand and silt strata (Washburn, 1979). The Mackenzie delta of Canada supports a high concentration of 1450 pingos, many of which are the largest in the world (Mackay, 1979).

Pingos of the Mackenzie delta evolve from slow seasonal freezing of shallow lakes rich in sand-sized bottom sediments (Mackay, 1962, 1979). Hence, unfrozen ground initially exists for a while beneath freezing lakes (Figure 8.9, top). With the progress of the winter season, however, the permafrost expands downward from the lake and from lake margins into the unfrozen sediments (Figure 8.9, center). The permafrost expanding in this manner creates a cryogenic pressure in the closed-system of sediment and water that squeezes the pore water from sediments and generates a strong uplift force to updome, and sometime rupture, the surface of evolving form. The expelled water migrates in the direction of force release, and freezes a little beneath the updomed surface as *intrusive ice* of the pingo hills (Figure 8.9, bottom). Pingos, however, cease to grow when permafrost permanently encroaches the lake area.

The saturated sediments from which the pingos rise provide only a small volume of pore water for each cycle of lake freezing. Hence, the closed-system pingos evolve slowly with time. The pingos of Mackenzie delta have thus developed at a slow rate of 0.3 to 0.5 m 1000 year^{-1} (Mackay, 1972). ^{14}C dates on two large pingos of the region suggest that they are respectively only 4000 to 7000 and 10,000 years old. Hence, the permafrost here is of recent origin and in *dynamic equilibrium* with the thermal stress of the environment.

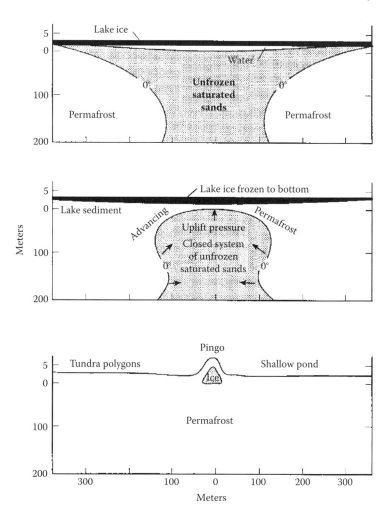

FIGURE 8.9 Stages in the development of a closed-system or Mackenzie-type pingo. A vertical exaggeration of five times is introduced to show the lake. (Source: Figure 15 in Mackay, J. R., "Pingos of the Pleistocene Mackenzie Delta Area," *Geogr. Bull.* 18 [1962]: 21–63. With permission from Geological Survey of Canada.)

Open-System Pingos

Open-system pingos are usually small-sized isolated forms of ground ice on hillslopes. They evolve in areas where surface water from an elevated outside source penetrates a thin section of the *discontinuous permafrost*, reaches *talik*, accumulates as a local body of confined subsurface water, ascends the intrafrost zone with sufficient *hydraulic head*, and freezes near the surface as *intrusive ice* (Figure 8.10). Thus, continuous flow of subsurface water is necessary to evolve a sufficiently thick ice core and a strong uplift force to updome and rupture impervious strata (Müller, 1968).

Evolution of open-system pingos by the artesian flow of water under a hydraulic head is debated (Washburn, 1979). Some researchers even doubt the existence of

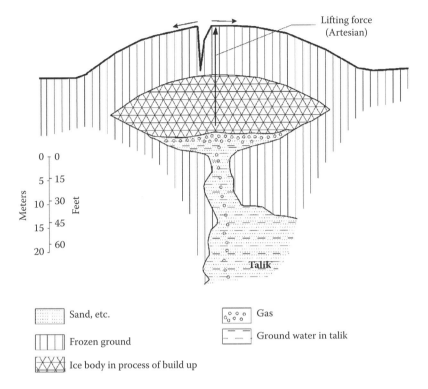

FIGURE 8.10   Hypothetical cross section through a pingo of the open system or Greenland type in the discontinuous or thin permafrost. (Source: Figure 2 in Müller, F., "Pingos, Modern," in *The Encyclopedia of Geomorphology*, ed. R. W. Fairbridge [New York: Reinhold Book Corp., 1968], 845–47. With permission from Kluwer Academic Publishers.)

artesian water in the system. Rastogi and Nawani (1976), however, report up to 10 m high egg-shaped pingos along the spring line of Tso Kar (Kar Lake) in the Kashmir Himalayas, recording a winter minimum of −40°C and a 30 m thick frozen ground beneath the lake. A section through one of the pingos reveals that artesian water at isothermal temperature of −1°C ascends the intrafrost zone with a hydraulic head of 90 cm, and freezes near the surface as intrusive ice. The freezing of water in this manner generates a sufficient uplift force to updome the surface, evolving the pingo form.

Fossil Pingos

Fossil pingos in parts of Asia, Europe, and North America are reminiscent of past cold environments. Their origin, however, remains suspect, as many forms bear semblance to the features of *dead ice*, *wind deflation*, and *solution processes* (Washburn, 1979).

Palsas

Bogs of the discontinuous permafrost environment of more than 300 degree-days year[−1] of thaw index differentially *frost heave* by the growth of ice lenses in the winter season (Harris, 1982), evolving up to 10 m high, 150 m long, and 30 m wide ice-cored ridges and mounds called palsas (Brown and Kupsch, 1974; Derbyshire,

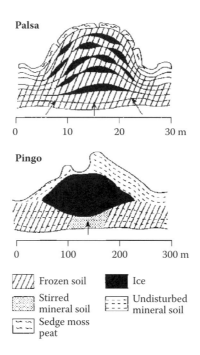

FIGURE 8.11 Schematic sections through active palsas and pingos, illustrating their relative dimensions and the distribution of ground ice. (Source: Figure 7-15 in Derbyshire, E., "Periglacial Environments," in *Applied Geomorphology*, ed. J. R. Hails [Amsterdam: Elsevier Science Publishing Co., 1977], 227–76. With permission.)

1977). Palsas are similar to but an order of magnitude smaller than average-sized pingo mounds (Figure 8.11).

Palsas are an active, stable, and relict form of ground ice in the subarctic environment of long severe winters. *Active palsas* evolve when bogs rapidly cool in the winter season and reach *thermal conductivity* nearly four times that of the unfrozen bog (Derbyshire, 1977). At this critical condition, the water freezes within pore spaces of mire plants as ice lenses, evolving closed and linear eminencies at the surface of frost-heaved terrain. The differentially heaved surface opens up tension cracks at the surface, exposing the ice lenses. This ice rapidly thaws in the following summer season, by which the frost-heaved surface returns to its original position. Active palsas, thus, evolve and decay rhythmically between the winter and summer seasons. *Mature palsas* are identified with stable frost-heaved eminencies that are sustained by relatively cool summer temperatures. *Relict palsas* are possibly the relic of a colder phase of subarctic climate.

Organic terrain is inherently poor in hydraulic conductivity. Therefore, migration of water for freezing in the system and its subsequent drainage has attracted debate. One view holds that water moves laterally at a shallow depth of the organic terrain, while the other argues that macropores within mire plants are as efficient openings for the flow of water as are the intergranular pores in coarse-textured sediments. Siegel et al. (1995) observed that water flow changes direction with the change in

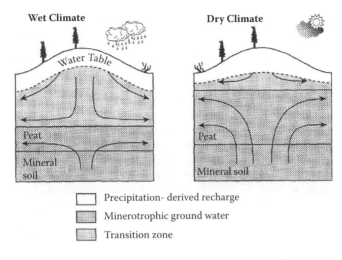

Precipitation- derived recharge

Minerotrophic ground water

Transition zone

FIGURE 8.12   Diagram showing changes in groundwater flow under a raised bog in the Lost River peatland during wet (left) and dry (right) years. Groundwater moves from high hydraulic head to low hydraulic head, measured as the elevation of the water in piezometers. During wet years, the hydraulic head of the water table mound is high enough to cause precipitation-derived recharge to flush mineral-rich groundwater from the peat column. During persistent drought, the water-table mound dissipates and groundwater in the under-lying mineral soil moves up through the peat to the surface of land. (Source: Figure 1 in Siegel, D. I., Reeve, A. S., Glaser, P. H., and Romanowicz, E. A., "Climate-Driven Flushing of Pore Water in Peat Lands," *Nature* 374 [1995]: 531–33. With permission from Macmillan Magazines Ltd.)

moisture status of *modern peats*, such that water drains laterally outward to the depth of peat bottom in wet periods but flows toward the peat surface in periods of long drought (Figure 8.12). A similar pattern of *lateral advective flow* of water is also expected in bogs of the discontinuous permafrost environment, in which water is drawn up the peat column in the dry winter season and utilized in the growth of *segregated ice*. The summer season thaw of ice lenses adds meltwater to the bog, which subsequently drains laterally outward to the peat bottom.

Ice Wedges

Ice wedges are vertical masses of pure ice in the *continuous permafrost* of −6°C or less mean annual temperature (Péwé, 1966) and thaw index of less than 100 degree-days year[-1] (Harris, 1982). They occupy some 2.6 million km[2] area in the Arctic of North America. Ice wedges also occur in the *discontinuous permafrost* environ-ment, which possibly suggests warming of the climate.

Ice wedges develop when the winter temperature suddenly drops sharply by sev-eral degrees, producing *thermal contraction* and *tensile stress* in the frozen ground (Lachenbruch, 1962). This stress is relieved by the opening of a millimeter wide and up to a meter deep vertical contraction crack in the frozen ground (Figure 8.13(a)). In the summer period, however, meltwater from the *active layer* flows into the crack and

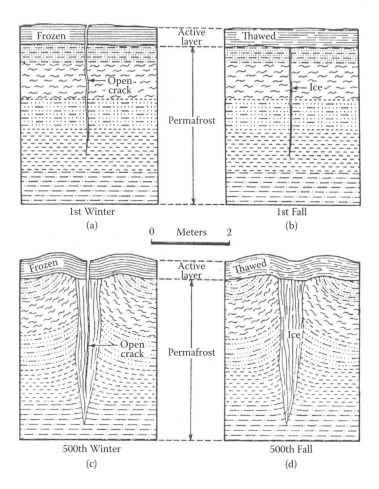

FIGURE 8.13 (a) Ice wedge evolves from the growth of ice vein that opens up in the permafrost by thermal contraction of the frozen ground by the sudden decline of air temperature in the periglacial environment. (b) The melting of seasonal permafrost in the following summer evolves an active layer that terminates the extension of vein ice through the mineral soil but provides small quantities of water to the vein each season for freezing and growth of the vein ice beneath. (c, d) Over a long period of freeze-thaw activity, the open vein ice expands sufficiently into an ice wedge. This expansive stress of growth deforms the sediments against the evolving ice wedge form.(Source: Figure 6 in Lachenbruch, A. H., "Permafrost," in *The Encyclopedia of Geomorphology*, ed. R. W. Fairbridge [New York: Reinhold Book Corp., 1968], 833–39. With permission from Kluwer Academic Publishers.)

freezes as *vein ice* (Figure 8.13(b)). The active layer freezes in the following winter, and the vein ice extends through it to the surface. The vein is a potential plane of weakness in the frozen ground. It, therefore, cracks open again in the following winter season. Thermally induced cracking of permafrost on an annual basis and freezing of water drawn from the active layer into the ice vein crack widens and deepens the vein ice to an *ice-wedge* form (Figure 8.13(c), (d)). The *thermal contraction process* of ice-wedge formation finds support from the observation of rhythmic crack opening along the same ice veins in the Arctic of Canada (Mackay, 1974). Steady expansion of the vein ice also deforms juxtaposed sediments in a way that they upturn against ice wedges. Ice wedges are about 1 m wide and 6 m deep in North America. However, they are extremely wide and nearly 60 m deep in the permafrost of Russia, requiring extreme deformation of adjacent sediments into nearly vertical columns (Washburn, 1979). Ice wedges present a *polygonal pattern* at the surface.

The thin active layer in the continuous permafrost releases only a small amount of water for the development of vein ice and the growth of an ice-wedge form. Hence, ice wedges of the Canadian Arctic grow slowly at the rate of less than 0.25 cm year^{-1} (Mackay, 1972). The growing ice wedges develop to an optimum size in equilibrium with the thermal stress of the environment, but then become inactive and thaw, forming fossil ice wedges called *ice-wedge casts*. Ice wedges also cease to grow beneath a cover of snow and gelefluction debris, which insulate the ground from thermal contraction. Ice-wedge casts in parts of Europe and the United States, however, are related to the warming of climate (Black, 1976, 1983).

Sand Wedges

Sand wedges are active and relict periglacial features that commonly comprise sand and silt-sized clastic fractions. They are an active feature of strong dry winds in the periglacial environment of Antarctica (Westgate, 1969). Inactive or fossil sand wedges exist outside the known periglacial limit, such as in the prairies of northern Canada (Figure 8.14). They suggest a dry and windier environment for the region in the Late Pleistocene period.

Thermokarst

Thermokarst refers to a group of *subsidence forms* as thaw lakes, pits, mounds, and dry valleys that evolve from uneven thaw of the ground ice. The ground ice can thaw from causes as diverse as the warming of climate, disturbance of the protective soil and vegetation cover from above the frozen ground, forest fires, and human activities, all of which alter the *thermal balance* of the frozen ground environment.

Observations suggest that the *Holocene* warming of climate has set the permafrost of *Pleistocene origin* in Russia to extensive degradation (Kachurin, 1962). Natural disturbance of the soil and vegetation cover in the perennially frozen ground environment also affects the thermal balance of permafrost due to which ice wedges begin to thaw even at temperatures as low as −11°C (Demek, 1978). The clearing of tundra vegetation from above ice wedges near Fairbanks, Alaska, has similarly thawed the ground ice, evolving thermokarst mounds of slumped structures (Price, 1972). Effects of human activities on the thaw of ground ice are pronounced in and around the Arctic settlements of Alaska, Canada, and the former republics of Soviet Russia.

PERIGLACIAL PROCESSES AND LANDFORMS

Frost, mass movement, nivation, and wind activities are major *mechanical processes* of the cold region environment (Table 8.1). *Frost action* refers to the stress

FIGURE 8.14 Fossil sand wedges confined between an upper gray and a lower brown till in a gravel pit near Edmonton, Canada. Note the orientation of stones in sand wedges in which the sediments are highly stained. The wedges display well-developed roots, which penetrate some 1.5 m into the till.

TABLE 8.1
Periglacial Processes and Associated Major Landforms

Process	Mechanism	Major Landforms
Frost action	Frost shattering	Talus, blockfields, rock glaciers
	Frost cracking	Ice wedges
	Frost heave	Pattern ground, involutions
	Needle ice	No major landform
Mass movement	Frost creep	No major landform
	Gelefluction	Gelefluction lobes, stone streams
Nivation	Erosion, freeze-thaw, hydration shattering, meltwater erosion, and mass wasting	Nivation hollows, solifluction terracettes
Wind activity	Not qualifiable	Sand wedges; ventifacts

activity of frost shattering, frost cracking of permafrost, frost heave, and growth of needle ice. *Mass movement* involves gelefluction and frost creep activities, which redistribute sediments from higher to lower ground without substantial loss of mass. *Nivation* is the typical process of erosion and snowmelt transport beneath immobile patchy snow covers of sizable extent. Finally, *wind activity* is particularly strong toward the polar direction of frozen ground environments. By comparison, *chemical weathering* is a minor stress activity, which possibly pulverizes clastic sediments in cold region environments. Many aspects of periglacial processes, however, are inadequately understood.

FIGURE 8.15   Frost-shattered granite fragments near Rohtang Pass (3978 m), Himachal Pradesh, India. At these seasonally cold altitudes in the Central Himalayas, rocks of other lithologic composition are also equally frost wedged into many angular fragments.

FROST SHATTERING

Seasonally and perennially frozen ground environments generate frost-shattered debris in bulk (Figure 8.15). The intensity of periglacial frost shattering, in general, depends on rock strength, thermal regime, and moisture content in the environment (Tharp, 1987; Matsuoka, 1991).

Rocks frost disintegrate by the freeze and thaw of water in the confined space of 5 to 25 cm deep cavities in rocks (French, 1994). The freezing of water generates *expensive stress* within the cavity space. This stress exceeds the *tensile strength* of most rocks (Chapter 3), by which they loosely partition and eventually disintegrate about geologically weak structures. The intensity of frost shattering, however, varies directly with the frequency of freeze-thaw cycles rather than with the magnitude of absolute temperature fluctuation about the freezing point (Washburn, 1979). The frequency of freeze-thaw is between forty and sixty cycles per year in the subarctic alpine environment: it decreases toward the frigid thermal regime of the Arctic and the warmer climatic regime of lower latitudes. Hence, the *subarctic environment* is typical of frost disintegration on the earth. Even in environments favorable to freeze-thaw, a snow cover of slight thickness (Harris, 1974; Ødegård et al., 1995) and subsurface snowmelt (Hall, 1988; Ødegård et al., 1995; Carey and Woo, 1998) greatly dampen the freeze-thaw activity. Observation from a part of the Baspa Valley in the north-western Himalayas, India, suggests that the air in rock cavities is warmer by 2°C than the ambient air temperature (Kaul and Singh, 1997). Hence, air-temperature-based arithmetic of freeze-thaw cycles and aspects of frost shattering require reassessment.

Water freezing rapidly expands by 9% at 0°C and occupies 13.5% more space at −22°C, the temperature at which ice is lowest in density (Davidson and Nye, 1985). Therefore, growth and deformation of ice, transmission of the stress of ice deformation normal to cavity space, and dilation of the confined space loosens and finally detaches rocks about the fissure pattern. The effective frost shattering, however, depends on *crack geometry* expressed by the ratio of maximum aperture of the crack to its length. Theoretically, pointed cracks are more susceptible to frost disintegration than cracks of other shape attributes (Tharp, 1987). Talus, block fields, and rock glaciers are major forms of frost disintegration activity.

Talus

Talus is the surface of frost-shattered angular rock fragments against the face of steep slopes and cliff faces. The talus debris originates on higher exposed slopes, and travels by rolling, sliding, and free falling to the site of accumulation, evolving a concave-up slope along the radial profile. The concavity of longitudinal profile depends on the manner of debris release from the upland source and debris accumulation pattern against the slope (Carson and Kirkby, 1972).

Blockfields

A bedrock surface in cold region environments of at least 200 degree-days year^{-1} of thaw index frost splits *in situ* into 2 m or more across angular rock blocks called blockfields, or *felsenmeer* (Harris, 1982). Such mega-sized rock fragments are regarded active and relict forms of frost disintegration. *Active blockfields* are freshly split, unstable clastic fragments on level to gently sloping surfaces, and *relict blockfields*, as in mid-latitude zones of severe climatic regime, are smooth in appearance and covered with endolithic lichen.

Rock Glaciers

Rock glaciers are large-sized active feature of the *continuous permafrost environment*, and active, inactive, and relict forms of *cold and dry alpine regions*. They appear similar to *valley glaciers* but are commonly built of either talus or moraine-derived debris containing ice lenses and ice cores of *segregated ice* (Martin and Whalley, 1987; Barsch, 1996; Haeberli and Mühel, 1996; Mitchell and Taylor, 2001). A few rock glaciers are also built of clastic debris, permafrost of periglacial origin, and ice core of glacier origin (Whalley et al., 1995; Johnson, 1995). Thus, rock glaciers are diverse in origin, material composition, and dynamic activity. These varied aspects of rock glaciers are highlighted in the following section with specific case studies.

Active rock glaciers in a part of the central Andes exist at subfreezing air temperature, and the inactive ones, suggesting a recent climate change for the region, occupy the limit of 0°C mean annual temperature (Payne, 1998). Ice-cored active rock glaciers in a part of the Osörna Valley in Kazakhstan are comprised of supraglacial debris of receding valley glaciers (Schröder et al., 2005). The ice-cored Nautárdalur rock glacier in Iceland (Whalley et al., 1995) and the ice-cored Naradu glacier in a part of Baspa Valley in the Central Himalayas of northwestern India (Kaul and Singh, 1997), however, are *relics* of thin, decaying valley glaciers carrying an overburden of thick *supraglacial debris* (Figure 8.16). The Naradu glacier is situated in an area where the temperature for the recorded period of 1993 to 1997 varied from 2.5 to 6°C in summer and from −29 to −21°C in the winter. This glacier has receded at the average rate of 4.05 m year^{-1}, leaving 6000- to 8000-year-old *recessional moraines*

FIGURE 8.16 Ice-cored Naradu glacier in a part of the Baspa Valley, Himachal Pradesh, India. It carries a thick overburden of supraglacial debris in the manner of decaying valley glaciers. (Source: Photo courtesy Dr. M. N. Kaul.)

FIGURE 8.17 Recessional moraines and a lateral moraine near the snout of Naradu glacier in a part of the Baspa Valley, Himachal Pradesh, India. (Source: Photo courtesy Dr. M. N. Kaul.)

at the terminus of ice (Figure 8.17). Thus, the ice core of many rock glaciers represents either the ice of valley glaciers set to decay by the climate change (Whalley and Martin, 1992; Evans, 1993; Barsch, 1996; Haeberli and Mühel, 1996; Elconin and LaChapelle, 1997; Mitchell and Taylor, 2001) or remains of an ice and detritus load of dynamically inactive glaciers (Shroder et al., 2000; Mitchell and Taylor, 2001). By comparison, the talus and moraine-type rock glaciers in a part of Wales, UK, are the relic of a periglacial environment (Harrison and Anderson, 2001).

Talus-type rock glaciers comprise frost-shattered ice-cemented debris, and *moraine-based* rock glaciers are built of ice and supraglacial debris. The talus-type rock glaciers suggest periglacial disintegration of rocks, and the moraine-type rock glaciers a climate change (Whalley and Martin, 1992; Evans, 1993; Barsch, 1996;

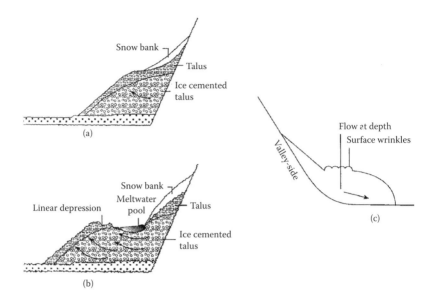

FIGURE 8.18 Schematic cross section of the incipient modification of talus slopes by plastic flow and compressive flow mechanisms. (a, b) Ice-cemented talus debris behave as a cohesive clastic mass that deforms by rotational sliding at the depth of the rock rubble, evolving long frontal lobes against the backing high-angled cliff face and surface depression immediately behind the lobe. (c) The talus debris held together by the segregated ice in the mass may also undergo compressive deformation by the stress of the weight of clastic debris that more or less behaves as a cohesive mass. This form of movement evolves steep frontal lobes and transverse ridges at the surface as an expression of the deceleration of flow through the depth of rubble. (Source: Figures 8 and 9 in Swett, K., Hambrey, M. J., and Johnson, D. B., "Rock Glaciers in Northern Spitsbergen," *J. Geol.* 88 [1980]: 475–82. With permission from University of Chicago Press. Figure 10 in Johnson, P. G., "Rock Glacier Formation by High-Magnitude Low-Frequency Slope Processes in Southwest Yukon," *Ann. Assoc. Am. Geogr.* 74 [1984]: 408–19. With permission from Routledge Publishers.)

Haeberli and Mühel, 1996; Mitchell and Taylor, 2001). This association of rock glaciers appears far too complex for the northwestern Himalayas, India, where talus- and moraine-type rock glaciers exist in proximity to each other and to the present-day glaciers (Mitchell and Taylor, 2001).

  Talus and moraine-based rock glaciers are narrow at the head, but progressively widen downslope and terminate in frontal lobes. *Talus-type rock glaciers* possibly flow by plastic deformation (Swett et al., 1980) or by compressive deformation at depth (Johnson, 1984). The *plastic deformation model* postulates that the detrital load held together by the ice content deforms by rotational sliding in the manner of ice flow in *cirque glaciers* (Figure 8.18(a), (b)). This form of deformation produces long frontal lobes. The *compressive deformation model* argues that rock glaciers flow by the stress of their own weight, evolving steep fronts and transverse ridges at the surface as an expression of the deceleration of flow through the depth of rock rubble (Figure 8.18(c)). *Moraine-based rock glaciers* are built of cohesive sediments that are susceptible to various forms of *slope failure* (Johnson, 1984, 1995; Barsch,

1996). They possibly deform by avalanche, slab, and slide failure of sediments (Figure 8.19). The *avalanche model* postulates rapid successive failure of the overriding mass of moist cohesive debris, evolving a typical frontal lobe and a shallow depression behind the terminus. The *slab failure model* perceives conditional instability of finer sediments along shear planes. This form of failure evolves small-sized rock glaciers of steep frontal lobe. The *slide failure model* postulates the presence of an impermeable substratum above the detritus load of rock glaciers, such that it interferes with the percolation of snowmelt through the sediment cover. Hence, the sediment above becomes saturated, loses shear strength, and slides at the interface, evolving long lobes and a moderately high front of rock glaciers.

Rock glaciers deform slowly at a rate at least an order of magnitude smaller than that of valley glaciers. Rock glaciers in mid-latitude mountains of the United States deform at the rate of 83 cm year^{-1} and of the mountains of subarctic Alaska and Canada at the rate of 50 to 70 cm year^{-1} (Price, 1972). Comparatively young rock glaciers in a part of Yukon, Canada, however, deform at a higher rate of 1 m year^{-1} (Johnson, 1995). The Nautárdalur rock glacier in Iceland has receded at the rate of 2 to 25 cm year^{-1} in 16 years of observation (Whalley et al., 1995), while the similar type of Naradu rock glacier in the Central Himalayas of northwestern India had retreated at an estimated rate of 12 m year^{-1} between 1922 and 1944 (Kaul and Singh, 1997). The studies referred to above suggest that deformation in rock glaciers varies considerably from one situation to the other.

FROST CRACKING

A sharp decline of air temperature in the *continuous permafrost environment* cracks open the frozen ground on an annual basis. Observations suggest that *true permafrost* cracks open when the air temperature rapidly declines by 4°C, and *dry permafrost* splits at a steeper temperature drop of 10°C (Lachenbruch, 1968). The frost cracking of true permafrost evolves vein ice and ice-wedge forms, discussed earlier in this chapter.

FROST HEAVE

Frost heave is the phenomenon in which mineral soils of the active layer expand by 100% or more in volume by the freezing of water into segregated ice. Freezing of all the pore water in the medium theoretically expands the soil some 9% in volume, but frost heave requires additional water for freezing to produce the given effect. Experimental studies suggest that this water is derived from the frozen ground itself, but processes of moisture transfer for freezing into segregate ice are far from understood.

In general, silt-textured soils are *frost susceptible* and clay-sized soils are least susceptible of all. In qualitative terms, frost heave is variously a function of particle size, heat capacity, and thermal conductivity of soils, and moisture retention capacity and hydraulic conductivity of the soil-water system. In test conditions, the dynamics of frost heave is understood in terms of one or the other aspect of soils and the soil-water system.

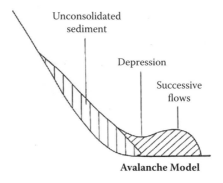

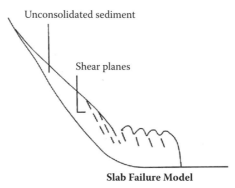

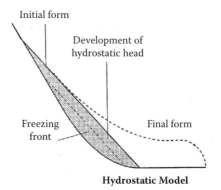

FIGURE 8.19 Compressing flow mechanisms and their surface expression in debris-type rock glaciers. (Source: Modified from Figures 4, 7, and 9 in Johnson, P. G., "Rock Glacier Formation by High-Magnitude Low-Frequency Slope Processes in Southwest Yukon," *Ann. Assoc. Am. Geogr.* 74 [1984]: 408–19. With permission from Routledge Publishers.)

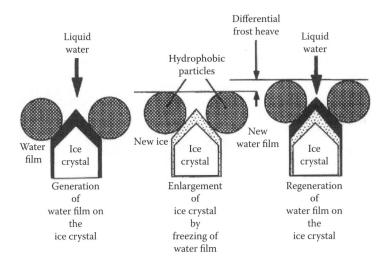

FIGURE 8.20   Differential frost heave in hydrophobic soil system. (Source: Figure 1 in Sage, J. D., and Porebska, M., Frost Action in a Hydrophobic Material, *J. Cold Regions Eng.* 7 [1993]: 99–113. With permission from American Society of Civil Engineers.)

Frost-susceptible soils imbibe moisture directly from the frozen ground (Mageau and Morgenstern, 1980). This moisture adsorbs as a film of water on hydrophobic soil grains and freezes into segregated ice (Anderson et al., 1978). Frost heave, thus, varies directly with hydraulic conductivity and moisture retention properties of the soils (Yang and Goodings, 1998). It is also likely that moisture adsorbs and freezes directly onto the interstitial ice, generating sufficient upthrust force for the heave of soils (Figure 8.20). Hence, the magnitude of heave depends on the volume of moisture available for freezing, and the number of adsorption sites in the soil system (Sage and Porebska, 1993). Clays, in comparison, are poor in hydraulic conductivity. They, therefore, are the least frost susceptible. Eigenbrod et al. (1996), however, suggest that clay-textured soils frost heave possibly by the migration of pore water in the direction of a hydraulic- and pore-water-deficit gradient, and freezing of this water a little below the freezing front as segregated ice.

Effects of Frost Heave

Volumetric expansion of frost-susceptible soils generates pull and push forces in the active layer. The frost pull and push ejects pebble- and gravel-sized stones vertically and horizontally through the mineral soil, evolving a *patterned ground* at the surface (Washburn, 1979). The mineral soils of the active layer also comprise sedimentary layers that differ in grain size, moisture retention, and thermal conductivity attributes. They, therefore, freeze at unequal rates and deform as *periglacial involutions* within the seasonal permafrost.

The *frost-pull mechanism* holds that stones firmly held in the freezing active layer lift by the like amount of frost heave. In the period of seasonal thaw, however, the saturated cohesive sediments quickly *collapse* at the base of stones and prevent their return to the previous position (Figure 8.21(a)). The *frost-push mechanism* envisages

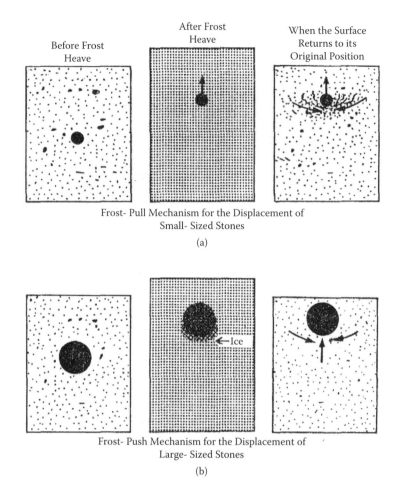

FIGURE 8.21 Illustration of the generalized processes of frost pull (a) and frost push (b) in frost-susceptible soils.

greater heat conductivity of stones than that of the soil matrix in which they are held. Hence, ice crystals form and melt first at the base of stones during seasonal freeze and thaw of the active layer. The space thus vacated by the melting of ice beneath stones is rapidly occupied by *lateral flow* of saturated sediments, which process pushes stones from their prethaw position (Figure 8.21(b)). The rhythmic *freeze-thaw cycles* in the active layer thus eject stones to the surface.

Stones expelled from the active layer appear either size sorted or nonsorted at the surface. The reason for this sorting of stones by size is not understood, but observations suggest that stones are better sorted in the permafrost of higher than lower frequency of freeze-thaw cycles (Corte, 1963). A sorted pattern is associated with the active layer of discontinuous permafrost of 200 degree-days year[-1] of thaw index, and that of the nonsorted pattern with the active layer and thermal cracking of the continuous permafrost of 100 degree-days year[-1] of thaw index (Harris, 1982).

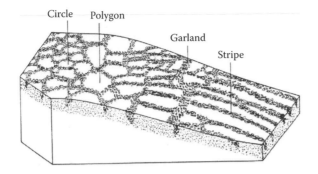

FIGURE 8.22 Polygons, nets, and strips showing the influence of ground slope on the development of patterned ground phenomena. (Source: Adapted from Figure 5 in Sharpe, C. F. S., *Landslides and Related Phenomena* [New York: Columbia University Press, 1938].)

The stones also possibly size sort by thermal convection in the active layer (Hallet et al., 1988).

PATTERNED GROUND

Stones ejected from the active layer present a symmetrical pattern of circles, polygons, nets, and stripes, in which the pattern shape depends entirely on the ground slope (Figure 8.22). Circles, and polygons develop on flat to nearly flat surfaces, garlands on 2 to 6° steep slopes, and stripes on surfaces up to 30° steep. Patterned ground, however, fails to establish on surfaces more than 30° steep. The morphodynamics of patterned ground is explained by hypotheses based on thermal cracking of the frozen ground, frost heave of the active layer, and thermal convection in the active layer.

POLYGONS

Polygons are a frost crack and active layer feature of cold region environments. *Frost crack polygons* are mud polygons in deep soils of seasonally cold alpine regions (Figure 8.23), wherein the polygon size is directly proportional to the magnitude of temperature drop below the freezing point. Polygons of the active layer are specific to areas of continuous and discontinuous permafrost. Those of the *continuous permafrost* present an extensive honeycomb pattern of nonsorted stones at the boundary of individual ice wedges, and of the *discontinuous permafrost* appear as high- and low-centered forms of sorted stones that tend to be coarser outward to the edge of the pattern (Figure 8.24). For reasons not understood yet, the stones of high-cenetred polygons reach the base of the active layer and of low-centered polygons remain confined to a shallow depth of the rim (Jahn, 1968).

Heat conductivity and temperature difference through the depth of the active layer possibly introduce a local system of *thermal convection* in the mineral soil. Thermal convection in the *asthenosphere* occurs at a *Rayleigh number* (Ra) between 10^6 and 10^7 (McKenzie and Richter, 1976) and in the active layer at a higher Ra value of 6 ×

FIGURE 8.23   Frost crack polygons near Banff, Alberta, Canada.

High-Centered Polygons

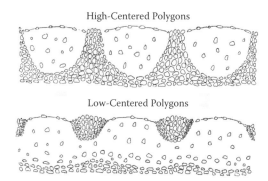

Low-Centered Polygons

FIGURE 8.24   Section of sorted polygons. Type A, stone borders reaching the base of the structures; type B, floating stone borders. (Source: Figure 2 in Jahn, A., "Patterned Ground," in *The Encyclopedia of Geomorphology*, ed. R. W. Fairbridge [New York: Reinhold Book Corp., 1968], 814–17. With permission from Kluwer Academic Publishers.)

10^{13} (Niemela et al., 2000). At this magnitude, the larger system of thermal convection breaks down into small-scale secondary motions. Thermal convection (Chapter 2), which accompanies coherent motion of the material in the direction of the thermal gradient, displaces and size sorts stones in the active layer (Figure 8.25).

INVOLUTIONS

Mineral soils of the active layer comprise sedimentary layers that individually distinguish in grain size, heat conductivity, and moisture content. Therefore, they freeze at an uneven rate in the active layer freezing downward of the surface and upward of the expanding permafrost table. Seasonal permafrost expanding in this manner exerts an enormous *cryostatic pressure* in the medium, and intensely deforms the laminae (Johnsson, 1962). These deformed sedimentary layers of the active layer are called periglacial involutions (Figure 8.26).

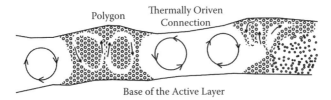

Frozen Ground

FIGURE 8.25 Schematic model for the thermally driven convection within the active layer showing the development of patterned sorted circles. (Source: Adapted from Hallet, B., Anderson, S. P., Stubbs, C. W., and Gregory, E. C., "Surface Soil Displacements in Sorted Circles in Western Spitzbergen," in *Proceedings, Fifth International Conference on Permafrost*, Trondheim, August 2–5, 1988, Senneset, K. ed. pp. 775–79, Tapir.)

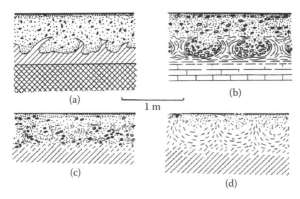

FIGURE 8.26 Various types of involutions: (a) injection of tongues of silt and clay into overlying sand and gravel; (b) frost pockets, frost kettles, or pillar involutions, with ascending tongues of finer sediment; (c) irregular or amorphous involutions; (d) festoons, with tongues of frost-shattered disintegrated rock rising into finer sediment. (Source: Figure 5.13a–d in West, R. G., *Pleistocene Geology and Biology*, 2nd ed. [London: Longman, 1977]. With permission from Pearson Education.)

NEEDLE ICE

In areas of nocturnal temperature fluctuation about the freezing point, the *soil water* migrates in the direction of thermal gradient and freezes near the surface as needle ice. The growth of needle ice stirs soil particles and displaces small-sized stones at the surface. This process aids *creep* of soils and rock fragments even on gentle slopes.

MASS MOVEMENT

Frost creep and gelefluction are principal mass movement activities of the cold region environment. Frost creep is associated with heave and thaw of the active layer. Observations suggest that stones lying at the surface of frost-heaved soils laterally displace when the surface returns to its original position upon thaw, but the

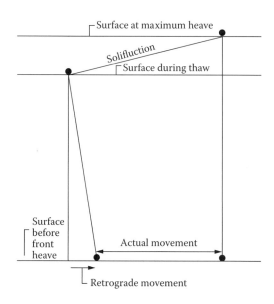

FIGURE 8.27   Illustration of retrograde movement in frost-susceptible soils. The particle affected by frost creep on the frost-heaved surface does not settle vertically but at an interme- diate position due to cohesion among soil grains. Therefore, there is a slight upslope or back- ward movement of the particle immediately after the thaw of frost-heaved surface. (Source: Adapted from Figure 5 in Washburn, 1967. Instrumentational observation of mass wasting in Mester's Vig District, Northeast Greenland. Meddelelser om Grønland, Bd. 166, No. 4, 296 pages.)

net displacement is somewhat less than in other environments (Figure 8.27). This *retrograde movement* possibly results from the *cohesive force* among saturated sedi- ments, which tends to hold back the displacement of individual grains somewhat (Washburn, 1979).

Cohesive soils of cold region environments rapidly saturate from ground thaw and seasonal snowmelt and, therefore, deform by viscous flow called gelifluction (Chapter 4). Gelifluction best develops on 5 to 20° steep slopes, but moist soils additionally stressed by the overburden of snow deform even on gentler slopes. Gelifluction dis- turbs the *thermal balance* of frozen ground environments and, thereby, hastens the thaw of ground ice.

NIVATION

Nivation is the process of erosion by the deformation of snow in the manner of ice flow, frost shattering, meltwater erosion, weathering, and mass movement around and beneath sufficiently thick snow covers of alpine and subpolar regions. Frost shat- tering, however, remains a debated nival process. Snow not only insulates the ground from the effect of air temperature fluctuations, but also interferes with the two-way exchange of heat between the ambient air and permafrost. In addition, the refreez- ing of meltwater in soils and subsurface rock cavities releases the *heat of fusion* that slightly warms the temperature there and alters conditions for freeze and thaw beneath snow covers of sizable extent. Observations also suggest that subsurface warming of temperature at the rate of 2°C min^{-1} generates a strong enough thermal stress in dry permafrost to split rocks (Ødegård et al., 1995).

French (1994) argues that the large volume of frost-shattered debris observed in the periglacial environment is not due alone to the frequency of *freeze-thaw cycles* but also to the mechanical stress of *hydration shattering* beneath lingering snow covers of limited extent. The rhythmic hydration expansion and dehydration contraction of rock-forming minerals splits rocks along mineral boundaries (Chapter 3). Geomorphologists opine that the expansive force of mineral hydration nearly equals the stress magnitude resulting from the *growth of ice crystals* in rock cavities (White, 1976).

Chemical weathering is a little known and rather little understood activity in periglacial environments. Walder and Hallet (1985) and Tharp (1987), however, suggest that decomposition pulverizes frost-shattered debris in cold region environments. This activity may perhaps be related to the atmospheric carbon dioxide and windborne salts locked in the snow cover.

Snowmelt and deformation of thick snow cause *abrasive wear* of the surface. Strong dustier winds in the periglacial environment are known to evolve *sand wedges* in permafrost and *ventifacts* in ice-free tracts of Antarctica.

SUMMARY

Periglacial refers to processes and landforms of the frozen ground environment. A periglacial or cold region environment is conveniently identified and classified by the magnitude and range of subfreezing mean annual temperature. The frozen ground is continuous, discontinuous, and sporadic by the amplitude of subfreezing mean annual air temperature, true and dry by the additional variable of moisture content in the subsoil, and seasonal by the subfreezing winter temperature of otherwise warmer climates. The surface layer of the frozen ground that thaws in summer and freezes in winter also constitutes the seasonal permafrost. This surface layer of seasonal permafrost produces active layers of loose mineral soil. Nearly 26% of the earth has frozen ground, much of which lies within the Arctic and subarctic zones. The extent of frozen ground in data-sparse Antarctica, however, is not known in certain terms.

The depth of frozen ground is in a state of delicate balance with the magnitude of subfreezing air temperature, heat conductivity of the subsurface medium, time given for the permafrost to develop, hydrologic and vegetation aspects of terrain, and geothermal heat loss from the ground. Seasonal temperature fluctuations affect the temperature of the frozen ground to a depth of 20 to 30 m from the surface. Down this depth, the permafrost temperature remains unaffected by variations in the air temperature. This interface is called the level of zero amplitude. At this depth, the frozen ground of North America is at a temperature of −15°C. Interactions between the cooler air temperature conducting through the surface and geothermal heat escaping the ground balance each other out at some depth. Therefore, unfrozen ground called talik appears at some depth in the ground.

Pingos, palsas, ice wedges, and thermokarst are major forms of ground ice that evolve above the level of zero amplitude in the frozen ground. Pingos, the ice-cored conical hills in continuous and discontinuous permafrost environments, are closed and open system in type with respect to the source of water and manner of water movement for freezing into ice. Pingo processes, however, are debated. Palsas are frost-heaved ice-cored ridges and mounds in the bogs of discontinuous permafrost. They possibly evolve from advective flow and seasonal freezing of water in organic terrain otherwise known for its poor thermal conductivity. Ice wedges are a typical feature of the continuous permafrost. They evolve from thermal cracking of the frozen ground on an annual basis, and develop from the vein ice nourished by moisture in the active layer. The thaw of permafrost due to natural causes and human activities produces a variety of subsidence forms called thermokarst.

Frost shattering, mass wasting, nivation, and wind activities excel in the periglacial environment. Activities like frost disintegration of rocks, thermal cracking of the frozen ground, and heave of frost-susceptible soils, however, are known in outline and theory only. Therefore, the evolutionary mechanism of major forms of frost activity as block fields, rock glaciers, and patterned ground remains inconclusive. Gelefluction is a slow flowage of cohesive sediments, and the process of altiplanation in cold region environments. Nivation is the process of hydration shattering, meltwater erosion, and mass wasting around and beneath immobile snow covers of sizable extent. The dynamics of many nival processes is also debated. Periglacial areas are known for dry, strong, and dustier winds. Sand wedges and ventifacts are some notable forms of wind activity in cold region environments.

REFERENCES

Anderson, D. M., Pusch, R., and Penner E. 1978. Physical and chemical properties of frozen ground. In *Geotechnical engineering for cold regions*, ed. O. B. Andersland and D. M. Anderson, 37–102. New York: McGraw-Hill.

Barsch, D. 1996. *Rock glaciers*. Berlin: Springer-Verlag.

Black, R. F. 1954. Permafrost—A review. *Bull Geol. Soc. Am.* 65:839–55.

Black, R. F. 1976. Periglacial features indicative of permafrost: Ice and soil wedges. *Quat. Res.* 6:3–26.

Black, R. F. 1983. Pseudo ice-wedge casts of Connecticut, northeastern United States. *Quat. Res.* 20:74–89.

Brown, R. J. E. 1960. The distribution of permafrost and its relation to air temperature in Canada and the U.S.S.R. *Arctic* 13:163–77.

Brown, R. J. E. 1970. *Permafrost in Canada*. Toronto: University of Toronto Press.

Brown, R. J. E., and Kupsch, W. O. 1974. *Permafrost terminology*. National Research Council of Canada Technical Memorandum III. National Research Council of Canada.

Carey, S. K., and Woo, Ming-ko. 1998. Snowmelt hydrology of two subarctic slopes, northern Yukon, Canada. *Nordic Hydrol.* 29:331–46.

Carson, M. A., and Kirkby, M. J. 1972. *Hillslope form and processes*. Cambridge: Cambridge University Press.

Corte, A. E. 1963. Relationship between four ground patterns, structure of the active layer, and type and distribution of ice in the permafrost. *Biul. Peryglac.* 12:7–90.

Davidson, G. P., and Nye, J. P. 1985. The photoelastic study of ice pressure in rock cracks. *Cold Regions Sci. Technol.* 11:141–53.

Dean, K. G., and Morrissey, L. A. 1988. Detection and identification of arctic landforms: An assessment of remotely sensed data. *Photogrammetric Eng. Remote Sensing* 54:363–71.

Demek, J. 1978. Periglacial geomorphology: Present problems and future prospects. In *Geomorphology: Present problems and future prospects*, ed. C. Embleton, D. Brunsden, and D. K. C. Jones Oxford, 139–53. Oxford: University Press.

Derbyshire, E. 1977. Periglacial environments. In *Applied geomorphology*, ed. J. R. Hails, 227–76. Amsterdam: Elsevier Science Publishing Co.

Eigenbrod, K. D., Knutsson, S., and Sheng, D. 1996. Pore-water pressures in freezing and thawing fine-grained soils. *J. Cold Regions Eng.* 10:77–92.

Elconin, R. F., and LaChappelle, E. R. 1997. Flow and internal structure of a rock glacier. *J. Glaciol.* 43:238–44.

Embleton, C. 1980. Nival processes. In *Process in geomorphology*, ed. C. Embleton and J. Thornes, 307–24. New Delhi: Arnold Heinemann.

Embleton, C., and King, C. A. M. 1968. *Glacial and periglacial geomorphology*. London: Edward Arnold.

Evans, D. J. A. 1993. High latitude rock glaciers: A study of form and processes in the Canadian arctic. *Permafrost Periglacial Processes* 4:17–35.

French, H. M. 1994. Freeze-thaw cycle. In *The encyclopaedic dictionary of physical geography*, eds. Goudie, A., Simmons, I. G., Atkinson, B. W., and Gregory, K. J., p. 193. 2nd ed. Oxford: Blackwell.

Haeberli, W., and Vonder Mühel, D. 1996. On the characteristics and possible origins of ice in rock glacier permafrost. *Z. Geomorph.* 104:43–57.

Hall, K. 1988. A laboratory simulation of rock-breakdown due to freeze-thaw in a maritime Antarctic environment. *Earth Surface Processes Landforms* 13:369–82.

Hallet, B., Anderson, S. P., Stubbs, C. W., and Gregory, E. C. 1988. Surface soil displacements in sorted circles in western Spitzbergen. In *Proceedings, Fifth International Conference on Permafrost*, Trondheim, August 2–5, 1988, Senneset, K. ed., pp. 775–79.

Harris, C. 1974. Autumn, winter and spring soil temperature in Okstindan, Norway. *J. Glaciol.* 13:521–33.

Harris, S. A. 1982. Identification of permafrost zones using selected permafrost landforms. In *Proceedings of the 4th Canadian Permafrost Conference*, Calgary, Alberta, March 2–6, 1981, pp. 49–58.

Harris, S. A., and Brown, R. J. E. 1982. Permafrost distribution along the Rocky Mountains in Alberta. In *Proceedings of the 4th Canadian Permafrost Conference*, Calgary, Alberta, March 2–6, 1981, pp. 59–67.

Harrison, S., and Anderson, E. A. 2001. Late Devensian rock glacier in the Nantlle Valley, north Wales. *Glacial Geol. Geomorph.*, rp01.

Jahn, A. 1968. Patterned ground. In *The encyclopedia of geomorphology*, ed. R. W. Fairbridge, 814–17. New York: Reinhold Book Corp.

Johnson, P. G. 1984. Rock glacier formation by high-magnitude low-frequency slope processes in southwest Yukon. *Ann. Assoc. Am. Geogr.* 74:408–19.

Johnson, P. G. 1995. Debris transfer and sedimentary environments: Alpine glaciated areas. In *Steepland geomorphology*, O. Slaymaker, 27–44. Chichester: John Wiley & Sons.

Johnsson, G. 1962. Periglacial phenomena in southern Sweden. *Geogr. Annlr.* 44:378–404.

Kachurin, S. P. 1962. Thermokarst within the territory of the U.S.S.R. *Biul. Peryglac.* 11:49–55.

Kaul, M. N., and Singh, B. P. 1997. *Fluctuation of climate and glacier in Naradu Basin and its geoecological significance*. New Delhi: DST Project ES/91/03/91.

Lachenbruch, A. H. 1962. *Mechanics of thermal contraction cracks and ice-wedge polygons in permafrost*. Geological Society of America Special Paper 70. Geological Society of America.

Lachenbruch, A. H. 1968. Permafrost. In *The encyclopedia of geomorphology*, ed. R. W. Fairbridge, 833–39. New York: Reinhold Book Corp.

Lachenbruch, A. H., Brewer, M. C., Greene, G. W., and Marshall, B. V. 1962. Temperature in permafrost. In *Temperature—Its measurement and control in science and industry*, 791–803. Vol. 3, Pt. I. New York: Reinhold Publishing Corp.

Lachenbruch, A. H., and Marshall, B. V. 1969. Heat flow in the arctic. *Arctic* 22:300–11.

Mackay, J. R. 1962. Pingos of the Pleistocene Mackenzie delta area. *Geogr. Bull.* 18:21–63.

Mackay, J. R. 1972. The world of underground ice. *Ann. Assoc. Am. Geogr.* 62:1–22.

Mackay, J. R. 1974. Ice-wedge cracks, Garry Island, Northwest Territories. *Can. J. Earth Sci.* 11:1366–83.

Mackay, J. R. 1975. The closing of ice-wedge cracks in permafrost, Garry Island, Northwest Territories. *Can. J. Earth Sci.* 12:1668–74.

Mackay, J. R. 1979. Pingos of the Tuktoyaktuk Peninsula area, North-West Territories. *Géographie Physique Quaternaire* 23:3–61.

Mageau, D. W., and Morgenstern, N. R. 1980. Observations on moisture migration in frozen soils. *Can. Geotech. J.* 17:54–60.

Martin, H. E., and Whalley, W. B. 1987. Rock glaciers. Part I. Rock glacier morphology, classification and distribution. *Progress Phys. Geogr.* 11:260–82.

Matsuoka, N. 1991. A model of the rate of frost shattering: Application to field data from Japan, Svalbard and Antarctica. *Permafrost Periglacial Proc.* 2:271–81.

McKenzie, D. P., and Richter, F. 1976. Convection currents in the earth's mantle. *Sci. Am.* 235:72–89.

Mitchell, W. A., and Taylor, P. J. 2001. Rock glaciers in the northwestern Indian Himalaya. *Glacial Geol. Geomorph.*, rp02.

Müller, F. 1962. Analysis of some stratigraphic observations and radiocarbon dates from the two pingos in the Mackenzie delta area, N.W.T. *Arctic* 15:278–88.

Müller, F. 1968. Pingos, modern.In *The encyclopedia of geomorphology*, ed. R. W. Fairbridge, 845–47. New York: Reinhold Book Corp.

Niemela, J. J., Skrbek, L., Sreenivasan, K. R., and Donnelly, R. J. 2000. Turbulent convection at very high Rayleigh numbers. *Nature* 404:837–40.

Ødegård, R. S., Etzelmüller, B., Vatne, G., and Sollid, J. L. 1995. Near-surface spring temperatures in an arctic coastal rock cliff: Possible implications for rock breakdown. In *Steepland geomorphology*, ed. O. Slaymaker, 89–102. Chichester: John Wiley & Sons.

Owens, E. H., and Harper, J. R. 1977. Frost-table and thaw depths in the littoral zone near Pearl Bay, Alaska. *Arctic* 30:155–68.

Payne, D. 1998. Climatic implications of rock glaciers in the arid western cordillera of central Andes. *Glacial Geol. Geomorph.*, rp03.

Peltier, L. 1950. The geographical cycle in periglacial regions as it is related to climatic geomorphology. *Ann. Assoc. Am. Geogr.* 40:214–36.

Péwé. T. L. 1966. Palaeoclimatic significance of fossil ice wedges. *Biul. Peryglac.* 15:65–73.

Péwé, T. L. 1969. Periglacial environment. In *The periglacial environment*, ed. T. L. Péwé, 1–9. Montreal: McGill-Queen's University Press.

Price, L. W. 1972. *The periglacial environment, permafrost, and man.* Resource Paper 14. Washington, DC: Commission on College Geography.

Rastogi, S. P., and Nawani, P. C. 1976. Permafrost areas in Tso-kar basin, a short note. In *Geological Survey of India Symposium*, Lucknow, November 21–23, 1976, Vol. III, Paper III.7.

Sage, J. D., and Porebska, M. 1993. Frost action in a hydrophobic material. *J. Cold Regions Eng.* 7:99–113.

Schmidlin, T. W. 1988. Alpine permafrost in eastern North America: A review. In *Proceedings, Fifth International Conference on Permafrost,* Trondheim, August 2–5, 1988. Senneset, K. ed., 775–79.

Schröder, H., Kokarev, A., and Harrison, S. 2005. Rock glaciers in the northern Tien Shan, Kazakhstan: New data on movement rates and distribution. *Glacial Geol. Geomorph.*, rp01.

Sharpe, C. F. S. 1938. *Landslides and related phenomena.* New York: Columbia University Press.

Shroder, J. F., Bishop, M. P., Copland, L., and Sloan, V. F. 2000. Debris-covered glaciers and rock glaciers in the Nanga Parbat Himalaya, Pakistan. *Geogr. Annlr.* 82A:17–31.

Shumskii, R. A. 1964. Ground (subsurface) ice. *Nat. Res. Coun. Can. Tech. Trans.* 1130.

Siegel, D. I., Reeve, A. S., Glaser, P. H., and Romanowicz, E. A. 1995. Climate-driven flushing of pore water in peat lands. *Nature* 374:531–33.

Steurer, P. M., and Crandell J. H. 1995. Comparison of methods used to create estimates of air-freezing index. *J. Cold Regions Eng.* 9:64–74.

Swett, K., Hambrey, M. J., and Johnson, D. B. 1980. Rock glaciers in northern Spitsbergen. *J. Geol.* 88:475–82.

Tharp, T. M. 1987. Conditions of crack propagation by frost wedging. *Bull. Geol. Soc. Am.* 99:94–102.

Thie, J. 1974. Distribution and thawing of permafrost in the southern part of the discontinuous permafrost zone in Manitoba. *Arctic* 27:189–200.

Thorn, C. E. 1992. Periglacial geomorphology: What, where? when? In *Periglacial geomorphology*, ed. J. C. Dixon and A. D. Abrahams, 1–30. Chichester: John Wiley & Sons.

Walder, J., and Hallet, B. 1985. A theoretical model of fracturing during freezing. *Bull. Geol. Soc. Am.* 97:336–46.

Washburn, A. L. 1967. Instrumentational observation of mass wasting in Mesters Vig District, Northeast Greenland. Meddelelser om Grønland, Bd. 166, 296 pages.

Washburn, A. L. 1979. *Geocryology*. London: Edward Arnold.

West, R. G. 1977. *Pleistocene geology and biology*. 2nd ed. London: Longman.

Westgate, J. A. 1969. *The Quaternary geology of the Edmonton area, Alberta*, 129–51. Contribution 465. Department of Geology, University Alberta.

Whalley, W. B., Hamilton, S. J., Palmer, C. F., Gordon, J. E., and Martin, H. E. 1995. The dynamics of rock glaciers: Data from Tröllaskagi, north Iceland. In *Steepland geomorphology*, ed. O. Slaymaker, 129–45. Chichester: John Wiley & Sons.

Whalley, W. B., and Martin, H. E. 1992. Rock glaciers. Part II. Models and mechanisms. *Progress Phys. Geogr.* 16:127–86.

White, S. E. 1976. Is frost action really only hydration shattering? *Act. Alp. Res.* 8:1–6.

Yang, D., and Goodings, D. J. 1998. Predicting frost heave using FROST model with centrifuge models. *J. Cold Regions Eng.* 12:64–83.

9 Aeolian Environment and Landforms

An aeolian environment is characteristic of 20 to 25% of the continental area in diverse geomorphic environments of strong dry winds, moisture scarcity, scant vegetation cover, and loose surface sediments. Dry regions of such qualifications are *ideal areas*, and areas of abundant supply of sediments along coasts and glacier margins and in cultivated lands are *suitable sites* for the activity of wind (Livingstone and Warren, 1996). Dry regions are broadly identified with hot and cold deserts and areas of semiarid environment.

BASES OF AEOLIAN ACTIVITY

Friction between the wind and rough surfaces produces shear in the wind. The *wind shear*, which depends directly on the wind speed and surface roughness elements and varies with thermal effects, generates a near-bed turbulence in the wind. The turbulence increases drag of the wind on the surface, reducing the mean velocity for the given pressure gradient. The wind shear or shear stress on a horizontal surface of air is reasonably given by the assumption

$$\tau = \mu \, (du/dz)$$

in which τ is the shear stress at the bed, μ is the dynamic viscosity, and du/dz is the rate of change of wind speed with height above the surface. The velocity gradient in the vertical, du/dz, is independent of viscosity but depends entirely on the density of air (ρ), the surface stress (τ), and height (z) above the surface.

The *wind shear velocity*, or friction velocity as it is sometimes called, varies with the shear stress at the bed and the density of air. The wind shear velocity is derived from the relationship

$$u_* = (\tau/\rho)^{1/2}$$

where u_* is the wind shear velocity, τ is the surface stress of the wind, and ρ is the density of air. This relationship also provides an estimate of the surface stress in terms of the wind shear velocity as

$$\tau = \rho \, u_*^2$$

By the *Karman–Prandtl equation*, the wind shear velocity is related to the logarithmic velocity profile as

$$u = u_*/\kappa \ln (z + c)$$

in which u is the wind velocity at a height z above the surface, κ is von Karman's constant with an experimental value 0.41, and c is the constant for surface conditions. The wind velocity profile is easy to measure in field and laboratory conditions.

The above equation suggests that the wind velocity increases logarithmically with height, producing a certain intercept, Z_0, for the straight-line relationship (Figure 9.1). This datum nearly equals the depth of a minutely thin layer of air hugging the deformable bed in which the flow is nearly stationary. This viscous sublayer obtains from the capacity of the wind to resist shear due to its property of *viscosity*. The depth of viscous sublayer is nearly one-thirtieth of the mean height of surface roughness elements and the size of surface particles (Lancaster and Nickling, 1994).

The wind shear provides energy for the aeolian activity of erosion, transportation, and sedimentation in the direction of wind flow. In essence, the three processes are interdependent and considerably merge with each other (Warren, 1979).

THRESHOLDS OF MOTION

Bed particles of a given size at rest are set in motion at a certain minimum shear force called the threshold shear force. The threshold shear force is difficult to measure. Therefore, the threshold of motion is conveniently determined by the drag (or shear) velocity for particles of a given size composition.

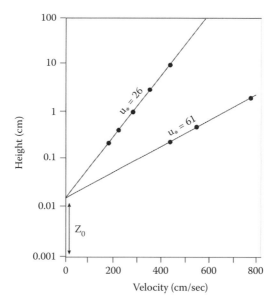

FIGURE 9.1   Variation of wind velocity with height plotted logarithmically for two different velocities. (Source: Figures 15A and B in Bagnold, R. A., *The Physics of Blown Sand and Desert Dunes* [London: Methuen, 1941]. With permission from Routledge.)

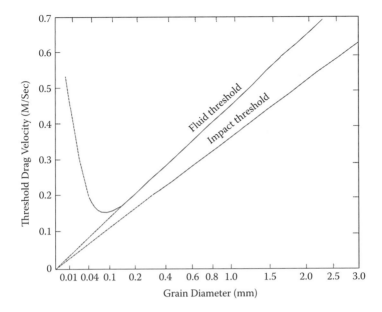

FIGURE 9.2 Relationship between drag velocity and grain diameter showing the fluid and impact thresholds. (Source: Figure 28 in Bagnold, R. A., *The Physics of Blown Sand and Desert Dunes* [London: Methuen, 1941]. With permission from Routledge.)

The movement of sediments at rest on a mobile bed of sand is governed by fluid (or static) and impact (or dynamic) thresholds of shear velocity (Figure 9.2). The *fluid threshold* defines that critical velocity at which grains of a given size first begin to move by the drag and lift forces. The grains slide or roll at the bed by the drag of the fluid force of the wind that produces a slight difference of pressure between the up- and downwind faces of particles at rest. The particles lift by the effect of a slight decrease of air pressure at high wind speeds. Observations suggest that drag and lift forces operate in a narrow zone above the mobile bed of sand (Werner, 1990).

The fluid threshold is a function of the bed material size, surface roughness elements, cohesion among grains, density of air, and wind turbulence (Bagnold, 1941). In general, *bed particles* of finer size composition that also make the surface smooth require a higher fluid threshold velocity for initiation of movement than coarser particles producing a relatively rough bed surface (Nickling, 1988). *Pore water* binds particles by the force of cohesion, which makes the aggregate more resistant to the shear force. Hence, a higher fluid threshold velocity is required to initiate displacement of particles with even 2 to 14% moisture content in the medium than would be needed for a similar-sized grade of dry particles at rest on the mobile bed of sand (Namikas and Sherman, 1995). *Temperature* of the environment affects the density of air, which in a global context is relatively dense and heavier in frigid climates. Therefore, same-sized bed particles of identical moisture content require a higher fluid threshold to initiate movement in colder climates than in other climatic regimes. *Wind turbulence* initiates deflation. Deflation alters the bed condition and initiates a widespread cascading effect on subsequent movement of an increasing number of bed particles (Williams et al., 1990; Lancaster and Nickling, 1994). A similar effect is also obtained in the initial stages of the movement of bed particles of varying size composition (Nickling, 1988).

Once moved, the particles are entrained in the wind flow for continuous transport at a velocity lower than the fluid threshold. This is the *dynamic threshold velocity*, at which grain movement is maintained by the energy derived from collision among grains. In test conditions, the dynamic threshold velocity is nearly 80% of the fluid threshold velocity (Anderson and Haff, 1988).

SEDIMENT TRANSPORT

The power of the wind across a deformable bed of sediments sets grains in motion. The wind transports sediments by saltation, surface creep, reptation, and suspension processes (Figure 9.3). Of these, saltation is the dominant process of aeolian transport system and a major control of other processes of sediment transport. In general, nearly three-fourths of the sediment moves in saltation mode, one-fourth in surface creep, and the remaining insignificant amount in reptation and suspension styles.

SALTATION

Saltation is the process of a leap and bounce movement of fine to medium-sized sand grains in the direction of wind flow. Experimental studies suggest that particles on a mobile bed of sand initially roll 30 to 40 cm before lift-off into the saltation mode (Bagnold, 1941), in which transition from rolling to lift stage lasts only a few seconds (Greeley and Marshall, 1985). The particles lift nearly 30 cm from the surface, and describe a nearly stable *parabolic trajectory* despite the turbulent wind structure in the near-bed region. On harder surfaces and in areas of wind acceleration due to topographic effects, however, the particles bounce to a height of 1.5 m from the ground (Laity, 1995). The height and length of saltation trajectory are related, such that the length is generally ten to fifteen times the bounce of particles. As a rule, particles in higher trajectories gain time for acceleration and, thereby, cover longer distances downwind before returning to the bed.

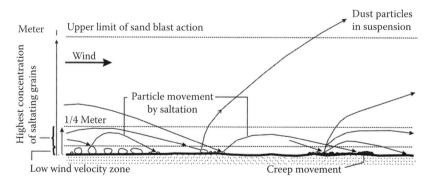

FIGURE 9.3   Models of grain movement by saltation, surface creep, and suspension.

Experimental studies suggest that particles in saltation mode launch at an angle of 30 to 40° from the horizontal, gain momentum during lift-off, and reach a maximum speed just before returning to the bed at an impact angle of 10 to 15° from the horizontal (Greeley et al., 1984). The particles landing at the bed generate an impact velocity of about 4 m s^{-1}, which ejects other grains from the bed and sets in motion a stream of saltating grains (Anderson and Haff, 1988). Constant *bombardment* of the bed by saltating grains, however, hardens the surface and lowers the *bed shear* that restricts per unit time the number of grains available for entrainment in the wind flow. In time, however, the *feedback* between the wind and sand flow rate reaches a *steady state* for sediment transport rate, in which the saltating grains eject an equal number of particles from the bed. This state is reached within 1 to 2 s (Anderson and Haff, 1988) to as long as 9 s from (McEvan, 1993) the initiation of grain movement.

CREEP

Creep is an imperceptibly slow near-bed movement of coarser bed particles. Surface creep results from the knocking forward of bed particles by saltating grains returning to the bed, and from the failure of a few grains in saltation mode to rebound from the bed after completion of their trajectory. The high-speed impact of saltating grains against the bed dislodges bed particles six times the diameter of impacting grains in saltation mode (Bagnold, 1941). These coarse-grained creeping particles contribute to the size sorting of sand deposits.

The creep occurs at the base of the saltation layer. Therefore, it is not only difficult to observe, but also a less well understood activity. Experimental observations suggest that creep is associated with a loss of forward momentum in creeping particles, such that the loss is more pronounced for plate-shaped particles than rounded grains (Willetts and Rice, 1986). For other conditions equal, therefore, the rounded grains travel farther in creep than the plate-shaped ones. The test data on creep further suggest (Willetts and Rice, 1986) that 0.335 to 0.6 mm sized bed particles move as a group at the wind shear velocity of 0.48 m s^{-1}. These particles travel at the speed of 0.005 m s^{-1} but soon disperse within 3 min of the initiation of movement by the effect of greater speed of a few creeping grains in the group. Livingstone and Warren (1996) suggest that creep is not the only near-surface activity of sediment transport, as there are other, less well understood processes also at work at the mobile bed of sand. Hence, much needs to be learned of creep and related processes in the near-bed region.

REPTATION

Reptation is also a near-bed aeolian transport activity. It is described as the effect of high-energy *splash impact* of saltating grains against the bed particles (Anderson and Haff, 1988). Observations suggest that the splash impact of each saltating grain against the bed ejects some ten particles in low hops in almost all directions to the wind flow. Therefore, reptation differs fundamentally from the manner of grain movement in saltation and surface creep activity. The number of grains ejected in reptation from the bed varies with the *wind shear velocity* (Livingstone and Warren, 1996).

SUSPENSION

Turbulence in the wind generates eddy motions, which keep particles of certain size composition afloat so long as the *inertial force* due to turbulence in the wind exceeds the *viscous force* due to the weight of particles. Suspension, thus, is defined as the condition in which grains follow turbulent motion above the saltation layer. This qualification also differentiates suspension from the saltation mechanism of sediment transport (Livingstone and Warren, 1996).

Particles of up to 0.07 mm size composition commonly pass in suspension. These dust particles have several sources, of which *crushing* of particles by collision among grains in the saltation stream and *bombardment* of the bed by the saltating grains returning to the bed generate a significant amount of smaller-sized particles in the environment. These particles are held in suspension above the saltation layer. Dampening of turbulence in the saltation layer, however, prevents a few particles that could have passed in suspension from escaping the stream of saltating grains (Livingstone and Warren, 1996). At times, stronger turbulence across undulatory topography, such as that of sand dunes, however, helps raise even particles coarser than 0.07 mm in suspension.

Particles entering suspension become *size sorted* during transport, such that the coarsest fractions drop out of the system first and finer fractions are carried farther downwind for sedimentation. In general, coarser fractions travel only a few tens of kilometers and finer fractions reach hundreds of kilometers before falling out of suspension. These particles also raise *dust storms*. The *loess*, which is composed of 0.02 to 0.07 mm sized particles, has been deposited in this manner. Particles finer than 0.002 mm in size, however, possess negligible *fall velocity*. They, therefore, always stay in suspension as the *atmospheric haze*.

SEDIMENT TRANSPORT RATE

The aeolian sediment transport rate is defined by the mass of sediment passing per unit time through a plane of unit width and infinite height perpendicular to the wind. It is measured in kg (m-width)$^{-1}$ s^{-1}. The transport rate, however, is difficult to measure directly, as it is affected in many different ways by the wind shear velocity, near-bed turbulence, surface encrustation, topography, vegetation, and cohesion among grains. Further, the wind shear velocity is hard to measure and near-bed processes likewise are less well understood. These conditions preclude a straightforward relationship between the sediment flow rate and the wind velocity.

Most researchers agree that the wind shear velocity for ideal surfaces and ideal wind conditions provides the best estimate for the sediment flow rate. Ideal surfaces are perceived to be horizontal and totally covered with well-sorted dry sand, and ideal wind conditions refer to a steady wind with a semilogarithmic velocity profile. These conditions, though, are rarely satisfied in nature. Therefore, all relationships predicting sediment discharge from the shear velocity tend to yield values of maximum sediment flux possible in the environment (Livingstone and Warren, 1996).

Bagnold (1941) predicted that the sediment transport rate (Q) is directly proportional to the bed shear velocity (u$_*$) and particle size (d) as

$$Q \propto u_*{}^3 \, d^{1/2}$$

suggesting that a small increase in the wind shear velocity and, thereby, the wind speed increases the sediment flux many times. In terms of the wind speed 1 m above the ground, a 16 m s^{-1} wind moves as much sand in 24 h as does a wind of 8 m s^{-1} in 3 weeks. Similarly, the relationship between the sediment transport rate and wind shear velocity, given as (Livingston and Warren, 1996)

$$Q \propto u_*{}^a \, (u_* - u_* t)^b$$

also provides a reasonable estimate for the sand flow rate. In the above expression, Q is the sand flow rate, u$_*$ is the wind shear velocity, u$_*$t is the fluid threshold velocity, and a and b are constants with a combined value of 3.

DIRECTIONAL VARIABILITY OF SEDIMENT FLUX

The wind is known by the direction from which it blows and the speed of its movement. The frequency of wind from each compass direction as a percentage of total observations describes the *consistency of the wind*. A consistent wind is the most frequent wind. It is also called the prevailing wind. The magnitude of the wind in a given direction (or vector) is obtained by combining the direction and the wind speed for each respective quadrant. This cumulative vector is called the *resultant wind*.

Winds of different magnitude blow from different directions and, thereby, transport sand and dust from as many directions. This directional variability of the sediment transport is a function of the wind speed. The maximum amount of sediment that the wind can transport in a year is called the *drift potential* (DP). It is expressed in vector units (VU). The net amount of sediment moved by the wind from all cardinal directions is the *resultant drift potential* (RDP) in a resultant drift direction.

Fryberger (1980) determined the potential sand drift by the wind energy as

$$Q \propto u^2 \, (u - u_* t) \, t$$

where Q is the potential sand drift, u is the wind velocity in knots (1.994 knots = 1 ms^{-1}), u$_*$t is the fluid threshold velocity, and t is the duration of wind speed as percentage of total time.

The drift potential and resultant drift potential are useful measures of the *wind climate* (Fryberger, 1980). The drift potential is a measure of the wind energy, providing for a classification of the wind into high (DP 401 VU or more), intermediate (DP 200 to 400 VU), and low (DP 199 VU or less) energy wind environments. A *high-energy wind environment* dominates parts of the deserts of Saudi Arabia, Kuwait, and Libya; an *intermediate-energy wind environment* prevails in parts of the deserts of Australia, Mauritania, Russia, Namibia, and Saudi Arabia; and a *low-energy wind environment* is typical of the Kalahari, Sahel, Gobi, Thar, and Takla Makan deserts. The ratio between the resultant drift potential and drift potential (RDP/DP) is a measure of the *directional variability* of the wind. This ratio varies between 0 and 1. The wind is unimodal for RDP/DP 1 and complex for RDP/DP 0.

WIND EROSION

Erosion by the wind is governed by the erosivity of the wind and the erodibility of impacted surfaces. The *erosivity* of the wind owes to the wind energy and duration of the wind. The *erodibility* of the surface is related to the shear resistance of unconsolidated sediments and the strength of rocks in the aeolian environment.

Abrasion, deflation, attrition, and aerodynamic erosion are interdependent processes of wind erosion. *Abrasion* is the impact process of airborne sediments against surfaces, evolving pits, grooves, facets, and patination to a height generally not exceeding 1 m above the surface. *Deflation* is a turbulence-controlled lifting and removal of loose or loosely held dry surface particles in areas particularly devoid of vegetal cover. It produces stone pavements, and hollows in unconsolidated sediments over the vast expanse of flat-lying soft rock surfaces. *Attrition* is the mutual wear of particles while in the wind transport. This process is rather less important to the evolution of landforms. *Aerodynamic erosion* is associated with the interfacial flow in contact with topographic surfaces. This process evolves aerodynamic forms of wind erosion.

Abrasion

Abrasion is the process of rubbing, pitting, and scouring of clasts and rock surfaces by the sand blast action, and friction of dust-sized particles with surfaces in contact with the interfacial flow. Abrasion, in general, is a function of the wind velocity, angularity of particles in the wind flow, and relative hardness of impacting mineral grains and impacted surfaces. Experimental observations suggest that saltating grains abrade the impact surface at an *impact velocity* of 15 m s⁻¹, and that saltating grains plaster onto the surface and protect the target from abrasion at a velocity lower than this critical velocity (Greeley et al., 1985). The *shape* of particles in saltation is also significant to the abrasion of surfaces (Greeley et al., 1984). Observations suggest that the impact of rounded particles against rocks develops surface indentation 10 to 30% larger than the diameter of impacting grains. The angular particles in saltation mode, however, induce inelastic deformation of rocks at the point of contact, opening vertical cracks at the surface of rocks. These cracks soon close in the unloading period of applied stress, and yield to the development of lateral vents. Cycles of loading and unloading stress thus loosen the impact surface along intersecting vertical and lateral cracks by which the rock eventually chips off (Figure 9.4). The *impact angle* of saltating trajectory with the impacted surface is also a variable of significance to abrasion, such that impacted surfaces are gouged at low and crushed at high angles of the particle impact (McKee et al., 1979). The friction of the dust-sized particles in the *interfacial flow* with surfaces evolves flutes and burnished surfaces (Whitney, 1983). The interfacial flow is discussed in the section on aerodynamic erosion.

The susceptibility of surfaces to abrasion (S_a) varies with the *kinetic energy* of impacting grains as (Greeley et al., 1984)

$$S_a \propto D^3 V_p^2$$

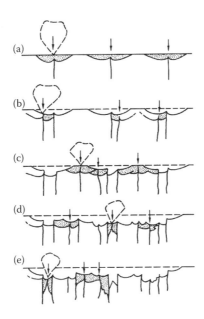

FIGURE 9.4   Possible fracturing and chipping sequence for surfaces impacted with angular particles. (a) Normal loading of the surface by impacting pointed grains generates a zone of inelastic deformation, which is precursor for crushing and/or plastic deformation of surfaces at the point of physical contact of sand grains. (b) At some threshold stress, the inelastic deformation initiates a vertical crack or median vent which, with increasing stress, propagates downward into the body of rock. (c) Chipping results when the lateral vent reaches the surface of the indented specimen. The length of surface trace due to the vertical crack varies directly with the depth of the vent. Upon unloading, the median vent closes but does not heal. (d) It leads to the development of a lateral vent in the zone of inelastic deformation. (e) The impacted surface chips off unevenly along the zone of intersecting median and lateral vents in the body of rock (e).(Source: Figure 28 in Greeley, R., Williams, S., Pollack, J., Marshall, J., and Krinsley, D., *Abrasion by Aeolian Particles: Earth and Mars*, NASA CR-3788 [NASA, 1984]. With permission.)

suggesting that the rate of abrasion increases with the particle size diameter (D) and particle velocity (V_p) in the wind flow.

DEFLATION

Deflation is the process of selective removal usually of 0.1 to 1.0 mm sized loose particles by the wind from bare rough surfaces. Deflation scoops up depressions of variable size, leaves a surface cover of stone pavements from the winnowing of finer sediments, and raises dust storms.

AERODYNAMIC EROSION

Aerodynamic erosion is the process of abrasion caused by the *interfacial flow* on leeward slopes and flanks of topographic surfaces (Whitney, 1978). Interfacial flow

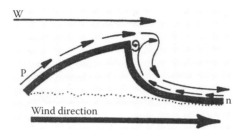

FIGURE 9.5 Diagram illustrating the components of aerodynamic erosion. W refers to the wind direction, P to the positive flow, and n to the negative flow. (Source: Figure 1 in Whitney, M. I., "Role of Vorticity in Developing Lineation by Wind Erosion," *Bull. Geol. Soc. Am.* 89 [1978]: 1–18. With permission from Geological Society of America.)

is initiated when the principal wind accelerates at the upwind face of obstacles, loses contact with the surface, and develops a relatively low pressure at the flow separation. This low pressure attracts a weak interfacial flow beneath the principal wind, which while ascending the surface from a direction opposite that of the principal wind peters out as a vortex near the apex of obstacles initiating flow separation in the first place (Figure 9.5). The pressure difference between the up- and downwind faces of the obstacle also attracts a secondary flow along the flanks of obstacles.

The weak interfacial flow transports dust particles, which groove and burnish the surface. These obstacles evolve the *aerodynamic shape* by the fluid force of the wind (Whitney, 1983). Thus, nearly circular objects increase in the effective length by flank erosion and deposition of sediments in the lee of objects. By comparison, objects that are longer than wide reach the aerodynamic shape by abrasion at the upwind and erosion at the downwind face of topographic surfaces.

DUST STORMS

Dust storm activity is common to semiarid regions of about 200 to 500 mm mean annual rainfall amount. These areas of frequent droughts of protracted duration, sustained above-average maximum temperature, high evaporation rate, reduced vegetation cover, and acute soil moisture deficit are hot spots of accelerated wind erosion and dust production activity (McTainsh et al., 2001, 2005). In essence, much of the global dust is derived from such erodible surfaces in semiarid regions. However, the manner of climate control on dust storm activity is far from understood. Bao et al. (2007) observe that rainfall variations on a centennial rather than decadal timescale statistically correlate well with the frequency of dust storm activity. McTainsh et al. (2005), however, suggest that dust storm activity follows a period of vegetation withering during severe climatic droughts of protected length. Therefore, the absence of vegetation, and drought combine to yield the highest intensity of wind erosion. Other studies observe a direct relationship between land use practices in semiarid regions and production of dust and dust storm activity. Tsoar and Pye (1987) suggest that dust has its source in the erosion of cultivated lands in semiarid regions. Huang and Bao (2001) observe that spatial variations in the frequency of dust storm activity correlate

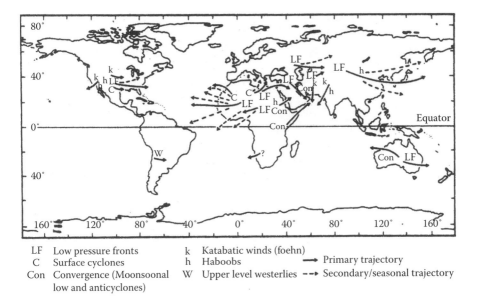

LF Low pressure fronts   k Katabatic winds (foehn)
C   Surface cyclones   h Haboobs   → Primary trajectory
Con Convergence (Moonsoonal W Upper level westerlies --→ Secondary/seasonal trajectory
      low and anticyclones)

FIGURE 9.6 Global map of main synoptic meteorological conditions associated with deflation or dust events. (Source: Figure 6 in Middleton, N. J., Goudie, A. S., and Wells, G. L., "The Frequency and Source Areas of Dust Storms," in *Aeolian Geomorphology*, ed. W. G. Nickling [Boston: Allen & Unwin, 1986], 237–59. With permission.)

critically with the type of land use in the Maowuse Desert, Mongolia. McTainsh et al. (2001) similarly conclude that the intensity of wind erosion in semiarid zones of Australia varies in direct proportion to the extent of cattle grazing activity. This relationship between land use and deflation rate is highlighted in Chapter 12.

Sources of Dust

The dust is commonly distributed between 0.1 and 0.3 mm sized particles. It is produced within the saltation layer, generated from the earth materials, and released in volcanic eruptions (Livingstone and Warren, 1996). The *saltation layer* creates much fine dust from the crushing of coarser particles and blasting of fine clays off the surfaces of quartz grains. *Unconsolidated sediments* in playas, pans, and dry lakes, such as in parts of Tunisia, Algeria, Chad, Nevada in the southwestern United States, and the Lake Eyre region of Australia, are prolific sources of dust in arid and semiarid environments. Fluvial and outwash sediments and the sand overlying emergent continental shelves likewise are rich sources of dust in the environment. *Volcanic eruptions*, however, release a large volume of dust directly into the atmosphere.

Dust-Generating and Dust-Yielding Systems

The dust-generating potential of a site or region depends on atmospheric conditions and erodibility of the earth materials (Middleton et al., 1986). Low-pressure weather fronts, surface cyclones, converging cold air masses, anticyclones, convectional cells, and upper air westerlies are major weather systems that raise dust from surfaces (Figure 9.6). By the comparative potential of dust release, the deserts of Central Asia and Sahara are *primary*, and the deserts of Australia, the southwestern United States, southern South America, and southwestern Africa are *secondary* areas of global dust production.

Playas, pans, outwash, other unconsolidated sediments, overgrazed land, and land cleared of vegetation for agriculture and habitation are ideal dust-yielding surfaces. Observations suggest that the dust-yielding potential of playas and pans varies with the moisture rhythm (McTainsh and Strong, 2007), such that the deflation rate is highest in the driest period of aridity and near zero in the period of pan inundation (Bryant, 2003). Changes in the frequency and extent of natural inundation of such large surfaces thus lead to significant fluctuations in the loading of dust on a seasonal basis. Human activities in semiarid regions also enhance the dust-yielding potential of surfaces at a local and regional scale (Middleton and Goudie, 2001), a few aspects of which were referred to in an earlier section of this chapter. Wiggs et al. (2003) similarly observed that the diversion of surface flow to the Aral Sea has exposed a new ground of several thousand square kilometers to deflation, and enhanced the dust-raising potential of the region by a few orders of magnitude. Optimum conditions for the release of dust, though, are best realized when dust-generating and dust-yielding systems are favorably placed in respect to each other, generating dust at the rate of about 18 to 20 billion t year^{-1} (Livingstone and Warren, 1996).

WIND DEPOSITION

The wind deposits its entrained load by sedimentation, accretion, and encroachment processes. *Sedimentation* occurs when particles drop usually from suspension with insufficient force to induce either saltation or surface creep. Most dust settles in this manner. *Accretion* is the process of accumulation of particles due to loss of forward momentum in creeping grains. *Encroachment* takes place when sufficiently high topographic obstructions ahead of the wind obstruct the movement of some particles.

FEATURES OF WIND EROSION

The earth materials are susceptible in different degrees to abrasion, deflation, and aerodynamic erosion. These wind processes evolve a variety of small- to large-sized forms of erosion in aeolian environments (Table 9.1). Many forms of wind erosion, however, are *polygenetic* or have *inheritance* in the activity of past processes.

GROOVES

Grooves are small- to large-sized features of abrasion and deflation in competent and incompetent materials alike. Small-sized grooves are a 1 mm or so deep and up to 1 m long expression of the *abrasive wear* of rock surfaces by dust particles held in the interfacial flow. Larger grooves exist as active and relict forms of deflation on

TABLE 9.1
Landforms of Wind Erosion

Feature	Size-Class	Dominant Processes	Major Characteristics
Grooves	Small to large	Abrasion and deflation	Small grooves related to the ballistic impact and aerodynamic erosion; large grooves probably caused by deflation
Ventifacts	Small to medium	Abrasion and aerodynamic erosion	Faceted stones and faceted cohesive sediment lumps
Yardangs	Medium to large	Abrasion, deflation, and aerodynamic erosion	Streamlined forms in rocks and cohesive sediments
Pans	Small to large	Deflation	Inland drainage basins in semiarid climate
Deflation depressions	Small to large	Deflation, solution, fluvial erosion, mass movement	Feature of arid lands; closed depressions or blowouts with parabolic dunes; at the foot of pedestal rocks and in the lee of stones at the surface of sand sheets; larger depressions are polygenetic
Stone pavements	Variable	Deflation, freeze-thaw, thermal expansion and contraction, capillary salts	Lag deposits at the surface of loose sediments in arid environments
Pedestal rocks	Small to medium	Deflation and fluvial erosion	Columnar forms of variable shape

cohesive sediments (Warren, 1979). Active grooves occur in areas of intense aridity, such as in parts of Iran, Sahara, and Peru. Inactive grooves are relatively common to the *indurated crust* of northwestern Europe and the Great Plains of the United States. This crust, which is believed to have evolved in the drier phase of the Late Pleistocene by the precipitation of *capillary salts* in sediments laid down by intermittent floods, was subsequently scoured by strong winds of the period. The wind-eroded grooves now support minor valleys in central Kansas.

VENTIFACTS

Ventifacts are up to 3 m long abraded and polished multifaceted stones (Figure 9.7). They are common to arid, periglacial, and polar environments of strong winds, abundant supply of sand, and absence of vegetation cover (Whitney and Splettstoesser, 1982). Ventifacts also evolve in permeable cohesive sediments of moist temperate climate in Nova Scotia, Canada (Laity, 1994), and in playa sediments of the arid environment in California (Laity, 1995).

Ventifacts are explained as a feature of sandblast action and aerodynamic erosion. The *sand blast hypothesis* argues that the upwind face of surface stones lying

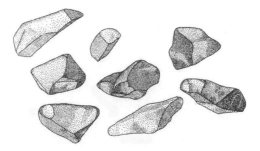

FIGURE 9.7   Ventifacts. (Source: Figure 10.3 in Warren, A., "Aeolian Processes," in *Process in Geomorphology*, ed. C. Embleton and J. Thornes [London: Edward Arnold, 1979], 325–51. With permission from Edward Arnold Ltd.)

undisturbed for a long enough period of time recedes sharply above the base, and develops a flat face by the abrasive wear of impacting grains. The stones subsequently roll or rotate in periods of exceptional flood or strong wind, presenting a new impact face to the wind. This surface similarly recedes and develops a flat face. The rotational mechanism for the faceting of stones finds support from the observation that less mobile large stones in California are single faceted and mobile smaller stones are commonly multifaceted (Laitly, 1995). Besides, the intersecting palimpsests of grooves on ventifacted stones also suggest that stones have rotated in the past (Whitney and Splettstoesser, 1982). Evidence to the contrary, however, suggests that immobile embedded stones are faceted and faceted stones of the same area present no evidence of variable wind conditions (Livingstone and Warren, 1996). Hence, evolution of ventifacted stones by rotation remains suspect.

The *aerodynamic model* proposes that faceting develops from the abrasive wear of dust particles held in the interfacial flow (Whitney and Dietrich, 1973; Whitney and Splettstoesser, 1982). Observations suggest that the *interfacial flow* undercuts leeward margins of stones and generates an upwind moving gentle vortex at the interface with surfaces that eventually destroys the stability of stones (Figure 9.8). The stones, therefore, tumble, presenting a new face to the wind.

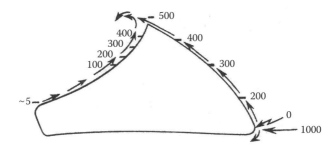

FIGURE 9.8   Wind velocities measured in feet per minutes around a ventifact. Impact velocity was approximately 1000 feet per minute, and the stagnation point is indicated with an O. (Source: Figure 6 in Whitney, M. I., and Dietrich, R. V., "Ventifacts Sculpture by Windblown Dust," *Bull. Geol. Soc. Am.* 84 [1973]: 2561–82. With permission from Geological Society of America.)

Ventifacts also possibly develop from the combined activity of sandblast action and aerodynamic erosion (Breed et al., 1989; Laity, 1994). The sandblast action evolves facets, and the abrasive wear of particles in the interfacial flow develops grooves on faceted stones.

The ventifacted stones and other features of aeolian erosion carry a bright surface coating of the oxides of iron and manganese. It is called *desert varnish*. The constituents of desert varnish have sources possibly in minerals moved in the *capillary suction* (Engel and Sharp, 1958) and those released by *biochemical processes* associated with the vegetation of an upwind area (Dorn, 1986). The desert varnish is datable, providing for the age of ventifacts and interpretation of the environmental change (Dorn and Oberlander, 1982; Dorn, 1995). *Radiometric dating* of the desert varnish on ventifated stones of *valley moraines* in Sierra Nevada suggests a windier and dustier climate for the region 65 ka to 14 ka BP (Bach, 1995; Dorn, 1995).

YARDANGS

Yardangs are narrow asymmetric ridges of rock, clay, or cemented sand in deserts of a *hyperarid environment*. These ridges, which appear similar to inverted ship hulls, occur in fields as a streamlined form of a high upwind blunt face and a long downwind tapering tail aligned to the wind (Figure 9.9). In general, mature yardangs present a perfect *aerodynamic shape*, but those in various stages of development deviate much from this ideal shape attribute (Halimov and Fezer, 1989; Livingstone and Warren, 1996). Notable yardang fields include Rogers Lake in the Mojave Desert, Turkestan in China, Kalut in Iran, Namib Desert in Namibia, the coastal desert of Peru, and the Western Desert of Egypt (Livingstone and Warren, 1996). Yardangs in China, Peru, and Iran are tens of kilometers long and more than 100 m high winderoded ridges (Ward and Greeley, 1984). Yardangs cut in lakebed sediments and cohesive materials are called *mudlions*.

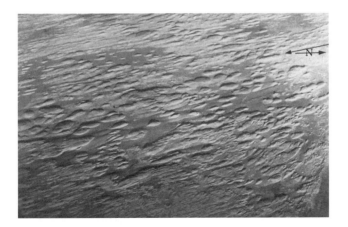

FIGURE 9.9   Yardangs on the limestone plateau of Egyptian Kharga. (Source: Photo courtesy Dr. M. I. Whitney.)

Yardangs evolve from sandblast action or abrasion, deflation, and aerodynamic erosion of the earth materials that vary much in lithology, structure, and climatic history of the region (Laity, 1994). The mudlions of extinct Pleistocene Rogers Lake, California, possibly evolved from abrasion and deflation of moderately consolidated sediments. Ward and Greeley (1984) attribute the initiation of yardang ridges to abrasion and of the aerodynamic shape to deflation and abrasion of sediments. Whitney (1983, 1985), however, observes that grooves along the flanks and windward face of the feature in rock and consolidated sediments do not survive the sandblast action. Therefore, yardangs evolve from aerodynamic erosion of surfaces aligned to the wind. The *aerodynamic model* argues that relatively low pressure over ridges and high pressure within corridors separating yardangs generates an interfacial flow encircling ridges. The dust particles in this flow evolve grooves on the body of yardangs and produce an overall burnished surface of the ridges. The streamlined form results from the fluid force of the wind against obstacles.

Pans

Most pans are periodically inundated shallow depressions of semiarid lands. They commonly occupy sites of interior drainage and comprise soft sediments that are susceptible to *deflation* (Goudie, 1991; Laity, 1994). Livingstone and Warren (1996) note that the pan size is between 0.0004 and 100 km^2 in southwestern Australia and between 0.05 and 30 km^2 in South Africa.

Similar depressions along the eastern coast of the United States, and in parts of the High Plains of Texas, Colorado, and western Kansas, may be either an active or relict feature of former semiarid climate of the area. The deflation depressions in soluble rocks of the arid southern High Plains of the United States, which appear similar to sinkholes, are *polygenetic* in nature (Sabins and Holliday, 1995). The best-known pans of deflation, however, exist along a 1100 km stretch of the Atlantic coast of the United States between Maryland and northern Florida. They have inheritance in the Late Pleistocene climatic aridity of strong dry winds and paucity of vegetation cover (Livingstone and Warren, 1996).

Wind deflation of pan surfaces yields sand and clay sediments and datable carbonate. These clastic and chemical sediments are deposited immediately downwind of the pans as crescent-shaped dunes called *lunettes*. The dated lunettes from a part of southern Australia suggest a history of desiccation before 45 ka and sustained aridity of the region from 17 ka to the present (Bowler, 1976; Oberlander, 1994).

Deflation Depressions

Deflation of a surface of unconsolidated sediments in arid lands produces small- to large-sized deflation depressions. Medium-sized deflation depressions (or blowouts) are integral to the evolution of *parabolic dunes*, and small-sized deflation depressions evolve at the foot of *pedestal rocks* and downwind of large-sized stones held in *sand sheets*. Large-sized closed depressions are polygenetic in nature.

A few hundred kilometers across and tens of meters deep large-sized depressions, such as the P'ang Kiang Hollows in Mongolia, the Big Hollow in the United States, large north African basins, and others of similar nature in Central Asia, are *polygenetic*. They have evolved from the activity of fluvial erosion, solution, salt weathering, and deflation of surface sediments (Livingstone and Warren, 1966). The Qattara Depression of the western Egyptian highlands that covers some 192,000 km^2 area over a part of the old Nile delta, and is 100 m below the msl (mean sea level), is also polygenetic. This gigantic depression, which initially occupied a stream valley, has developed into its present form by solution, mass movement, and deflation processes (Albritton et al., 1990).

Stone Pavements

A mantle of angular to subrounded stones overlying unconsolidated sediments in hot deserts is called a stone pavement. The stone pavements are locally called gobi in Central Asia, hammada in the Sahara, gibber in Australia, and desert armor in the United States. These pavements are a feature either of wind deflation or of process combinations with deflation as the dominant process. The classic hypothesis on stone pavements revolves around deflation of finer sediments from surfaces, which concentrates coarser particles at the surface as lag deposits. Later studies, however, observe that the evidence for deflation is not conclusive in all situations, and that deflation is but one of many possible processes of stone pavements. In Atacama Desert, and in California and in Israel, loose fine material at the surface is much in deficit for deflation to be the effective process of stone pavements. Geomorphic conditions in these deserts strongly indicate removal of fine sediments from above the stone pavements by surface wash activity (Cooke, 1970). The stone pavement in a part of the Mojave Desert is similarly a surface of volcanic debris from which the dust has been washed down by running water (McFadden et al., 1987). Surface disturbance due to thermal expansion and contraction, freeze-thaw activity, and crystallization of capillary salts in arid environments are other possible processes for the evolution of stone pavements (Cooke, 1970; Cooke et al., 1993).

The clastic fragments of stone pavements are coated with desert varnish. The radiometric dates on desert varnish suggest that stone pavements are indeed very old. The stone pavement of Colorado Plateau is more than 300 ka old (Knauss and Ku, 1980). The desert varnish also provides regional data on periods of initiation and cessation of aeolian activity, which suggests that stone pavements are *inherited* from the activity of past processes (Dorn, 1986).

Pedestal Rocks

Pedestal rocks are vertical eminences of a protective cap rock of relatively resistant lithology above a pedestal of a lithologically weak formation. They are explained as a feature of deflation in arid, and deflation and fluvial erosion in semiarid environments. Pedestal rocks in the *badlands* of South Dakota and in southern Alberta, Canada, are called hoodoos, and in the French Alps demoiselles.

FEATURES OF WIND DEPOSITION

Deposition from the wind evolves characteristic small-sized forms of ripples and large-sized features of sand dunes, sand sheets and loess plains. Ripples evolve from the perturbation of a mobile bed of sand, and large-sized features of deposition develop from deformation of the principal wind and interaction between the wind and the mobile bed of sand.

Sand Ripples

Ripples are wave-like asymmetric *bed forms* on dry sandy surfaces. They are aligned perpendicular to the wind. The usual ripples are 0.1 to 5.0 cm high, 0.5 to 2.0 m in wavelength, and approach an ideal 13° of the upwind and 30° of the downwind slope. In test conditions, the ripple height and spacing between successive ripples depends directly on the particle size and the wind shear velocity (Seppälä and Lindé, 1978; Livingstone and Warren, 1996). Ripples, thus, are higher and more widely spaced in coarser than finer sands and are larger in wavelength on gentler than steeper surfaces. In certain wind and topographic conditions, however, the ripples are more than 50 cm high and up to 25 m in wavelength. They are called mega-ripples. Mega-ripples are shaped by the wind of high directional variability (Greeley and Iversen, 1985) and take years to establish into stable bed forms (Ellwood et al., 1975).

Sand ripples evolve from the disturbance of a mobile bed of sand. This bed can be perturbed by Helmholtz waves, creep of sand, and reptation activity. The *Helmholtz waves* are a wave-like phenomenon of unstable wind produced by the relative motion of a denser layer of saltating grains against a less dense sand-free air above the surface. The proponents of the Helmholtz waves hypothesis for the rippled bed presume that ripples evolve by the mobilization and regrouping of bed particles beneath the higher-velocity descending waveform. The existence of Helmholtz waves, however, has not been established so far.

The *creep model* for the evolution of ripples is based on the observation that bed particles of heterogeneous size composition do not travel an equal distance in unit time. Hence, they evolve a rhythmic wave-like pattern by the *steady-state* migration of sand grains across the crest of rippled bed form (Ellwood et al., 1975). The creep-related evolution of ripples, though, remains suspect, as the relationship between saltation trajectory and ripple spacing has not been established so far (Livingstone and Warren, 1996).

The *reptation model* holds promise for the explanation of rippled bed form morphology. Livingstone and Warren (1996) summarize experimental observations on the relationship between ripples and the reptation process. These studies demonstrate that the ripple spacing is in close agreement with the bed shear velocity, and that ripples are spaced six times the mean reptation length. The above relationship, however, falls short of providing a direct process link to the evolution of sand ripples.

Obstacle Dunes

Obstacle dunes evolve from local deformation of the wind by compression ahead of sufficiently high topographic obstructions in areas of shifting sand. They, therefore, are a class apart from classic sand dunes that evolve from interaction between the wind and the mobile bed of sand.

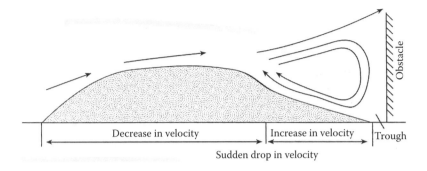

FIGURE 9.10   Development of an obstacle dune.

Compression of the wind follows deceleration of the wind flow. Therefore, the wind sheds a part of its entrained load as echo dunes and climbing dunes some distance upwind of obstacles to the wind flow. The *echo dunes* represent a stable sand ridge parallel to, and some distance from, high-angled cliff faces. The *climbing dunes* are a migratory form of sand deposit at the ramp of obstacles initiating deformation of the wind flow pattern.

Experimental data suggest that the wind sheds a part of its load in the form of echo dunes, where it begins to decelerate and loses energy against cliff faces at least 50° steep (Tsoar, 1983). In test conditions, the trapped sand builds up to a height 0.3 to 0.4 times the height of the obstacle, and at a distance 0.4 to 2.0 times the obstacle height (Figure 9.10). The downwind extension of this deposit, however, is restricted by the reverse eddy flow initiated at the cliff face. Further, any addition of sand to the equilibrium dune form is carried beyond the trough by the vortex for sedimentation at the ramp of obstacles as climbing dunes. Experimental observations suggest that climbing dunes also evolve in situations where eddies poorly develop for low-angled obstacle faces. A few climbing dunes are also identified with the downwind extension of major sand deposits, such as the *outwash* (Seppälä, 1993). The lee-side accumulation of fine sand derived from a *glaciofluvial deposit* in Kevojoki Valley near the Arctic Circle in eastern Finland is one such case in point.

SAND DUNES

An aeolian dune is a 0.3 to 400 m high and between 1 and 500 m wide sand deposit shaped by deflation and deposition into a bed form (Livingstone and Warren, 1996). All dunes pass through processes of initiation, growth, replication, and development of slip faces, the morphodynamic aspects of which are discussed in this section of the chapter.

Interaction between the wind and the mobile bed of sand evolves morphologically distinct crescentic, longitudinal, and complex types of basic dunes, which also distinguish in their alignment to the wind (Table 9.2), wind regime, and the drift potential. Each basic type also comprises a family of dune forms that evolve by deformation of the wind in a particular manner by conditions external and internal to the wind flow.

TABLE 9.2

Basic Dunes and Their Characteristics

Type	Morphology	Relation to Wind
Crescentic (transverse, barchan, sigmoidal, and parabolic)	Heights up to 100 m	All perpendicular to wind: barchan has horns downwind; parabolic has horns upwind. Transverse dunes have relatively straight crest
Longitudinal (seif, parallel, and whaleback)	As much as 100 m high, 1 to 2 km wide and 20 to 200 km long; linear, sinuous; cross-winds increase height and width; form spectacular "windrows" in some continental deserts; may be straight or slightly sinuous; whalebacks are flat-topped	Extend downwind roughly parallel to average wind vector
Complex (pyramidal, hooked, dune "massifs" heaps, star, and reversing)	Many and varied but pyramidal dunes to 200 m or more	Variable, shifting winds plus merging dune forms
Sand sheets	Flat to gently undulating; can cover wide areas	Accretion probably predominates over avalanche

Source: Table 11-8 in Pettijohn, F. J., Potter, P. E., and Siever, R., *Sand and Sandstone* (Berlin: Springer-Verlag, 1972). With permission from Springer-Verlag.

CRESCENTIC DUNES

Crescentic dunes are transverse, barchan, and parabolic in type. They are aligned perpendicular to the wind (Figure 9.11). The crescentic dunes occur in humid temperate zones through to arid tropical regions of low drift potential and low directional variability of the wind (Figure 9.12). They are both an active and relict form of wind deposition.

Transverse Dunes

Transverse dunes are up to 1 km long and 70 m high sinuous asymmetric ridges perpendicular to the wind in humid temperate and arid tropical regions of strong winds and abundant supply of sand. However, they become unstable in the interior of deserts of low sand supply and transform into isolated *barchans* and *seif dunes*.

Transverse dunes are possibly initiated by the affect of *Helmholtz waves* on a mobile bed of sand. Helmholtz waves are produced by a slight change in *surface roughness* or a rare *temperature inversion* in arid environments (Warren, 1979). The supporters of the Helmholtz wave model for transverse dunes hold that the descending wave mobilizes sand grains at the bed, which regroup as a low ridge transverse to the wind (Figure 9.13(a)). In time, the migrating sand achieves *steady-state equilibrium* with the wind climate, cascades into the leeward sheltered zone,

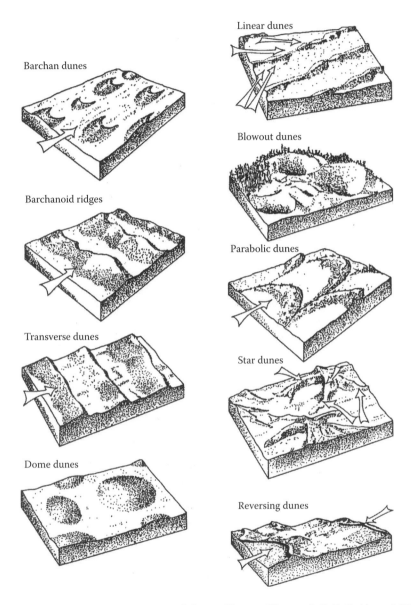

FIGURE 9.11 Some common types of dunes. (Source: Figures 3–5, 7, 9, 11, and 12 in McKee, E. D., "Introduction to a Study of Global sand Seas," in *A Study of Global Sand Seas*, ed. E. D. McKee, U.S. Geological Survey Professional Paper 1052 [U.S. Geological Survey, 1980b], 1–19. With permission from U.S. Geological Survey.)

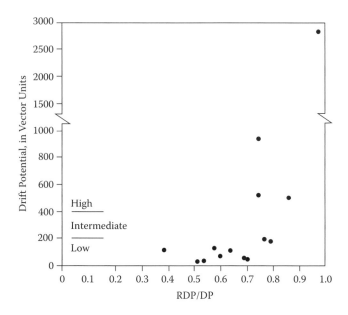

FIGURE 9.12   Wind energy environment of crescentic dunes. Drift potential (DP) versus RDP/DP for fourteen stations represented by dots suggests that transverse (barchanoid) dunes occur in wind environments of low directional variability. Barchanoid dunes in environments with high drift potential (>399 VU) are associated with higher RDP/DP (less directional variability) than are barchanoid dunes in environments with low drift potentials (<200 VU). (Source: Figure 98 in Fryberger, G., "Dune Forms and Wind Regime," in *A Study of Global Sand Seas*, ed. E. D. McKee, U.S. Geological Survey Professional Paper 1052 [U.S. Geological Survey, 1980], 137–69. With permission from U.S. Geological Survey.)

and establishes a downwind sequence of straight parallel ridges perpendicular to the wind (Figure 9.13(b)). The drag of strong wind over rough surfaces, however, disturbs its flow pattern and generates *eddy motion* in the lee of ridges. The oscillatory drag of divergent and convergent eddy flow respectively evolves barchanoid (concave downwind) and linguoid (convex downwind) segments of the ridges (Figure 9.14). The very existence of Helmholtz waves, however, is suspect.

Barchan Dunes

Barchans are crescent-shaped dunes with horns pointing downwind. They are generally 20 to 30 m high and 200 to 600 m wide at the base, with 10 to 15° of a longer convex-up segment facing the wind and 32 to 35° of a shorter concave-up leeward slope down the wind. Barchans are common to tropical deserts of limited supply of sand and unidirectional gentle wind regime.

Barchans are initiated from chance deposition of sand from a sand-laden wind (Bagnold, 1941). This patchy accumulation of greater surface roughness than the surrounding grows by trapping sand from the wind, which on reaching a height of 30 cm from the surface increases the *wind shear* and affects stability of the wind flow over the sand heap. Initially, the wind shifts sand from thin margins of this patchy

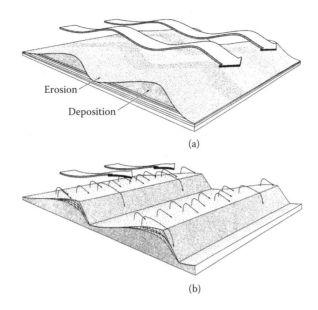

FIGURE 9.13 Oscillatory aerodynamic drag model of a transverse dune formation. (a) A wave-like pattern in the wind erodes and deposits alternate ridges and hollows. (b) Slip faces develop. (Source: Figure 10.10 in Warren, A., "Aeolian Processes," in *Process in Geomorphology*, ed. C. Embleton and J. Thornes [London: Edward Arnold, 1979], 325–51. With permission from Edward Arnold Ltd.)

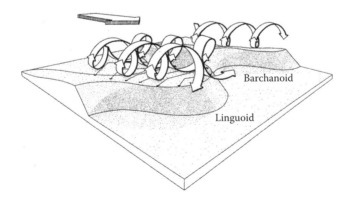

FIGURE 9.14 Distortion of a transverse dune ridge by a longitudinal vortex pair. (Source: Figure 10.11 in Warren, A., "Aeolian Processes," in *Process in Geomorphology*, ed. C. Embleton and J. Thornes [London: Edward Arnold, 1979], 325–51. With permission from Edward Arnold Ltd.)

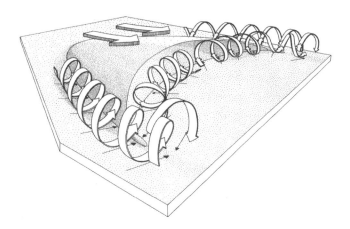

FIGURE 9.15   The pattern of flow around an isolated barchan dune. The flow pattern is reconstructed from the field data collected near Salah, Algeria, by P. Knott. (Source: Figure. 10.13 in Warren, A., "Aeolian Processes," in *Process in Geomorphology*, ed. C. Embleton and J. Thornes [London: Edward Arnold, 1979], 325–51. With permission from Edward Arnold Ltd.)

mass, evolving downwind horns about the body of sand. This *embryo barchan* also disproportionately increases the wind shear more over the main body of sand mass than in the region of horns. The horns, therefore, lengthen downwind into the sheltered leeward zone as a compensation for unequal distribution of the wind shear across the dune. In time, the downwind extension of horns is arrested when the wind shear at horns becomes equal to the wind shear on the main body of the dune. The embryo dune finally evolves into a barchan when sand grains migrating over the upwind segment cascade down the crest, forming the *slip face*.

A vortex system that develops in the lee of the dune explains the steep leeward slope (Figure 9.15) and cross-bedded sedimentary structure of the barchan form (McKee, 1980a). The migration of sand along dunal flanks similarly provides for the growth of new dunes behind slip faces and an echelon arrangement of dunes in the barchan field. Active barchans migrate across the desert terrain, in which the rate of migration depends on the rate of sand supply. Barchans of the Thar Desert, India (Kar, 1993), and of the Salton Sea near California (Haff and Presti, 1995) migrate at the rate of 27 to 32 m year^{-1}.

Parabolic Dunes

Parabolic dunes are partly stabilized U-shaped deposits of sand that characteristically enclose a *deflation hollow* within upwind pointing horns. The dunes rise gently from the blowout and continue beyond the crest as a steep convex leeward slope. The sand of parabolic dunes is derived from sources as diverse as coastal sand, obstacle dunes, sand shields, and outwash.

Parabolic dunes are common to the vegetated margins of deserts of strong winds and a perennial supply of sand. They have been described from parts of South Asia, Saudi Arabia, Namibia, and southwestern United States, and reported from the cold climate vegetated areas east of the Cordillera of western Canada, central United States, Poland, Baltic Sea coast of Latvia and Estonia, Atlantic coast of France, and

FIGURE 9.16   Small-sized coalescent parabolic dunes near the northeastern edge of Thar Desert, India.

coastal Colombia and Venezuela. The parabolic dunes are larger along coastal deserts of strong winds and a perennial supply of sand than toward the interior of deserts of low wind strength and diminishing sand flux, where they join in a rake-like fashion (Figure 9.16). The extensive parabolic dunes of the Thar Desert of India and Pakistan are, on average, 2.6 km long, 2.4 km wide, and tens of meters high, but the dunes along the Colombia-Venezuela border are perhaps the largest, with arms 12 km long (Breed et al., 1980).

Parabolic dunes in the interior of deserts and in temperate zones are *relics* of the Early Holocene climate. Such dunes in the Thar Desert of South Asia evolved in a short humid phase of intensely arid postglacial climate some 5000 years BP as an *anchored form* of deflation (Verstappen, 1970). The Early Holocene dune-building activity in temperate zones of North America and Europe coincided with a major dune-building episode shortly after the retreat of the Pleistocene ice that left extensive areas of glaciofluvial sediments at the surface. This period has been known for strong winds and slow recovery of vegetation cover (Livingstone and Warren, 1996). The material of coastal parabolic sand dunes along the shore of Lake Michigan, however, has been derived from the sand of preexisting wave-cut dunes. These dunes, therefore, suggest a more recent environmental change in the region some 2900 to 2500 calendar years before present (Arbogast et al., 2000; Harman and Arbogast, 2004).

Parabolic dunes are an *anchored form* of deflation around the stabilizing effect of vegetation at the thin margins of sand heaps. Ranwell (1975) discusses a model

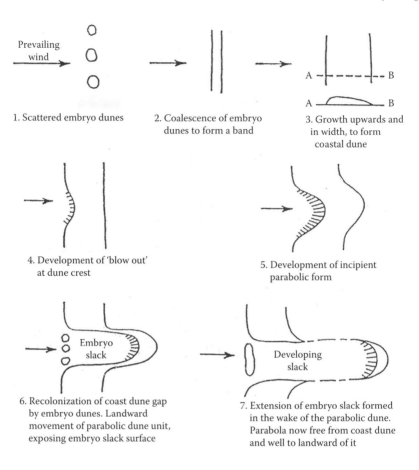

FIGURE 9.17 The formation and development of a parabolic dune unit in a coastal environment. (Source: Figure 50 in Ranwell, D. S., *Ecology of Salt Marshes and Sand Dunes* [London: Chapman & Hall, 1975]. With permission from Kluwer Academic Publishers.)

for the evolution of parabolic dunes in coastal environments of adequate supply of sand and strong winds, suggesting that the sand-laden wind initially sheds a part of its entrained sand as isolated heaps. The heaps coalesce by trapping sand from the wind, forming a ridge transverse to the wind flow. The ridge ceases to grow further on reaching an equilibrium height with the sand flow rate, at which the wind energy is consumed in *deflation* at the bare crest, evolving a blowout and horns pointing upwind (Figure 9.17). The zone of dead sand ahead of dune horns affects sedimentation from the wind, which similarly generates a sequence of sand heap deposition through to fully developed parabolic dunes.

LONGITUDINAL DUNES

Longitudinal dunes are linear, seif, and whaleback bed forms aligned to the wind. *Linear dunes* appear in parallel to subparallel ridges notably in the Kalahari and

Simpson deserts, *seif* are short, sharp-crested sinuous ridges conspicuous in the deserts of Sahara and Arabia, and flat-topped *whaleback* dunes are noteworthy by their height of up to 400 m in the deserts of Sahara and Namibia. In general, longitudinal dunes are some 30% of all sand deposits and 50 to 60% of all sand dunes worldwide.

Linear Dunes

Linear dunes are simple, compound, and complex in morphologic form (Breed et al., 1980). *Simple* linear dunes extend uninterrupted for many kilometers in the direction of dominant wind as nearly equi-spaced single and bifurcated ridges of sharp and subdued crests. On average, linear dunes of the Kalahari Desert are 26 km long and 300 m wide at the base, and of the Simpson Desert are 20 to 25 km long, 10 to 35 m high, and 150 to 300 m apart (Breed et al., 1980). The main ridge of many such dunes is attached to the upwind opening secondary ridges of a shorter length. *Compound* linear dunes carry smaller ridges at crests, and *complex* linear ridges comprise superimposed crescentic and other dunes aligned to the wind. Despite the morphologic variation, the mean wavelength of linear dunes is consistently twice the dune width (Figure 9.18).

Linear dunes evolve in a strong unidirectional wind regime. However, conditions that initiate and evolve such dunes are far from understood (Livingstone and Thomas, 1993). Linear dunes are erosional and depositional, and an active and relict feature of the aeolian landscape. The *erosional mechanism* holds that strong wind over a vast expanse of vegetation-free rough surface of dry loose sand breaks up into a system of isotropic eddies transverse to the wind flow (Folk, 1971; Warren, 1979). The eddy pair of divergent wind sweeps and of convergent wind scoops up sand from the bed, respectively evolving interdunal tracts and sand ridges parallel to the wind (Figure 9.19). Strong deflation by the divergent eddy flow, such as in the Simpson Desert, Australia, has swept the thin sand cover completely from the substratum. Therefore, the linear dunes here are an *active* feature of the present-day wind erosion (Folk, 1971). In contrast, 3 ka to 200 ka old linear dunes of a part of the western Thar Desert of India are *relicts* of the aeolain landscape (Wasson et al., 1983; Kar, 1995). The ridges of convergent eddy flow develop characteristic cross-laminae dipping outward from the dune crest (Figure 9.20). The forked form of dune ridges, referred to earlier, develops when the eddy pair on either side of the dune ridge is replaced by a central cell circulation system.

The *depositional mechanism* for linear dunes is based on deflection, diversion, and funneling of the sand-laden wind ahead of sufficiently high topographic obstacles by which the wind loses *competence* for the transport of a given load (Twidale, 1972). The wind, therefore, sheds its load in the lee of obstacles as lee-linear dunes. The lateral and downwind extension of lee-linear dunes, however, is restricted by *horseshoe vortices* that develop from the interaction between the principal wind and the interfacial flow (Figure 9.21).

Linear dunes become unstable at a critical *biomass density*. Observations suggest that the dune vegetation traps dust from the wind, holds moisture in the sand, binds the surface sand, and generates a silty crust resistant to deflation (Tsoar and Møller, 1986; Thomas and Tsoar, 1990). Algae and lichen on the dune sand likewise provide

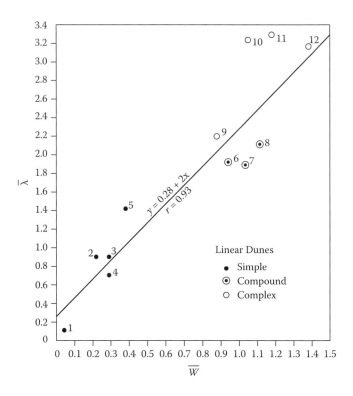

FIGURE 9.18 Scatter diagram showing the correlation of mean dune wavelength (λ) and mean dune width (W) in linear dunes in relation to dune form. Measurements are in kilometers. (Source: Figure 172 in Breed, C. S., Fryberger, S. G., Andrews, S., McCauley, C., Lennartz, F., Gebel, D., and Horstman, K., "Regional Studies of Sand Sea, Using LANDSAT (ERTS) Imagery," in *A Study of Global Sand Seas*, ed. E. D. McKee, U.S. Geological Survey Professional Paper 1052 [U.S. Geological Survey, 1980], 305–97. With permission from U.S. Geological Survey.)

a *biological crust* resistant to the deflation of sand (Yair, 1990). Hence, destruction of biomass by grazing and cultivation practices, and depletion of vegetation in the period of climatic drought weaken the *intrinsic threshold* of dune stability. Livingstone and Thomas (1993) observe that the biomass density varies more with the seasonal and interannual variability of rainfall than with variations in the mean annual rainfall total. Hence, episodic variations in climate are significant to the sustainability of the dune form.

SEIF

Seifs are up to a few kilometer long sharp-crested sinuous sand ridges. They are 12 km long, 660 m wide, and 1.58 km apart from crest to crest in the western Sahara (Breed et al., 1980) and less than 5 km long, 90 to 130 m wide at the base, and 20 to 50 m high in a part of the Paroli Plain along the coast of the Arabian Sea, Pakistan (Verstappen, 1970).

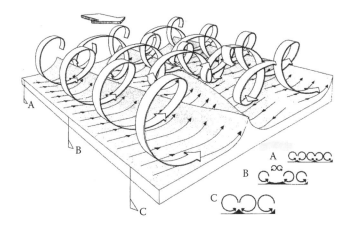

FIGURE 9.19 Helical vortices or Langmuir circulation that may form a linear dune. The central vortex pair rises buoyantly and is replaced from either side to form a Y-junction. (Source: Figure 10.14 in Warren, A., "Aeolian Processes," in *Process in Geomorphology*, ed. C. Embleton and J. Thornes [London: Edward Arnold, 1979], 325–51. With permission from Edward Arnold Ltd.)

FIGURE 9.20 An idealized internal structure of a longitudinal dune. (Source: Figure 1 in Tsoar, H., "Internal Structure and Surface Geometry of Longitudinal (Seif) Dunes," *J. Sediment. Petrol.* 52 [1982]: 823–31. With permission from Society for Sedimentary Geology.)

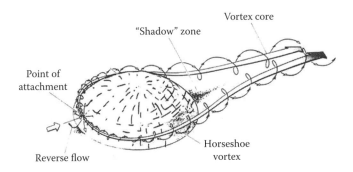

FIGURE 9.21 Horseshoe vortex formed around a low topographic obstruction. Deposition occurs in the shadow zone and on the outside margin of the trailing vortices. Erosion is pronounced beneath the vortex cores. (Source: Figure 6 in Greeley, R., Aeolian landforms: Laboratory simulations and field studies. In *Aeolian Geomorphology*, ed. W. G. Nickling [Boston: Allen & Unwin, 1986], 195–211. With permission from Allen & Unwin.)

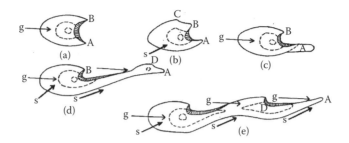

FIGURE 9.22 Transition from a barchan to a seif dune form due to a bidirectional wind regime. (a) An idealized barchan transverse to a unidirectional gentle wind, g. (b) One of the two dune horns, A, shifts downwind of an occasional strong wind, s. (c–d) The horn affected by the wind of this type moves rapidly downwind, and enters the area of sand being moved also off the horn B. (e) The barchan is transformed into a seif by the sand moved off the two horns by the bidirectional wind regime. (Source: Bagnold, R. A., *The Physics of Blown Sand and Desert Dunes* [London: Methuen, 1941]. With permission from Routledge.)

Seifs are widely recognized as a modified form of barchan dunes, and are also justified to have evolved from parabolic dunes. Bagnold (1941) proposed that seifs develop from the modification of barchans by a *bidirectional* wind regime (Figure 9.22). Accordingly, one of the horns (A) of the barchan aligned transverse to the gentle wind (g) shifts downwind of an occasional strong wind (S) and enters the area where the sand is also moving off the other horn (B) of the barchan. The horn (A), therefore, grows rapidly by the wind from both directions and transforms the barchan into a seif dune. The evolution of seifs from barchans is widely accepted (Warren, 1979; Tsoar, 1982). Tsoar (1984), however, argues that the barchan horn is elongated by both gentle and strong winds, and not by the strong wind alone.

The morphology and internal structure of seif dunes are governed by the angle of bidirectional wind approach to the dune crest (Tsoar, 1982). Wind measurements across a seif suggest that the wind arriving at an angle of up to 40° to the dune crest

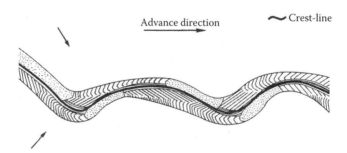

FIGURE 9.23 Model of the internal structure of a longitudinal dune. Lines represent the traces of the laminae on the surface, and dots represent the main depositional areas. Arrows show the wind direction. (Source: Figure 13 in Tsoar, H., "Internal Structure and Surface Geometry of Longitudinal (Seif) Dunes," *J. Sediment. Petrol.* 52 [1982]: 823–31. With permission from Society for Sedimentary Geology.)

deflects parallel to the leeward face at a higher velocity, causing flank erosion. For a higher angle of the incident wind, however, the deflected wind decelerates along the lee flank and sedimentation occurs from the wind. Erosion and sedimentation thus simultaneously occur on both faces of the dune (Figure 9.23). In essence, the wave-like morphology of the seif mimics the effect of incident wind relative to the dune crest.

Seifs such as those along the Arabian Sea coast, Pakistan, and in a part of southeastern Arabian Peninsula, however, are recognized as a complete blowout of the parabolic dunes (Verstappen, 1968). Such dunes in the Thar Desert of South Asia evolved in a *dry phase* of the Holocene climate about 4000 to 5000 years before the present (Verstappen, 1968, 1970).

Complex Dunes

Dome, star, and reversing dunes are major types of complex dunes (see Figure 9.11). *Dome dunes* are up to 10 m high semicircular to elliptical sand mounds in warm-humid coastal areas and continental deserts (Breed and Grow, 1980) of a wind environment similar to that of barchans (Fryberger, 1980). They establish upwind of barchans in unobstructed sites of possibly strong and buoyant wind that interferes with the normal growth of barchans (McKee, 1980a). Dome dunes occur in coastal Brazil, and in parts of Libya, Saudi Arabia, Oman, Yemen, Algeria, the United States, and Central Asia. *Star dunes* comprise three or more arms radiating outward from a high central mass. The dune arms, which possibly reflect sedimentation from the shifting wind, grow by trapping sand from the complex wind regime of moderate to high *drift potential* (Fryberger, 1980). Star dunes are a more than 100 m high, imposing form of the desert landscape in hundreds of square kilometer area in Grand Erg Oriental and Grand Erg Occidental of Algeria (Breed and Grow, 1980). *Reversing dunes* resemble *transverse dunes*, except that the dune crest about a firm base shifts position in response to opposing seasonal winds of equal strength and duration of time (McKee, 1980b). Observations from a part of South Africa suggest that the wind arriving at the reversing dune compresses and decelerates about the dune base and accelerates near the dune crest (Burkinshaw et al., 1993). The accelerated wind erodes the windward face and causes its seasonal displacement downwind. Reversing dunes commonly occur in parts of Central Asia, Namibia, the United States, Brazil, and Antarctica.

Sand Sheets

Many surfaces of thick sand of sizable extent that are without recognizable dune forms are called sand sheets (Breed et al., 1980). Sand sheets carry a surface layer of coarse sand or gravel, or both, as *lag deposits*, which possibly prevent the mobilization of sand into dunes (Breed et al., 1987). They, therefore, are regarded erosional remnants of the previous sand sea (Fryberger et al., 1984; Livingstone and Warren, 1996). The Selima Sand Sheet, covering a 10^5 km^2 area in southern Egypt, Libya, and northern Sudan, is the largest in the world (Livingstone and Warren, 1996).

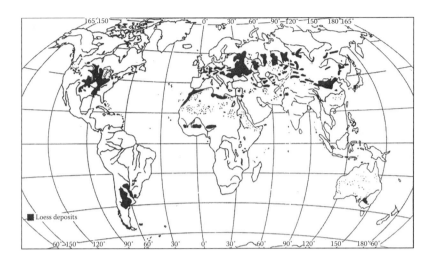

FIGURE 9.24 Principal loess-covered areas in the world. (Source: Figure 4.14 in Livingstone, I., and Warren, A., *Aeolian Geomorphology: An Introduction* [London: Longman, 1996]. With permission from Pearson Education.)

LOESS

Loess is a homogeneous and nonstratified deposit of *dust* that had been carried a long distance from the area of its first suspension and deposited on land and in water bodies. The dust on land was transformed into lightly cemented loess by *digenesis,* involving dissolution of carbonates from the dust (Livingstone and Warren, 1996). Some loess sequences have been either affected by mass movement activity or redeposited by streams.

Loess is distributed in 0.01 to 0.05 mm sized mineral grains of primary origin, and carries chemical sediments of secondary origin. Among the *primary minerals*, quartz is the dominant suite. It is derived from the dust-producing earth materials. The secondary minerals are *authogenic.* They are mostly carbonates and clay minerals, which vary in composition and concentration with the climatic regime. In general, the concentration of carbonates is high in dry and low in humid climatic regimes. The loess of low- and high-latitude climates is rich in kaolinite and illite clay minerals, respectively. The limonite clay mineral, which is comprised of hydrous iron oxide, however, is universal to the loess. It provides a characteristic gray to buff color to the deposit.

Loess occupies some 10% of land area in semiarid and temperate zones between 24 and 55° north and between 30 and 45° south latitudes (Figure 9.24). It is broadly classified as a high-latitude and peri-desert deposit of the wind (Livingstone and Warren, 1996). The *high-latitude loess* is thick and extensive in parts of Central Europe, north-central United States, Argentina, Central Asia, and northern China. The loess thickness, in general, varies directly with the expanse of Pleistocene ice (Shroder et al., 1989), vegetal density downwind of the dust release (Livingstone and Warren, 1996), and pattern of airborne dust sedimentation (Singhvi and Wagner, 1986). The high-latitude loess has diverse sources. The *Quaternary* loess of the Canary Islands (Coudé-Gaussen and Rognon, 1993) and 200 to 300 m thick *Late Pleistocene* loess of China (Li and Zhou, 1993) are derived from the sand overlying emergent continental shelves. The *interglacial* loess of Britain has a source in the outwash (Livingstone and Warren, 1996) and of Morocco in the rocks of the northern Sahara and southern Atlas Mountains (Coudé-Gaussen and Rognon, 1993).

The *peri-desert loess* occupies areas in low latitudes, where it occurs in sequences of reworked thin deposits on mountain slopes and in intermontane basins. The peri-desert loess of the northwestern Himalayas in South Asia is Late Pleistocene to Early Holocene in age (Owen et al., 1992). The loess of Potwar Plateau (Rendell, 1989) and Kashmir Basin (Rendell et al., 1989) is little more than 8 m thick and of Thal Chotiali in Pakistan makes an extensive plain (Pascoe, 1950).

The loess of China is thermoluminescence (TL) dated to be between 2.5 and 3 million years old (Rolph et al., 1993; Livingstone and Warren, 1996). It is the longest surviving continental archive of the environmental change since the onset of the Great Ice Age. The TL-dated sequences of loess provide a chronology of the climate change comparable to that of the *oxygen isotope* record in the deep sea organisms (Mavidanam, 1996).

SUMMARY

The wind is a weak agent of change due primarily to the low density of air. However, it performs the geomorphic work with greatest of ease in bare areas of dry strong winds and abundantly available sand-sized fractions at the surface. The wind flow over rough surfaces is frictional, causing shear in the wind. The wind shear, which provides energy for the geomorphic work, is estimated by the Karman-Prandtl equation.

The wind entrains particles of a given size at a certain critical velocity called the fluid threshold velocity. Once entrained, the particles continue to be moved at a velocity much lower than the fluid threshold velocity. This is called the impact threshold velocity for given sized particles in motion. The wind entrains particles by lift and drag forces, of which drag is a more powerful force in the initiation of particle movement at and in the near-bed region. The lift results from a difference of pressure at and above the bed, and drag is caused by the effect of the fluid force of the wind against particles at rest on a mobile bed of sand.

The wind transports sediments by interdependent saltation, surface creep, reptation, and suspension processes. Saltation is a near-bed leap-and-bounce process of sand transport over a mobile bed of sand. Creep is an extremely slow process of the movement of 0.3 to 0.6 mm sized bed particles at the base of the saltation layer. Hence, creep is difficult to observe and instrument in field studies. Reptation is the process of splash impact of saltating grains against particles on the mobile bed of sand, in which each saltating particle returning to the bed ejects some ten reptating grains from the surface. The collision among grains in the saltation trajectory and bombardment of the bed by saltating particles produce dust-sized particles for transport in suspended mode. The dust-generating potential of a site or region, however, also depends on weather systems, geomorphic environment, and other stress conditions due to human activities at the surface of the earth. The mechanisms producing dust in sufficient quantity, however, are debated. The sediment transport rate depends on the wind climate. It is expressed by the indices of drift potential and resultant drift potential.

Abrasion, deflation, and aerodynamic erosion are interdependent processes of erosion by the wind. Abrasion results from the impact of saltating grains against surfaces and of the dust-sized particles in the interfacial flow with topographic surfaces. Deflation is a turbulence-controlled removal of loose particles of selective size from surfaces. The process also initiates a cascading effect on subsequent movement of the bed particles. Aerodynamic erosion generates the aerodynamic shape of obstacles, and causes their abrasive wear by the dust particles in the interfacial flow.

Grooves, ventifacts, yardangs, deflation hollows, rock pavements, and pedestal rocks are major features of wind erosion. The dynamics of erosional features is debated, but most erosional landforms are the product of more than one process activity. The radiometric dating of desert varnish on ventifacted stones and rock pavements enables interpretation of the environmental change.

Sand ripples are minor, and sand dunes, sand sheets, and loess are major features of deposition by the wind. Sand ripples evolve from the disturbance of a mobile bed of sand by the debated processes related to the Helmholtz waves, surface creep, and reptation. Basic sand dunes and their family of forms are commonly classified in relation to the wind direction as transverse, linear, and complex. The dunes are also recognized as anchored and free form, and as an active and relict feature of deposition by the wind. Sand sheets make extensive plains at the margin of sand seas. Loess is a thick deposit of dust on land, and the longest surviving record of the environmental change since the Pleistocene epoch.

REFERENCES

Albritton, C. C., Brooks, J. E., Issawi, B., and Swedan, A. 1990. Origin of Qattara Depression, Egypt. *Bull. Geol. Soc. Am.* 102:952–60.

Anderson, R. S., and Haff, P. K. 1988. Simulation of eolian saltation. *Science* 241:820–23.

Arbogast, A. F., Hansen, E. E., and Van Oort, M. D. 2002. Reconstructing the geomorphic evolution of large coastal dunes along the southeastern shore of Lake Michigan. *Geomorphology* 46:241–55.

Bach, A. J. 1995. Aeolian modification of glacial moraines at Bishop Creek, eastern California. In *Desert aeolian processes*, ed. V. P. Tchakerian, 179–97. London: Chapman & Hall.

Bagnold, R. A. 1941. *The physics of blown sand and desert dunes.* London: Methuen.

Bao, Y., Achim, B., Zhang, Z., Dong, Z., and Jan, E. 2007. Dust storm frequency and its relation to climate change in northern China during the past 1000 years. *Atmos. Environ.* 41:9288–99.

Bowler, J. M. 1976. Aridity in Australia: Age, origin and expression in eolian landforms and sediments. *Earth Sci. Rev.* 12:279–310.

Breed, C. S., Fryberger, S. G., Andrews, S., McCauley, C., Lennartz, F., Gebel, D., and Horstman, K. 1980. Regional studies of sand sea, using LANDSAT (ERTS) imagery. In *A study of global sand seas*, ed. E. D. McKee, 305–97. U.S. Geological Survey Professional Paper 1052. U.S. Geological Survey.

Breed, C. S., and Grow, T. 1980. Morphology and distribution of dunes in sand seas observed by remote sensing. In *A study of global sand seas*, ed. E. D. McKee, 253–304. U.S. Geological Survey Professional Paper 1052. U.S. Geological Survey.

Breed, C. S., McCauley, J. F., and Davis, P. A. 1987. Sand sheets of eastern Sahara and ripple blankets on Mars. In *Desert sediments, ancient and modern*, ed. L. E. Frostick and I. Reed, 337–60. Geological Society of London Special Publication 35. Oxford: Blackwell.

Breed, C. S., McCauley, J. F., and Whiney, M. I. 1989. Wind erosion forms. In *Arid zone geomorphology*, ed. D. S. G. Thomas, 284–307. New York: John Wiley & Sons.

Bryant, R. G. 2003. Monitoring hydrologic controls on dust emissions: Preliminary observations for Etosha Pam, Namibia. *Geog. J.* 169:131–41.

Burkinshaw, J. R., Illenberger, W. K., and Rust, I. C. 1993. Wind-speed profiles over a reversing transverse dune. In *The dynamics and environmental context of aeolian sedimentary systems*, ed. K. Pye, 25–36. Geological Society Special Publication 72. Geological Society.

Cooke, R. U. 1970. Stone pavements in deserts. *Ann. Assoc. Am. Geogr.* 60:560–77.

Cooke, R. U., Warren, A., and Goudie, A. S. 1993. *Desert geomorphology*. London: UCL Press.

Coudé-Gaussen, G., and Rognon, P. 1993. Contrasting origin and character of Pleistocene and Holocene dust falls on the Canary Islands and southern Morocco: Genetic and climatic significance. In *The dynamics and environmental context of aeolian sedimentary systems*, ed. K. Pye, 277–91. Geological Society Special Publication 72. Geological Society.

Dorn, R. I. 1986. Rock varnish as an indicator of aeolian environmental change. In *Aeolian geomorphology*, ed. W. G. Nickling, 291–307. Boston: Allen & Unwin.

Dorn, R. I. 1995. Alterations of ventifact surfaces at the glacier/desert interface. In *Desert aeolian processes*, ed. V. P. Tchakerian, 199–217. London: Chapman & Hall.

Dorn, R. I., and Oberlander, T. M. 1982. Rock varnish. *Progress Phys. Geogr.* 6:317–67.

Ellwood, J., Evans, P., and Wilson, I. G. 1975. Small-scale aeolian bedforms. *J. Sediment. Petrol.* 45:554–61.

Engel, C. G., and Sharp, R. P. 1958. Chemical data on desert varnish. *Bull. Geol. Soc. Am.* 69:487–518.

Folk, R. L. 1971. Longitudinal dunes of the northwestern edge of the Simpson Desert, Northern Territory, Australia. 1. Geomorphology and grain size relationships. *Sedimentology* 16:5–54.

Fryberger, G. 1980. Dune forms and wind regime. In *A study of global sand seas*, ed. E. D. McKee, 137–69. U.S. Geological Survey Professional Paper 1052. U.S. Geological Survey.

Fryberger, S. G., Al-Sari, A. M., Clisham, T. J., Rizvi, S. A. R., and Al-Hinai, K. G. 1984. Wind sedimentation in the Jafurah sand sea, Saudi Arabia. *Sedimentology* 31:413–31.

Goudie, A. S. 1991. Pans. *Progress Phys. Geogr.* 15:221–37.

Greeley, R., and Iversen, J. D. 1985. *Wind as a geological process*. Cambridge: Cambridge University Press.

Greeley, R., and Marshall, J. R. 1985. Transport of venusian rolling 'stones' by wind? *Nature* 313:717–73.

Greeley, R., Williams, S., Pollack, J., Marshall, J., and Krinsley, D. 1984. *Abrasion by aeolian particles: Earth and mars*. NASA CR-3788. NASA.

Greeley, R., Williams, S., White, B. R., Pollack, J., and Marshall, J. 1985. Wind abrasion on earth and mars. In *Models in geomorphology*, ed. M. J. Woldenberg, 372–422. Boston: Allen & Unwin.

Haff, P. K., and Presti, D. E. 1995. Barchan dunes of the Salton Sea region, California. In *Desert aeolian processes*, ed. V. P. Tchakerian, 153–77. London: Chapman & Hall.

Halimov, M., and Fezer, F. 1989. Eight yardang types in central Asia. *Z. Geomorph.* 33:205–17.

Harman, J. R., and Arbogast, A. F. 2004. Environmental ethics and coastal dunes in western lower Michigan: Developing a rationale for ecosystem preservation. *Ann. Assoc. Am. Geogr.* 94:23–36.

Huang, F., and Gao, Q. 2001. Climate controls on dust storm occurrence in eastern Australia. *J. Arid Environ.* 39:457–66.

Kar, A. 1993. Aeolian processes and bedforms in the Thar Desert. *J. Arid Environ.* 25:83–96.

Kar, A. 1995. Geomorphology of arid western India. *Geol. Soc. India Mem.* 32:168–90.

Knauss, K. G., and Ku, T. 1980. Desert varnish: Potential for age dating via uranium series isotopes. *J. Geol.* 88:95–100.

Laity, J. E. 1994. Landforms of aeolian erosion. In *Geomorphology of desert environments*, ed. A. D. Abrahams and A. J. Parsons, 506–35. London: Chapman & Hall.

Laity, J. E. 1995. Wind abrasion and ventifact formation in California. In *Desert aeolian processes*, ed. V. P. Tchakerian, 295–321. London: Chapman & Hall.

Lancaster, N., and Nickling, W. G. 1994. Aeolian sediment transport. In *Geomorphology of desert environments*, ed. A. D. Abrahams and A. J. Parsons, 447–73. London: Chapman & Hall.

Li, P.-Y., and Zhou, L.-P. 1993. Occurrence and palaeoenvironmental implications of the Late Pleistocene loess along the eastern coasts of the Bohai Sea, China. In *The dynamics and environmental context of aeolian sedimentary systems*, ed. K. Pye, 293–309. Geological Society Special Publication 72. Geological Society.

Livingstone, I., and Thomas, D. S. G. 1993. Modes of linear dune activity and their palaeoenvironmental significance. In *The dynamics and environmental context of aeolian sedimentary systems*, ed. K. Pye, 91–101. Geological Society Special Publication 72. Geological Society.

Livingstone, I., and Warren, A. 1996. *Aeolian geomorphology: An introduction.* London: Longman.

Mavidanam, S. R. L. 1996. Studied on physical basis of luminescence geochronology and its applications. PhD thesis, Nagpur University.

McEvan, I. K. 1993. Bagnold's kink: A physical feature of a wind velocity profile modified by blown sand. *Earth Surface Processes Landforms* 18:145–56.

McFadden, L. D., Wells, S. G., and Jercinovich, M. J. 1987. Influence of eolian and pedogenic processes on the origin and evolution of desert pavements. *Geology* 15:504–8.

McKee, E. D. 1980a. Sedimentary structures in dunes. In *A study of global sand seas*, ed. E. D. McKee, 83–134. U.S. Geological Survey Professional Paper 1052. U.S. Geological Survey.

McKee, E. D. 1980b. Introduction to a study of global sand seas. In *A study of global sand seas*, ed. E. D. McKee, 1–19. U.S. Geological Survey Professional Paper 1052. U.S. Geological Survey.

McKee, T. R., Greeley, R., and Krinsley, D. H. 1979. Simulated aeolian erosion of quartz. In *37th Annual Proceedings of the Electron Microscopy Society of America*, San Antonio, TX, pp. 624–25.

McTainsh, G. H., Chan, Y., McGowen, H., Leys, J., and Tews, K. 2005. The 23rd October 2002 dust storm in eastern Australia: Characteristics and meteorological conditions. *Atmos. Environ.* 39:1227–36.

McTainsh, G. H., Leys, J., and Tews, K. 2001. *Wind erosion trends from meteorological records.* Technical Paper Series (Land) 2. Canberra, Australia: State of the Environment.

McTainsh, G. H., and Strong, C. 2007. The role of aeolian dust in ecosystems. *Geomorphology* 89:39–54.

Middleton, N. J., and Goudie, A. S. 2001. Sahara dust: Sources and trajectories. *Trans. Inst. Br. Geogr.* 26:165–81.

Middleton, N. J., Goudie, A. S., and Wells, G. L. 1986. The frequency and source areas of dust storms. In *Aeolian geomorphology*, ed. W. G. Nickling, 237–59. Boston: Allen & Unwin.

Namikas, S. L., and Sherman, D. J. 1995. A review of the effects of surface moisture content on aeolian sand transport. In *Desert aeolian processes*, ed. V. P. Tchakerian, 269–93. London: Chapman & Hall.

Nickling, W. G. 1988. The initiation of particle movement by wind. *Sedimentology* 35:499–511.

Oberlander, T. M. 1994. Rock varnish in deserts. In *Geomorphology of desert environments*, ed. A. D. Abrahams and A. J. Parsons, 106–19. London: Chapman & Hall.

Owen, L. A., White, B. J., Rendell, H., and Derbyshire, E. 1992. Loessic silt deposits in western Himalayas: Their sedimentology, genesis and age. *Catena* 19:493–509.

Pascoe, E. H. 1950. *A manual of the geology of India and Burma*. 3rd ed., Vol. III. Calcutta: Government of India Press.

Pettijohn, F. J., Potter, P. E., and Siever, R. 1972. *Sand and sandstone*. Berlin: Springer-Verlag.

Ranwell, D. S. 1975. *Ecology of salt marshes and sand dunes*. London: Chapman & Hall.

Rendell, H. M. 1989. Loess deposition during Late Pleistocene in northern Pakistan. *Z. Geomorph.* 76:247–55.

Rendell, H. M., Gardner, R. A. M., Aggarwal, D. P., and Juyal, N. 1989. Chronology and stratigraphy of Kashmir loess. *Z. Geomorph.* 76:213–23.

Rolph, T. C., Shaw, J., Derbyshire, E., and Jingtai, W. 1993. The magnetic mineralogy of a loess section near Lanzhou, China. In *The dynamics and environmental context of aeolian sedimentary systems*, ed. K. Pye, 293–309. Geological Society Special Publication 72. Geological Society.

Sabins, Ty. J., and Holliday, V. T. 1995. Playas and lunettes on the southern High Plains: Morphometric and spatial relations. *Ann. Assoc. Am. Geogr.* 85:286–305.

Seppälä, M. 1993. Climbing and falling sand dunes in Finnish Laplands. In *The dynamics and environmental context of aeolian sedimentary systems*, ed. K. Pye, 269–74. Geological Society Special Publication 72. Geological Society.

Seppälä, M., and Lindé, K. 1978. Wind tunnel studies of ripple formation. *Geografiska Annaler* 60A:29–49.

Shroder, J. F., Khan, M. S., Lawrence, R. D., Madin, I. P., and Higgins, S. M. 1989. *Quaternary glacial chronology and neotectonics in the Himalayas of northern Pakistan*, 275–94. Geological Society of America Special Paper 232. Geological Society of America.

Singhvi, A. K., and Wagner, G. A. 1986. Thermoluminescence dating and its applications to young sedimentary deposits. In *Dating young sediments*, ed. A. J. Hurford, E. Jäger, and J. A. M. Ten Cate, 159–97. CCOP Technical Publication 16. Bangkok: CCOP.

Thomas, D. S. G., and Tsoar, H. 1990. The geomorphological role of vegetation in desert dune systems. In *Vegetation and erosion*, ed. J. B. Thornes, 471–89. Chichester: John Wiley & Sons.

Tsoar, H. 1982. Internal structure and surface geometry of longitudinal (seif) dunes. *J. Sediment. Petrol.* 52:823–31.

Tsoar, H. 1983. Wind tunnel modelling of echo and climbing dunes. In *Eolian sediments and processes*, ed. M. E. Brookfield and T. S. Ahlbrandt, 247–59. Amsterdam: Elsevier Science Publishing.

Tsoar, H. 1984. The formation of seif dunes from barchans—A discussion. *Z. Geomorph.* 28:99–103.

Tsoar, H., and Møller, J. T. 1986. The role of vegetation in the formation of linear sand dunes. In *Aeolian geomorphology*, ed. W. G. Nickling, 75–95. Boston: Allen & Unwin.

Tsoar, H., and Pye, K. 1987. Dust transport and question of desert loess formation. *Sedimentology* 34:139–53.

Twidale, C. R. 1972. Evolution of sand dunes in the Simpson Desert central Australia. *Trans. Inst. Br. Geogr.* 56:77–109.

Verstappen, H. Th. 1968. On the origin of longitudinal (seif) dunes. *Z. Geomorph.* 12:200–20.

Verstappen, H. Th. 1970. Aeolian geomorphology of the Thar and palaeo-climates. *Z. Geomorph.* 14:104–20.

Ward, A. S., and Greeley, R. 1984. Evolution of the yardangs at Rogers Lake, California. *Bull. Geol. Soc. Am.* 95:829–37.

Warren, A. 1979. Aeolian processes. In *Process in geomorphology*, ed. C. Embleton and J. Thornes, 325–51. London: Edward Arnold.

Wasson, R. J., Rajaguru, S. N., Misra, V. N, Aggarwal, D. P., Dhir, R. P., Singhvi, A. K., and Rao, K. K. 1983. Geomorphology, Late Quaternary stratigraphy and palaeoclimatology of the Thar dune field. *Z. Geomorph.* 45:117–51.

Werner, B. T. 1990. A steady-state model of wind-blown sand transport. *J. Geol.* 98: 1–17.

Whitney, M. I. 1978. Role of vorticity in developing lineation by wind erosion. *Bull. Geol. Soc. Am.* 89:1–18.

Whitney, M. I. 1983. Eolian features shaped by aerodynamic and vorticity processes. In *Eolian sediments and processes*, ed. M. E. Brookfield and T. S. Ahlbrandt, 223–45. Amsterdam: Elsevier Science Publishing.

Whitney, M. I. 1985. Yardangs. *J. Geol. Edu.* 33:93–96.

Whitney, M. I., and Dietrich, R. V. 1973. Ventifacts sculpture by windblown dust. *Bull. Geol. Soc. Am.* 84:2561–82.

Whitney, M. I., and Splettstoesser, J. F. 1982. Ventifacts and their formation: Darwin Mountains, Antarctica. In *Proceedings of the International Society of Soil Science: Aridic Soils and Geomorphic Processes*, Jerusalem, Israel, March 29–April 4, 1982. pp. 175–94.

Wiggs, G. F. S., O'Hara, S. L., Wegerdt, J., Van der Meer, J., Small, I., and Hubbard, R. 2003. The dynamics and characteristics of aeolain dust in dryland Central Asia: Possible impacts on human exposure and respiratory health in Aral Sea basin. *Geogr. J.* 169:142–57.

Willetts, B. B., and Rice, M. A. 1986. Collision of quartz grains with a sand bed: The influence of incident angle. *Earth Surface Processes Landforms* 14:719–730.

Williams, J. J., Butterfield, G. R., and Clark, G. D. 1990. Rates of aerodynamic entrainment in a developing boundary layer. *Sedimentology* 37:1039–48.

Yaalon, D. H. (ed.) 1982. Aridic soils and geomorphic processes. *Catena Suppl 1,* 103–115.

10 Karst

Geomorphologists first used the term *karst* for typical surface and subsurface features of *solution* in pure limestone of the Dinaric Alps and the plateau of Istaria bordering the Adriatic Sea. The term now applies to *solution processes* and *landforms* in chalk, limestone, and dolomite rocks that are largely comprised of *calcite*, a mineral readily soluble in the solution of carbon dioxide in water called *carbonic acid*. The solution effects, however, are more pronounced in areas where surface water is diverted to subterranean routes through selectively corroded conduits in sufficiently thick *limestone beds*. The subsurface solution enlarges hydraulically efficient joint and fissure openings, and creates many others within the rock, establishing an integrated network of vertical, horizontal, and diagonal passages to the depth of *phreatic water table* in karst. Hence, karst topography is diagnostic of a developed subsurface drainage and disrupted or depleted drainage at the surface.

Karst landforms evolve by the subcutaneous flow of solution and reaction of carbonic acid with the calcite mineral, requiring applications of the principles of hydraulics to the fluid flow in subterranean passages and of the chemistry of carbonic acid reacting with karst. These applications essentially integrate karst geomorphology with hydraulics and chemistry (Palmer, 1984).

TYPES OF KARST LANDSCAPE

Karst landscape is identified with the locus of solution activity, disposition of soluble rocks, and climatic regime. The karst at the surface is exhokarst and beneath is endokarst (Bögli, 1980). The *exhokarst* comprises the surface forms of solution and related near-surface features of dissolution activity, and the *endokarst* represents the subsurface forms of solution to the depth of karst phreatic zone and the precipitates of solute load in limestone caves.

The *stratigraphic position* of soluble and insoluble rocks distinguishes denuded and mantled karst. The *denuded karst* develops on exposed limestone beds, evolving sinkholes and blind valleys as characteristic form elements of the landscape. The *mantled karst*, which underlies impermeable strata, is typical of karst valleys and gorges.

Karst is true karst, fluviokarst (nival karst), tropical karst, and semiarid and arid karst by the regional climate (Sweeting, 1972). *True karst* represents an extensive landscape of closed and linear forms of solution depressions in denuded karst (Jennings, 1985; White, 1990). *Fluviokarst* evolves in up to 50% pure karst, where insoluble impurities clog solution openings in the limestone terrain and generate stronger fluvial than solution activity at the surface. *Glaciokarst* of clints and stepped pavements is the relic of a glacial (Bögli, 1980) and uncommonly of a periglacial modification of the existing karst landscape. The glaciokarst in a part of Tibet, however, is in equilibrium with the contemporary periglacial processes of the region (Sweeting et al., 1991; Zhang, 1998). *Tropical karst* is the imposing landscape of accentuated limestone residuals of cone karst and tower karst features. The karst of *semiarid and arid climates* is the relic of former humid environments. Sinkholes in Morocco (Ford, 1980); collapsed sinkholes, subterranean passages, and near-surface caves 30 to 32 ka old in Sahara and Arabia (Oberlander, 1994); and sinkholes and poljes in a part of the Indian Thar Desert (Kar, 1995) are among the relics of solution in semiarid and arid environments.

GROUNDWATER IN KARST

In *limestone rocks*, the surface flow is diverted to subterranean routes through up to 3 mm wide hydraulically active karst openings along joints, faults, and bedding plane separations. The subsurface solution openings progressively widen by dissolution and establish a widespread network of short and abruptly terminating vertical, lateral, and horizontal passages within the body of rock until, at a certain depth, the network integrates into a system of extensive horizontal passages about the level of water table in karst. The rock depth above the water table is the *karst vadose zone*, and that about the water table is called the *karst phreatic zone*. The gravity flow of water in the karst vadose zone is understandably *turbulent*, in which the rate and direction of flow depends on fissure frequency and fissure alignment, dip of limestone beds, and relief of the karst terrain. The karst phreatic zone represents water-filled passages and saturated pore spaces in the body of rock. The seasonal and aperiodic flux of surface water into the karst, though, temporarily raises the water table, producing an *intermediate karst zone* of water table fluctuation.

DISSOLUTION MECHANISM

Calcite ($CaCO_3$) and dolomite ($CaMg(CO_3)_2$) are two of the most significant karst minerals, which readily dissolve in carbonic acid (H_2CO_3) or the water solution of carbon dioxide. Carbon dioxide (CO_2) is derived directly from the *atmosphere* and imbibed from the *decomposition* of organic matter at the surface and within the body of limestone. In general, the *biogenic* (oxidative) source is the largest pool of CO_2 for the dissolution of calcite-rich rocks.

The dissolution of karst involves interrelated air (gas), water (aqueous), and rock (solid) phases of the chemical environment (Figure 10.1). The ordinarily insoluble calcite reacts with the carbonic acid to form water-soluble *bicarbonate* as

$$CaCO_3 \text{ (solid)} + CO_2 \text{ (gas)} + H_2O \text{ (aqueous)} \rightarrow Ca(HCO_3)_2 \text{ (aqueous)}$$

Calcite	Carbon dioxide	Water	Calcium bicarbonate

The reaction between calcite and carbonic acid is *reversible*, and proceeds in stages through interdependent equilibrium constants of each reaction state. The dissolution of calcite (Chapter 3) may be given in a simple form as

$$CO_2 + H_2O \leftrightarrow H_2CO_3$$

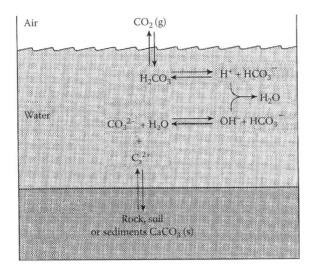

FIGURE 10.1   Reactions among three phases (air, water, rock) of the $CO_2/HCO_3^-/CO_3^{2-}$ system. (Source: Figure 8-1 in Baird, C., *Environmental Chemistry* [New York: W. H. Freeman and Co., 1995]. With permission from W. H. Freeman and Company.)

The carbonic acid dissociates into hydrogen (H^+) and bicarbonate (HCO_3^-) ions as

$$H_2CO_3 \leftrightarrow H^+ + HCO_3^-$$

The reaction between calcite ($CaCO_3$) and carbonic acid (HCO_3^-) liberates the ions of calcium (Ca^{2+}) and bicarbonate (HCO_3^-) as

$$CaCO_3 + CO_2 + H_2O \leftrightarrow Ca^{2+} + 2HCO_3^-$$

In this case, the molar concentration of ionic species stays in equilibrium with the carbon dioxide content in solution.

SOLUBILITY OF KARST

In general, the solubility rate of pure karst varies with the type of dissolution system and volume of carbon dioxide per unit volume of solution in the system. The solubility of karst depends on equilibrium or open and anaerobic or closed types of dissolution system states (Smith and Atkinson, 1976; Bögli, 1980; Ford and Drake, 1982; Palmer, 1990; White, 1990; Sheen, 2000). The *equilibrium solubility* results from continuous replenishment of the carbon dioxide being consumed in the dissolution of karst. Hence, the concentration of *ionic species* in karst solution remains in equilibrium with the solubility rate of karst. The *anaerobic solubility* refers to a closed-system interaction between the one-time input of carbon dioxide in the system and its reaction with the karst. The solubility rate of karst, therefore, progressively decreases as the content of carbon dioxide is consumed through time. Experimental

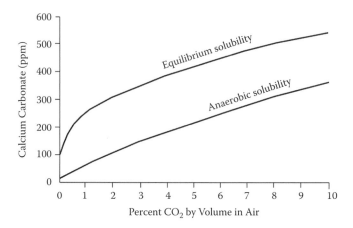

FIGURE 10.2 Calcium carbonate solubility at 10°C. (Source: Figure 13.1 in Smith, D. I., and Atkinson, T. C., "Process, Landforms and Climate in Limestone Regions," in *Geomorphology and Climate*, ed. E. Derbyshire [London: John Wiley & Sons, 1976], 367–410. With permission from John Wiley & Sons.)

data on the ionic concentration of calcium at 1 atm and 10°C solution temperature similarly suggests that the solubility rate of karst is higher for the equilibrium than anaerobic system state (Figure 10.2).

The solubility of karst depends on the volume of carbon dioxide held in karstic systems. The volume of carbon dioxide per unit volume of water varies with water temperature, partial pressure of carbon dioxide, and pH of the chemical environment. In general, the calcite-rich rocks are more soluble in cooler than warmer waters of the earth. All other conditions being equal, the solubility of calcite is 28% higher in solution at 0°C than at 10°C (Smith and Atkinson, 1976). Thus, diurnal and seasonal temperature lowering within 10 to 15 m depth of the *karst vadose zone* and cooling of *karst spring discharge* at the surface enhance the *open-system solubility* of calcite. Limestone caves in Budapest, Hungary (Bögli, 1980), and in parts of the Tibetan Plateau (Zhang, 1995), have evolved by the cooling effect of hot spring discharges within the Karst and attendant increase in the solution aggressiveness.

Carbon dioxide is only about 0.033% by volume in the atmosphere, which exerts a partial pressure (PCO_2) of 0.0003 atm at the sea level. The *biogenic release* of carbon dioxide from the decomposition of organic matter carried by the surface flow into the karst vadose zone, however, remarkably increases the volume of the CO_2 and PCO_2 in the system. In comparative terms, the *oxidation* of organic matter releases three hundred times more volume of carbon dioxide than is ordinarily present in the atmosphere (Moore, 1968) and generates up to three orders of magnitude higher PCO_2 in the karst vadose zone than at the surface of karst (Smith and Atkinson, 1976; Bögli, 1980; Ford and Drake, 1982). The subsurface solution thus always remains an aggressive solvent of karst.

The solubility of karst is *pH dependent*: it oscillates between the dissolution and depositional phases of calcite in equilibrium with the pH of chemical environment. The pH refers to the hydrogen ion activity in the aqueous phase (see the "Dissolution

Mechanism" section in this chapter). In a general sense, the increase in the concentration of bicarbonate ion in the solution decreases the pH, and the ions of calcium or magnesium, or both, in the solution raise it. Hence, a critical shift in the solution pH produces a large change in the direction and rate of process activity. The solution is *acidic* at pH less than 7, at which karst dissolves. However, the solution is *alkaline* at pH of more than 7, precipitating calcite in the environment.

Depth Variation

The concentration of ionic species remarkably changes in the karst solution as it moves from surface to the subsurface depth. This depth variation in the molar concentration of calcium and bicarbonate ions remarkably affects the pattern of dissolution and precipitation of calcite. The variation of solution activity with depth is related to the mixing effect of solutions of same or different hardness values, and CO_2 generating and PCO_2 enhancing subsurface decomposition of organic matter in karstic systems.

The solubility of karst is a nonlinear function of the partial pressure of carbon dioxide. Moore (1968) observed that mixing solutions of the same or different *hardness* invariably produces an *undersaturated solution* of varying karst dissolution potential (Figure 10.3). This effect is most pronounced in the *intermediate zone* of water table fluctuation, where the mixing of vadose and phreatic waters solutionally enlarges horizontal rather than vertical passages in the body of karst. The nonlinear relationship between calcite solubility and partial pressure of carbon dioxide is central to hypotheses on the evolution of certain karst landforms.

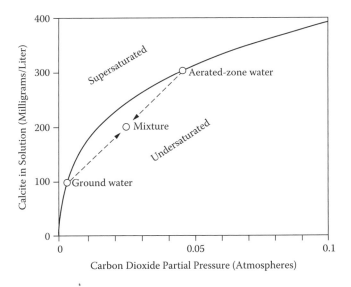

FIGURE 10.3 Solubility curve of calcite in solution at various partial pressures of carbon dioxide. (Source: Figure 3 in Moore, G. W., "Limestone Caves," in *The Encyclopedia of Geomorphology*, ed. R. W. Fairbridge [New York: Reinhold Book Corp., 1968], 652–53. With permission from Kluwer Academic Publishers.)

Organic matter is carried into the karst by the surface flow diverted to subterranean routes, where its decomposition releases *biogenic* carbon dioxide. Wood (1985) showed that the volume of biogenic carbon dioxide increases with the depth increase of particulate and dissolved organic matter in the karst vadose zone (Figure 10.4). Hence, continuously replenishing carbon dioxide in the system, higher PCO_2, and lower solution pH make the solution an *aggressive solvent* of the subsurface karst. The solution, however, becomes saturated with the solute load of calcium and another seventy-odd ionic species about the level of *limestone caves*. The solution reaches saturation by the escape of carbon dioxide from the flow in cave passages and from the dripping of solution in air-filled runway voids.

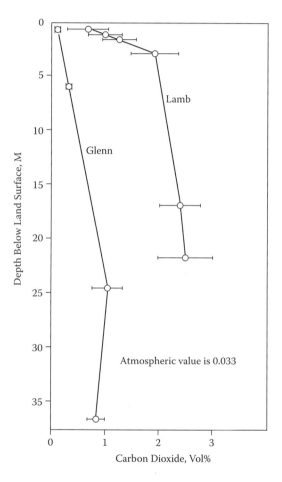

FIGURE 10.4 Observed CO_2 distribution in undersaturated zone from two sites in west Texas. Error bands represent one standard deviation of concentration based on eighteen samples at each depth, over a 1-year period. (Source: Figure 2 in Wood, W. W., "Origin of Caves and Other Solution Openings in the Undersaturated (Vadose) Zone of Carbonate Rocks: A Model for CO_2 Generation," *Geology* 13 [1985]: 822–24. With permission from Geological Society of America.)

CHEMICAL DENUDATION

The dissolution of karst yields ions of either calcium (Ca^{2+}) or magnesium (Mg^{2+}), or both, in solution, the molar concentration of which in stream flow and karst spring discharge is a measure of water hardness. The *water hardness* varies with the purity of rock, open- and closed-system solubility of karst, partial pressure of carbon dioxide in the system, water temperature, and flow volume. The water hardness thus also provides a reasonable estimate of the chemical denudation rate in karst drained by streams. The empirical equations $Q_m T/A_d$ (10^6 r) n and 4 Q_m/A_d 100 n, in which Q_m is the mean annual runoff, T is the mean hardness, A_d is the basin area, r is the density of soluble rocks, and n is the proportion of basin area in soluble rocks, are commonly used variables for estimating the *regional rate* of chemical denudation in karst terrain (Smith and Atkinson, 1976). At *local scales*, however, the chemical properties of soils strongly influence the dissolution rate. Observations on solution lowering of the hillslope of a *mantled karst* near Sheffield, England, suggest that the solution moving through *acidic soils* intensely dissolves the limestone beneath, but neutralizes in contact with *alkali soils* lower down the slope, affecting the net denudation rate in the system (Crabtree and Trudgill, 1987).

CLIMATE AND CHEMICAL DENUDATION RATES

The solubility of carbon dioxide was earlier observed to be higher at cooler than warmer water temperature. The systemic response of karst dissolution rate on climate, however, is debated. Sweeting (1972) argues that the solubility rate of karst progressively increases in the direction of warm moist climates, while Corbel (1959; Smith and Atkinson, 1976) holds that the solution is a more aggressive solvent of the karst in cooler climates of higher carbon dioxide holding capacity. Given the two opposing views, Smith and Atkinson (1976) statistically analyzed the grouped *solution hardness* data on spring and stream flow discharges in different climatic regimes and concluded that the solution hardness does not significantly differ between climatic regimes (Table 10.1). Analysis of solution hardness in stream flows similarly suggests that 50% variation in solution hardness and 97% variation in chemical denudation rate are governed by the magnitude of stream flow alone (Ford and Drake, 1982). Hence, the solubility rate of karst depends on the average stream discharge (Figure 10.5), suggesting further that differences in karst landscape characteristics are independent of the climatic regime (Zambo and Ford, 1997).

The karst landscape of humid tropics, however, is more accentuated than that of the temperate and cold region environments, requiring an explanation. Smith and Atkinson (1976) view regional variations in the karst landscape as due to long-term stability of the climate in the tropics rather than in other climatic provinces, suppression of the chemical activity in dry phases of the Pleistocene climate in temperate zones, and waterlogged conditions above the frozen ground in cold region environments. Ford and Drake (1982), however, explain the regional difference in karst landscapes as due to greater solubility of calcite in open- (coincident) than closed- (sequential) system states (Figure 10.6). Water hardness data without discrimination to the source in spring and stream discharges similarly suggest temperature-

TABLE 10.1

Mean Hardness and Distribution of Erosion in Relation to Climate

Climate	Mean Hardness (mg^{-1})	Coefficient of Variation (%)	Sample Size	Remarks
Arctic/alpine	82.6	42.4	43	Dissolution up to a depth of 10 cm in regolith; diffused CO_2 from atmosphere; general lack of biogenic CO_2
Temperate	210.5	41.1	154	Similar to that in the tropics but abundance of sinkholes and related features; interrupted climatic conditions; biogenic release of CO_2 and lower temperature conditions
Tropical	174.0	14.0	34	Dissolution distributed throughout the depth of karst; long uninterrupted climatic conditions; higher concentration of biogenic CO_2 and higher temperature conditions

Source: Abstracted in Smith, D. I., and Atkinson, T. C., "Process, Landforms and Climate in Limestone Regions," in *Geomorphology and Climate*, ed. E. Derbyshire (London: John Wiley & Sons, 1976), 367–410. With permission from John Wiley & Sons.

dependent variations in karst dissolution rates that are more intense in the tropics than in the temperate, and the arctic and alpine environments (Figure 10.7). The tropics are known to generate a large volume of carbon dioxide from the oxidation of organic matter and root respiration. Sheen (2000) also concludes that the *denudational potential*, given by the type of system state, soil carbon dioxide concentration, carbonate rock type, runoff, and air temperature amplitude, is higher in the tropics than in other climatic regimes.

MECHANICAL AND BIOCHEMICAL EROSION

Mechanical and biochemical erosional activities are rather insignificant to the evolution of karst landscape. Surface pits from *rain beat action* on denuded karst, and abrasive wear of the walls of solutionally enlarged conduits and passages carrying *suspended particles* in the flow are perhaps the only two observed micro forms of *mechanical erosion* in karst landscapes. The *biochemical erosion* manifests in the form of solution pans, and jagged spongy pinnacles called phytokarst that develop beneath the wet root zone of microscopic algae, lichen, and moss colonizing bare karst.

OTHER FACTORS

Soluble rocks yield to *karstification* in different degrees. *Chalk* and *reef limestone* possess low strength and large primary porosity, *dolomite* is a rock of high strength

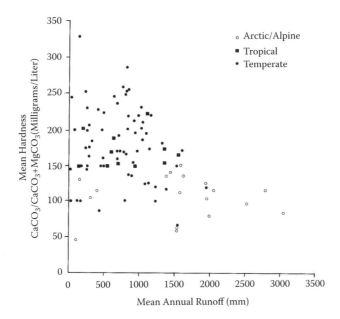

FIGURE 10.5   Mean hardness of spring and river waters plotted against mean annual runoff. (Source: Figure 13.6 in Smith, D. I., and Atkinson, T. C., "Process, Landforms and Climate in Limestone Regions," in *Geomorphology and Climate*, ed. E. Derbyshire [London: John Wiley & Sons, 1976], 367–410. With permission from John Wiley & Sons.)

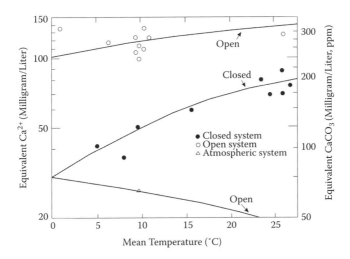

FIGURE 10.6   Lines for the equilibrium Ca^{2+} concentration for the coincident and sequential systems with PCO_2 given as a function of temperature. The data selected from literature represent groundwater hardness. (Source: Figure 7.9a in Ford, D. C., and Drake, J. J., "Spatial and Temporal Variations in Karst Solution Rates: The Structure of Variability," in *Space and Time in Geomorphology* [London: George Allen & Unwin, 1982], 147–70. With permission from George Allen & Unwin.)

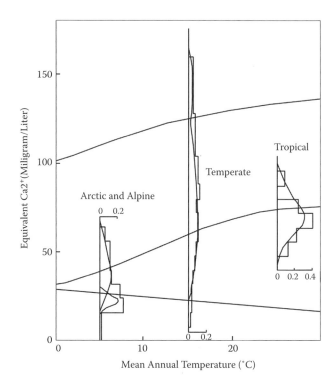

FIGURE 10.7 Plots of solute concentration from diverse sources of water in limestone in arctic/alpine, temperate, and tropical environments. The equilibrium concentration of calcite for the data from Smith and Atkinson (1976) is shown as histograms. The arctic and alpine data show equilibrium solubility of karst, probably resulting from the recharge through waterlogged soils. (Source: Figure 7.9c in Ford, D. C., and Drake, J. J., "Spatial and Temporal Variations in Karst Solution Rates: The Structure of Variability," in *Space and Time in Geomorphology* [London: George Allen & Unwin, 1982], 147–70. With permission from George Allen & Unwin.)

but of low secondary permeability, and *limestone* is a rock of high secondary permeability and adequate strength for the support of solutionally widened structures of subterranean passages along joint and bedding plane separations. Hence, sufficiently thick limestone rocks satisfying requirements of secondary permeability and sufficient head for the circulation of solution are ideal candidates for the development of karst.

KARST LANDFORMS

Karst is a various ensemble of small- to large-sized forms of dissolution at and near the earth's surface, and of dissolution and precipitation of calcite at about the level of phreatic water in karst (Table 10.2). The karst at or near the surface (or exhokarst) comprises linear and closed depressions and accentuated residual forms, and the subsurface karst (or endokarst) carries inactive and active cave systems.

TABLE 10.2
Classification of Karst Landforms

Locus	Scale	Principal Landforms	Dependency
Earth's surface	*Small*: Greatest dimension (e.g., length) commonly <10 m	Karren—Solution pits, channels, and grooves	Varies; types dependent and independent of groundwater systems exist
	Intermediate: Greatest dimension commonly 10 to 1000 m	Closed depressions— Dolines (sinkholes) and compound forms (uvala) Upstanding residuals— Karst cones and towers	Dependent; the existence of efficient system of groundwater discharge is essential
	Large: Greatest dimension commonly >1000 m	Polje, dry valleys, and gorges	Mantled karst
Underground	*All scales*	Cave systems	Not applicable

Source: Table 1 in Ford, D. C., "Threshold and Limit Effects in Karst Geomorphology," in *Thresholds in Geomorphology*, ed. D. R. Coates and J. D. Vitek (London: George Allen & Unwin, 1980), 345–62. With permission from George Allen & Unwin.

KARREN

The solution activity on bare and beneath the humus-rich soils in *steeply inclined* limestone beds develops small-sized linear forms of selective dissolution called karren or lapiés (Figure 10.8). The *surface solution* on bare rocks etches grooves and flutes, and enlarges joints in limestone pavements. The aggressive solution beneath *humus-rich soils*, however, rapidly enlarges joints in rock pavements. These widened joints, which are generally 1 to 4 m deep and 20 to 30 cm wide in bare bed and up to 50 cm wide in soil-covered limestone beds, are called grikes. The solution acts vertically and laterally beneath the humus-rich soils. The vertical and deep solution corrodes solution notches, called undercut karren, on inclined limestone beds (Figure 10.9 left). The solution acting laterally, however, evolves up to 15 cm deep steep-sided solution pans (Figure 10.9, right). The solution at the rootlet of boring plants like algae and moss vigorously etches the rock to the depth of root penetration, evolving up to 1.5 m high pinnacle called phytokarst. The solution beneath

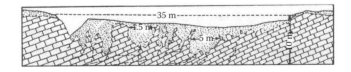

FIGURE 10.8 Lapiés formed in limestone covered by residual clay in a quarry near Generalski Sto on the railway line Zagreb-Rijeka in Croatia. A doline filled with clay forms when lapiés are destroyed. (Source: Figure 21 in Cvijić, J., "The Evolution of Lapis," *Geogr. Rev.* 14 [1924]: 26–49. With permission from American Geographical Society.)

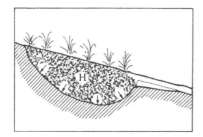 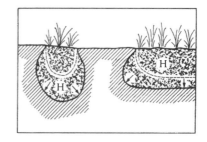

FIGURE 10.9   Cross section of undercut karren (left) and solution notch (right). Broken lines indicate the initial stage and H refers to humic soil. (Source: Figure 3.6 in Bögli, A., *Karst Hydrology and Physical Speleology* [Berlin: Springer-Verlag, 1980]. With permission from Springer-Verlag.)

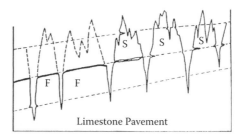

FIGURE 10.10   Morphology of pinnacles (S) and clints (F). (Source: Figure 3.8 in Bögli, A., *Karst Hydrology and Physical Speleology* [Berlin: Springer-Verlag, 1980]. With permission from Springer-Verlag.)

the vegetation cover of humid tropics is particularly aggressive. This solution rapidly enlarges intersecting joints, leaving sharp-crested low-relief pinnacles and their modified form of level-crested pinnacles called clints between them (Figure 10.10). The pinnacles in the *periglacial environment* of Tibet, however, are the relic of a former humid phase of a tropical climate (Sweeting et al., 1991; Zhang, 1996).

Sinkholes

Sinkholes are generally funnel-shaped shallow solution openings of an intermediate size at joints and bedding plane separations in limestone beds. A few, however, also indirectly develop as shaft-like forms of solution-related collapse of the endokarst structure. Sinkholes extend over several hundred square kilometers of terrain, and reach a high surface density (Figure 10.11). This density indeed exceeds 2500/km^2 in some limestone regions (Ford and Williams, 1989).

Sinkholes are a fundamental unit of the karst landscape. They are classified into five genetic types: solution dolines, subsidence dolines, collapse dolines, subjacent karst collapse dolines, and tropical cockpit (Figure 10.12). *Solution dolines* are funnel-shaped depressions that develop and grow in contact with the solution held in impervious impurities of limestone dissolution. In time, the insoluble impurities reach a critical thickness against the supporting sinkhole walls and become unstable. This impervious residue then gradually slumps into narrow funnel pipes, evolving

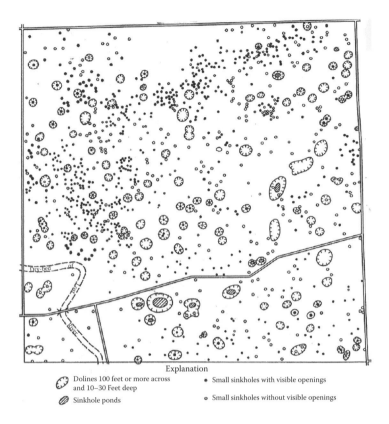

Explanation

⬭ Dolines 100 feet or more across   • Small sinkholes with visible openings
and 10–30 Feet deep

⬭ Sinkhole ponds   ○ Small sinkholes without visible openings

FIGURE 10.11   Map showing the number of sinkholes (1022) in 1 square mile (395 sinkholes per square kilometer), southwest of Orleans, Indiana. (Source: Figure 13.3 in Thornbury, W. D., *Principles of Geomorphology* [New York: John Wiley & Sons, 1969]. With permission from John Wiley & Sons.)

subsidence dolines. The plugged pipes interfere with normal drainage and, thereby, create *sinkhole ponds* within subsidence dolines. *Collapse dolines* are shaft-like depressions to the depth of subterranean channels. They develop when a part of the roof above near-surface subterranean passages collapses into the cavity. *Subjacent karst collapse dolines* develop and enlarge as subsurface sinkholes in mantled karst, causing collapse of the impervious strata of sufficient depth above the enlarged cavity. *Tropical cockpits* are star-shaped solutional depressions within the residual hills or hummocks called *cone karst.* They possibly develop at favorable sites on joints in limestone beds and grow to an optimum size. Williams (1972) similarly envisaged a sequence of cockpit evolution in horizontal crystalline limestone beds of evenly spaced intersecting joints (Figure 10.13(a), I). Accordingly, the initial centripetal drainage of the area converges at master joints to evolve optimum-sized solution depressions in equilibrium with the solution intensity (Figure 10.13(b), II). A second generation of solution depressions subsequently establishes on slightly less favorable joint sites, which similarly reach an equilibrium size, and merge with the first-generation solution depressions on master joints. Once all preferable joint sites have

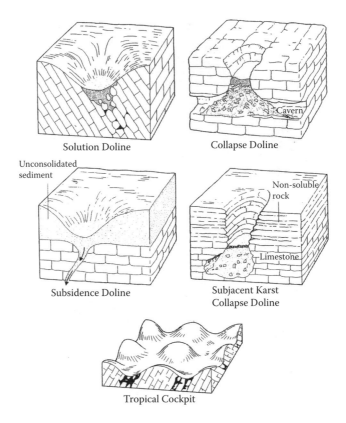

Solution Doline

Collapse Doline

Subsidence Doline

Subjacent Karst
Collapse Doline

Tropical Cockpit

FIGURE 10.12   Major types of dolines. (Source: Figure 12.3 in Goudie, A. S., *The Nature of the Environment*, 3rd ed. [Oxford: Blackwell, 1993]. With permission from Blackwell Publishers.)

enlarged into depressions, the least favorable joint sites are then similarly solutionally enlarged. They also coalesce with the depressions of earlier generations, evolving the composite morphology of cockpit depressions. The solutional depressions on joint systems leave an unfissured section of the limestone as *residual hills*.

Uvala

Uvalas are fairly deep and sufficiently long uneven depressions in karst. A few among such forms are identified with *coalesced sinkholes* (Jennings, 1985), and others represent a surface of collapsed roof above subterranean channels (Thornbury, 1969). The uvala floor of the former type supports as many solution openings into the karst as there are number of sinkholes comprising the feature.

Poljes

Poljes are isolated flat-floored depressions backed by steeply rising limestone hills. They are common to the *exposed karst* of the Mediterranean region and the *mantled*

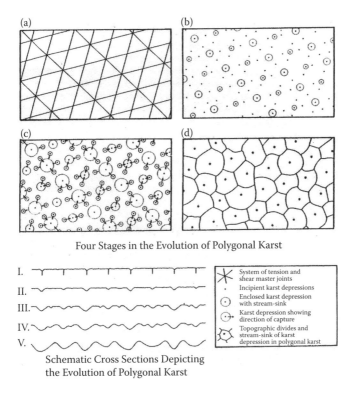

Four Stages in the Evolution of Polygonal Karst

I.
II.
III.
IV.
V.

Schematic Cross Sections Depicting
the Evolution of Polygonal Karst

	System of tension and shear master joints
.	Incipient karst depressions
⊙	Enclosed karst depression with stream-sink
⊙→	Karst depression showing direction of capture
	Topographic divides and stream-sink of karst depression in polygonal karst

FIGURE 10.13 Evolution of cockpit in New Guinea polygonal karst. (a) An idealized system of evenly-spaced intersecting joints in flat-bedded limestone terrain. (b) The surface drainage into intersecting master joints evolves incipient karst depressions of definable limits that grow to an optimum size. (c) Over a period of time, the surface drainage exploits less and less favourable joint openings that also reach a certain optimum-sized area of centripetal drainage. (d) Topographic divides as upstanding residuals and surface depressions of karst solution within them evolve cockpit depressions. (Source: Figure 18 in Williams, P. W., "Morphometric Analysis of Polygonal Karst in New Guinea," *Bull. Geol. Soc. Am.* 83 [1972]: 761–96. With permission from Geological Society of America.)

karst of Central Europe, and even occur in areas of tropical climates. Poljes can be dry or water-filled depressions at times.

The evolution of poljes is debated. They are regarded as surfaces of either solution planation or structural depressions. The proponents of solution planation take recourse to the potential of mixing effect of two solutions of similar or dissimilar hardness values (see Figure 10.3) on the solubility of karst. Sweeting (1965) holds that the mixing of stagnant surface water with the karst water table seasonally rising to above the polje floors produces an *undersaturated solution*, which uniformly dissolves and smoothens the depression floor. A few poljes, however, also extend along the *strike* of bedded limestone strata. They, therefore, are regarded as solutionally modified structural depressions in folded and downfaulted karst terrain (Thornbury, 1969; Bögli, 1980). Whether poljes are polygenetic (Gams, 1977) or a relic of former wet climates (Bögli, 1980) is also debated.

Residual Karst

Residual karst is an ensemble of intermediate-sized morphologic forms of low coni-cal hills called cone karst and vertical hills called tower karst. The *cone karst* rises from above the floor of cockpit depressions, and the *tower karst* abruptly rises from above the solution planated plains. These upstanding residuals are an explicit active form of limestone dissolution in the tropical realm of East Asia, Southeast Asia, and Central America. The two forms of residual karst, however, also uncharacteristi-cally occur outside the tropical zone as an active form of limestone dissolution in the periglacial environment of Canada, and as a relic of former hot-humid climatic conditions in temperate regions (Ford, 1980).

Cone Karst

Cone karst is an ensemble of up to 100 m high conical to hemispherical hills that occur at the periphery of *cockpits* in evenly jointed *horizontal* limestone beds. These hills represent a residual form of limestone dissolution in unfissured and case-hardened rocks. They also demarcate the drainage divide between adjacent cockpit depressions (Figure 10.14). The cone karst is called pepino and haystack hills in Puerto Rico, kegel karst in Java, and hum in the former republic of Yugoslavia.

Tower Karst

The tower karst (turm karst) rises abruptly from solution planated plains as up to 300 m high isolated precipitous hills in *steeply inclined* limestone rocks of high-purity content. Tower karst is the characteristic landscape of monsoonal southern China and parts of North Vietnam.

The evolution of tower karst is debated, and explained by possible mechanisms of uneven rates of calcite dissolution beneath sediments, solution sapping at the foot of hills, and modification of the cone karst. Jakucs (1977) reasoned that rates of limestone dissolution beneath a cover of noncalcareous sediments vary in direct pro-portion to the sediment thickness and, thereby, to the amount of residual moisture in sediments. Thus, a higher rate of limestone dissolution beneath thicker rather than thinner cover of sediments segments the rock into depressions and upstanding residuals of varying proportions (Figure 10.15 (I)). In the hot-humid tropical climate of high evaporation rate, the upstanding residuals also become mantled by the *sec-ondary calcite*. In time, the spalling off of this resistant calcite leaves upstanding residuals in vertical slopes (Figure 10.15 (II, III)). The *solution sapping mechanism* (Day, 1978) argues that the release of karst spring discharge at the foot of hills and its mixing with seasonally pounded surface water, produces an *undersaturated solu-tion*. This solution saps the basal slope at the line of spring discharge and causes its collapse, producing a vertical slope in hills. The tower karst also possibly evolves from the solution sapping of cone karst. Williams (1990) argues that the seasonal rise of the water table and its mixing with the surface water in cockpit depressions undercuts the base of conical hills. This process causes *slump failure* and, thereby, the steep-sided hill slopes. Tower karst, thus, is *polygenetic*, evolving from a direc-tional change in the solution activity from vertical to predominantly horizontal in cockpit depressions.

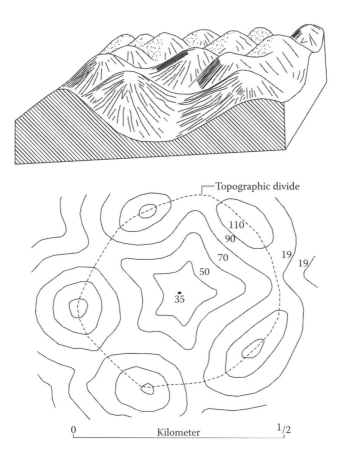

FIGURE 10.14   Cone karst with associated map of cockpit. (Source: Figure 6 II.9 in Williams, P. W., "The Geomorphic Effects of Groundwater," in *Water, Earth and Man*, ed. R. J. Chorley [London: Methuen & Co. Ltd., 1969], 269–94. With permission from Routledge.)

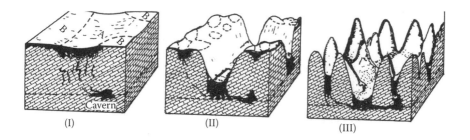

FIGURE 10.15   Three stages in the development of tower karst. (I) Preferential solution and debris accumulation in area A, compared with B. (II) Intense corrosion under the soil cover increases the disproportionate elevation of areas A and B. (III) The B areas are reduced to small, steep towers whose rate of erosion may be only one-tenth that of the intervening, chemically active lowlands. (Source: Figure 45 in Jakucs, L., *Morphogenetics of Karst Regions* [Bristol: Adam Hilger, 1977]. With permission from Institute of Physics Publishing.)

Dry Valleys and Gorges

Dry valleys are blind valleys, dry valleys in chalk, and karst valleys in limestone. *Blind valleys* develop as a normal valley in mantled karst only to disappear in down-channel *swallow holes*. Blind valleys are termed semiblind valleys when, at times, the surface flow extends the channel beyond swallow holes of smaller water draining capacity. These extended channels, though, remain dry for most of the year. *Dry valleys*, as in the *chalk* of southern England and northern France, are relics of a former *periglacial climate*. They had developed when the chalk was frozen and impermeable to infiltration. *Karst valleys* evolve as normal valleys in mantled karst. In time, surface channels trench to the depth of buried karst, by which the channel flow progressively disappears through solution openings in limestone terrain, evolving wide and deep valleys reminiscent of the higher flow volume of the past. Such wide and deep valleys with negligent stream flow are called *misfit streams*. *Gorges* generally develop from a concentrated flow in mantled karst.

Caves

Limestone caves are solutionally enlarged subterranean runway voids with distinct entrances, horizontal passages, and large-enough halls accessible to human explorers (White, 1988). They are vadose, phreatic, and mixed with reference to the groundwater system, and active and relict in relation to the contemporary solution activity. *Vadose caves* are active and inactive features of the endokarst. *Active vadose caves* are hydraulically efficient, steep and narrow voids with openings at spring discharge in karst, and *relict vadose caves* are dry and steep abandoned passages that had developed at the level of the higher karst water table of the past (Figure 10.16). As a rule, caves higher in topographic level had developed earlier than the caves lower down the slope. A few inactive cave passages, though, have been flooded by the *postglacial* rise of sea level. The vadose Bungonia Caves at the interface of limestone and *impervious substratum* in the Southern Highlands of New South Wales, Australia, however, are similar in morphology and hydraulic activity to the phreatic caves (Osborne, 1993). *Phreatic caves* develop at the level of the water table in karst. Their passages are nearly horizontal. They evolve independent of the geologic structure and cut across bedding and joint patterns in limestone rocks (Figure 10.17). The caves, however, randomly expand and contract, reflecting variations in the purity of limestone, irregularity of the cave floor, and escape of carbon dioxide from the circulating solution and that entering caves from the roof and walls. *Compound caves* evolve from frequent mixing of the vadose and phreatic waters in the karst vadose zone, and can be identified by their keyhole form (Zhang, 1995).

Vadose and phreatic caves are solutionally enlarged hydraulically active openings in limestone rocks. The *vadose caves* develop from the gravity circulation of solution in short and randomly interconnected vertical and horizontal karst passages. Many such horizontal passages, however, terminate in dead ends, but the one of sizable exit to the surface of karst terrain make vadose caves at the level of karst spring discharge. The Bungonia Caves, Australia, are an exception to this rule on account

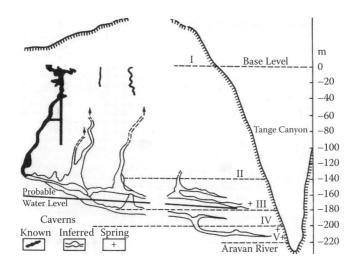

FIGURE 10.16 Schematic east-west section through the Tyuva Muyun ridge and valley of Aravan River. (I) Caverns—known. (II) Caverns—inferred. (III) Kok-Bulak springs in Tange canyon. (I–V) Base levels. (Source: Figure 5 in Chikishev, A. G., "Landscape-Indicator Investigations of Karst," in *Landscape Indicators*, ed. A. G. Chikishev [New York: Plenum Press, 1973]. With permission from Plenum Publishing Corporation.)

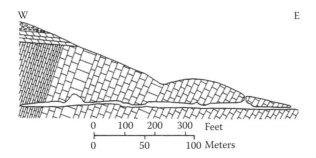

FIGURE 10.17 True water table cave passage networks are typically horizontal, as shown by the profile of Lehman Caves, Nevada. They are suggested to have formed just below a horizontal karst water table. (Source: Figure 2 in Moore, G. W., "Limestone Caves," in *The Encyclopedia of Geomorphology*, ed. R. W. Fairbridge [New York: Reinhold Book Corp., 1968], 652–53. With permission from Kluwer Academic Publishers.)

of the peculiar geologic setting of the region (Osborne, 1993). The *phreatic caves* are not attached to any of the geologic controls that govern the movement of solution in vadose caves. Phreatic caves are *water table caves* that appear in an amazing complex of nearly horizontal interconnected passages of sizable extent. Besides, the solution in phreatic caves moves only with a slight head and with least expenditure of energy in the system (Dreybrodt, 1990; Palmer, 1990, 1991).

Cave Deposits

The reaction between carbonic acid and karst releases diverse *ions* in the aqueous phase. Calcium, which is overwhelming in abundance in the solute load, precipitates from the solution reaching saturation in respect of the partial pressure of carbon dioxide and solution temperature (Thrailkill, 1971; Bögli, 1980). The ions of calcium and bicarbonate combine in a saturated solution, precipitating calcium carbonate in the medium as

$$Ca^{2+} + 2HCO_3^- \rightarrow CO_2\uparrow + H_2O + CaCO_3\downarrow$$

| Ions of calcium | Carbon | Water | Calcium |
| and bicarbonate | dioxide | | carbonate |

The precipitated calcium carbonate evolves a variety of cave deposits called *speleothems*.

Types of Cave Deposit

Cave deposits present a great diversity of forms that may be classified by the evolutionary style into gravity and eccentric forms (Bögli, 1980). The *gravity forms* are stalactites, stalagmites, sinter flags, curtains, flow stones and rimstones, and the *eccentric forms* represent helictite and heligmite as deposits of a saturated karst solution.

Stalactites hang vertically from cave roofs as mineral precipitates of slow-draining solution escaping fissures and interstices in the cave roof (Figure 10.18(a)). The evaporation and escape of carbon dioxide from the solution that runs down the incline of cave roof and walls forms massive calcite deposits, and a free column of the deposit at the terminus of solution rundown along the cave roof. This calcite column is called type 1 stalactite. The solute load, however, also precipitates from solution droplets that hang in the cave roof. These air-filled solution droplets, though, stick for a while to the surface of greater surface tension before burst, forming a small ring of the precipitated matter attached to the roof. Subsequent passage of the solution down the ring similarly precipitates the mineral matter around the ring by which it grows into 0.3 to 0.6 mm wide a *sinter tube* opening for the passage of solution (Moore, 1968). The mineral matter precipitated in this manner makes type 2 stalactites. *Stalagmites* grow vertically up from the cave floor as convex and splash-cup-type deposits of solution droplets falling to the surface (Figure 10.18(b)). The convex-type pillars depict smaller and splash-cup-type larger fall distance of the droplets. *Sinter flags and curtains* are massive deposits of the solution that avoid even minor surface irregularities, and adopt a winding path down the incline of the cave roof and wall projections (Figure 10.18(c)). Sinter

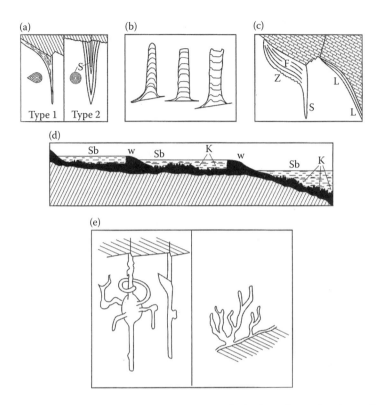

FIGURE 10.18 Common forms of speleothems. (a) Stalactites of type 1 and type 2. Sinter tube is shown as S in type 2 stalactite. (b) Stalagmites' longitudinal profiles, from left to right with increasing distance of fall of solution drops. (c) Sinter flag (F) with saw-toothed sinter (Z), stalactite of type 1, and sinter band (L). (d) Longitudinal section of a row of rimstone pools (Sb), rimstone bars (W), and calcite crystals (E). (e) Helictites and heligmites. (Source: Figures 13.16–13.20 in Bögli, A., *Karst Hydrology and Physical Speleology* [Berlin: Springer-Verlag, 1980]. With permission from Springer-Verlag.)

flags (F) comprise color-differentiated banded precipitates, such that each layer signifies a certain impurity in the solution of karst. Saw-tooth sinter flags (Z) result from the evaporation of solution hanging in the indentations of sinter flags. The sinter flags and curtains, as observed earlier in this section, continue at the free end as type 1 stalactite. *Flow stones and rimstones* are crystallized deposits of a saturated solution held in inclined ledges and an irregular cave floor (Figure 10.18(d)). They comprise rimstone bars (W), rimstone pools (Sb), and calcite crystals in shallow pools (K).

Helictites and *heligmites* are eccentric forms of mineral precipitation from the evaporation of seepage solution in caves. The *seepage solution* has greater surface tension than the force of gravity. Hence, precipitates from the evaporation of this solution orient randomly, and grow into millimeter-sized twisted dendritic forms around the initial stem without touching each other (Figure 10.18(e)). Helictites shoot outward from cave walls and heligmites grow upward from the cave floor.

Environmental Significance

Speleothems are a sensitive indicator of the chemical denudation rate of the karst. The *oxygen isotope* record ($^{18}O/^{16}O$) in stalagmites provides for the chemical lowering of the karst terrain at the rate of 0.1 to 10.0 cm year^{-1} and 350 ka age for the precipitates (Ford and Drake, 1982). Hence, limestone caves are at least as old.

SUMMARY

Chalk, dolomite, and limestone rocks are comprised of the mineral calcite, which is ordinarily insoluble in waters of the earth but readily dissolves in the water solution of carbon dioxide called carbonic acid. The solution processes and resulting landforms, called karst, ideally express in massive limestone rocks that have the advantage of efficient secondary permeability and sufficient inherent strength for the support of solutionally hollowed structure of the rock. These attributes are denied to the other two calcite-rich rocks.

The strength of carbonic acid and its potential to dissolve limestone rocks depend on the volume of carbon dioxide held in solution. The volume of carbon dioxide per unit volume of the water content depends on air temperature, partial pressure of carbon dioxide in karst systems, and solution pH. All other conditions being equal, the carbon dioxide holding capacity of water is low at higher and high at lower temperatures. Carbon dioxide is derived from the atmosphere and the biogenic source in the decay of organic matter at the surface of karst and that carried to its subsurface depth by the solution circulating in contact with the limestone rock. Carbon dioxide is only 0.003% by volume in the atmosphere, and a minor contributor to the solution strength. This source of carbon dioxide exerts a partial pressure of only 0.0003 atm at the sea level. By comparison, the decay of organic matter releases three hundred times more carbon dioxide than is available from the atmosphere and exerts three orders of magnitude higher partial pressure of carbon dioxide in the system. Hence, decaying organic matter is the principal source of carbon dioxide for the dissolution potential of solution. The dissolution of calcite is also contingent upon closed- and open-system solubility of calcite. The closed-system state defines solubility of calcite from a one-time input of carbon dioxide, and the open-system state is nourished by an uninterrupted supply of carbon dioxide in the dissolution system. The dissolution of calcite releases calcium and bicarbonate ions. The relative proportion of the molar concentration of either calcium or bicarbonate ion in solution governs the solution pH, such that the acidic solution with pH less than 7 dissolves and the basic solution with pH more than 7 precipitates calcite in the medium. It may be remembered that all acid-base reactions are reversible in nature. In sum, the chemical reactivity of carbonic acid depends on the biogenic release of carbon dioxide, open and closed solubility state of karst systems, air temperature, type of carbonate rock, and runoff volume in the karst terrain.

The solution of carbon dioxide exploits fissures, joints, and bedding plane separations at and beneath the surface of limestone rocks, evolving a variety of small- and intermediate-sized closed and linear features of dissolution, and phreatic caves of variable dimension at the depth of the water table in karst. The solution on bare and beneath the humus-rich soils on steeply inclined limestone beds evolves small-sized features called karren. The intermediate-sized sinkholes are a fundamental unit of karst dissolution at joint and bedding plane openings in limestone rocks. They are classified into five genetic types by the process of evolution. The collapse of the roof above shallow subterranean runway voids and coalesced compound sinkholes develop uvala. Poljes are intermediate-sized flat-floored depressions surrounded by steep limestone hills. They are either solutionally modified structural depressions in folded and faulted alpine terrain or the product of solution planation. Cone and tower karst are upstanding residuals of limestone dissolution in the humid tropics. These forms of accentuated relief perhaps reflect higher denudational potential of the solution in the tropics than in temperate and arctic climates. Dry valleys are an active feature of mantled karst and a relic of the chalk terrain. Karst valleys are a feature of mantled karst.

Limestone caves are natural voids in the vadose and phreatic karst zones. Vadose caves are short, steep, and generally dry solution passages in hydraulically efficient joint and bedding plane separations. They exit at the level of karst spring discharge or signify a former higher water table of the past. Phreatic caves represent the culmination of the solution process at the depth of the water table in karst. Such caves develop horizontal water-filled passages of sizable extent that are independent of the joint and fissure system in limestone rocks. The solution entering air-filled phreatic caves loses carbon dioxide and becomes saturated with the solute load of several ionic species, in which the calcium ion is by far in abundance. The solute load precipitates as gravity and eccentric forms called cave deposits or speleothems. The oxygen isotope record in speleothems enables the reconstruction of denudational rates of karst, the age of cave deposits, and by inference, the age of limestone caves.

REFERENCES

Baird, C. 1995. *Environmental chemistry*. New York: W. H. Freeman and Co.

Bögli, A. 1980. *Karst hydrology and physical speleology*. Berlin: Springer-Verlag.

Chikishev, A. G. 1973. Landscape-indicator investigations of karst. In *Landscape indicators*, ed. A. G. Chikishev, 48–63. New York: Plenum Press.

Crabtree, R. W., and Trudgill, S. T. 1987. Hillslope solute sources and solutional denudation on a Magnesian Limestone hillslope. *Trans. Inst. Br. Geogr.* 12:97–106.

Cvijić, J. 1924. The evolution of lapiés. *Geogr. Rev.* 14:26–49.

Day, M. J. 1978. Morphology and distribution of residual limestone hills (mogotes) in the karst of northern Puerto Rico. *Bull. Geol. Soc. Am.* 89:426–32.

Dreybrodt, W. 1990. The role of dissolution kinetics in the development of karst aquifers in limestone: A model simulation of karst evolution. *J. Geol.* 98:639–55.

Ford, D. C. 1980. Threshold and limit effects in karst geomorphology. In *Thresholds in geomorphology*, ed. D. R. Coates and J. D. Vitek, 345–62. London: George Allen & Unwin.

Ford, D. C., and Drake, J. J. 1982. Spatial and temporal variations in karst solution rates: The structure of variability. In *Space and time in geomorphology*, ed. C. E. Thorn, 147–70. London: George Allen & Unwin.

Ford, D. C., and Williams, P. 1989. *Karst geomorphology and hydrology*. Winchester: Unwin Hyman Ltd.

Gams, I. 1977. Towards a terminology of the poljes. In *Proceedings of the Seventh International Congress of Speleology*, September 10–17, 1977. Sheffield, pp. 201–2.

Goudie, A. S. 1993. *The nature of the environment*. 3rd ed. Oxford: Blackwell.

Jakucs, L. 1977. *Morphogenetics of karst regions*. Bristol: Adam Hilger.

Jennings, J. N. 1985. *Karst geomorphology*. Oxford: Blackwell.

Kar, A. 1995. Geomorphology of arid western India. *Geol. Soc. India Memoir* 32:168–90.

Moore, G. W. 1968. Limestone caves. In *The encyclopedia of geomorphology*, ed. R. W. Fairbridge, 652–53. New York: Reinhold Book Corp.

Oberlander, T. M. 1994. Global deserts: A geomorphic comparison. In *Geomorphology of desert environments*, ed. A. D. Abrahams and A. J. Parsons, 13–25. London: Chapman & Hall.

Osborne, R. A. L. 1993. A new history of cave development in Bungonia, N. S. W. *Austr. Geogr.* 24:62–74.

Palmer, A. N. 1984. Recent trends in karst geomorphology. *J. Geol. Edu.* 32:247–53.

Palmer, A. N. 1990. Groundwater processes in karst terranes. In *Groundwater geomorphology; the role of subsurface water in earth-surface processes and landforms*, ed. C. G. Higgens and D. R. Coates, 177–209. Geological Society of America Special Paper 252. Geological Society of America.

Palmer, A. N. 1991. Origin and morphology of limestone caves. *Bull. Geol. Soc. Am.* 103:1–21.

Sheen, S.-W. 2000. A world model of chemical denudation in karst terrains. *Prof. Geogr.* 52:397–406.

Smith, D. I., and Atkinson, T. C. 1976. Process, landforms and climate in limestone regions. In *Geomorphology and climate*, ed. E. Derbyshire, 367–410. London: John Wiley & Sons.

Sweeting, M. M. 1965. Denudation in limestone regions: A symposium. *Geogr. J.* 131:34–37.

Sweeting, M. M. 1972. *Karst landforms*. London: Macmillan.

Sweeting, M. M., Sheng, B. H., and Zhang, D. 1991. The problem of palaeokarst in Tibet. *Geogr. J.* 157:316–25.

Thornbury, W. D. 1969. *Principles of geomorphology*. New York: John Wiley & Sons.

Thrailkill, J. 1971. Carbonate deposition in Carlsbad Caverns. *J. Geol.* 79:683–95.

White, W. B. 1988. *Geomorphology and hydrology of karst terrains*. New York: Oxford University Press.

White, W. B. 1990. Surface and near-surface karst landforms. In *Groundwater geomorphology; the role of subsurface water in earth-surface processes and landforms*, ed. C. G. Higgins and D. R. Coates, 157–75. Geological Society of America Special Paper 252. Geological Society of America.

Williams, P. W. 1969. The geomorphic effects of groundwater. In *Water, earth and man*, ed. R. J. Chorley, 269–94. London: Methuen & Co. Ltd.

Williams, P. W. 1972. Morphometric analysis of polygonal karst in New Guinea. *Bull. Geol. Soc. Am.* 83:761–96.

Williams, P. W. 1990. Hydrological control and the development of cockpit and tower karst. In *Karst hydrogeology and karst environmental protection*, 281–87. International Association of Hydrology Sciences Publication 176. International Association of Hydrology Sciences.

Wood, W. W. 1985. Origin of caves and other solution openings in the undersaturated (vadose) zone of carbonate rocks: A model for CO_2 generation. *Geology* 13:822–24.

Zambo, L., and Ford, D. C. 1997. Limestone dissolution processes in Beke Doline Aggtelek National Park, Hungary. *Earth Surface Processes Landforms* 22:531–43.

Zhang, D. 1995. Geology, palaeohydrology and evolution of caves in Tibet. *Cave Karst Sci.* 21:111–14.

Zhang, D. 1996. A morphological analysis of Tibetan limestone pinnacles: Are they remnants of tropical karst towers and cones? *Geomorphology* 15:79–91.

Zhang, D. 1998. A mineralogical analysis of karst sediments and its implication to the middle-late Pleistocene climatic changes on the Tibetan Plateau. *J. Geol. Soc. India* 52:351–59.

11 Coastal Processes and Landforms

A coast is up to a few kilometers wide tract of interacting terrestrial and marine processes between the shoreline and landward limit of the first major change in terrain, usually in the form of a cliff or coastal dunes. Coastal geomorphology explains the contemporary processes and their landforms, translates the effect of global sea level changes of the Pleistocene and Holocene periods and of the tectonic instability of landmass on the evolution of coastal landscape, and interprets the chronology of Quaternary events for the present state of coastal development. In brief, coastal geomorphology is concerned essentially with the physics of wave motion, hydraulics, sediment transport mechanisms, and interpretation of the environmental change (Kidson, 1968).

GRADATIONAL PROCESSES

Coasts are areas of incessant motion of shallow seawater against the littoral tract that varies much in lithology, geologic structure, and resistance to erosion. The coastal water moves onshore and offshore by the effects of waves, tides, and currents—the three process domains of marine erosion and deposition. In a general term, waves are destructive in storm conditions, tides are known more for constructive than destructive work, and currents possess much less erosive potential. Tsunamis are a type of destructive waves, which develop by sudden dislocation of the seabed.

WAVES

Transfer of the wind energy onto the surface of a deep water body generates surface waves, such that the wave size depends directly on the wind speed, duration of time the wind blows from one direction, and extent of the open water, called *fetch* of the water surface, over which the wind blows prior to reaching a given observation point. These deep water waves are sinusoidal in form with interdependent length, height, amplitude, and period attributes. *Wavelength* (L) is the horizontal distance between successive wave crests or troughs, *wave height* (H) is the vertical distance between the crest and its adjacent trough, and *amplitude* is half the wave height (Figure 11.1). *Wave period* (T) is the interval of time between successive crests or troughs at a fixed point. *Deep water* refers to a water depth more than one-half of the wavelength.

DEEP WATER WAVES

Deep water waves originate in the area of strong wind, called sea, as a confused spectrum of overriding waves of different length, height, and period attributes. Among

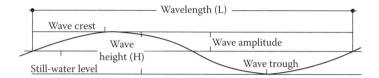

FIGURE 11.1   Basic components of a deep water wave.

them, the shorter waves reach maximum height quickly and then are destroyed. A few waves of different periods simultaneously present in the area of *sea* also cancel out by overlapping of the crests of one wave train with troughs of another wave train passing under the wind. The waves of longer period, however, soon segregate from the rest as a wave train of similar period and height called *swell* of a wave speed $C = L/T$, in which C is the celerity or wave phase velocity of the wave train in the spectrum of waves. *Wave speed* is independent of water depth and a function of wave period given as C = 3.03T, where C is in knots and T in seconds. *Wavelength*, in meters, is the product of celerity and *wave period* as $1.5CT^2$. The wave height and wind speed are also related. As a general rule, the wave height never exceeds half the wind speed (Bascom, 1959). Thus, a 6 km h^{-1} wind produces waves only up to 3 m high.

General observations suggest that waves move floating objects up and down and back and forth without transporting them for a sufficient distance, water moves horizontally in the direction of wave propagation at the crest and opposite to it in the trough, and that water returns to its original position upon passing of each wave. Thus, each water particle at the surface describes a circular orbit of a dimension equal to wave height. This regularity of the waveform obtains from transfer of energy from one orbiting water particle to the next, which places water particles in vertical motion through the depth of the water column (Bascom, 1959). The convergence of instantaneous velocity in these orbiting particles produces crests, and divergence crests in surface waves (Figure 11.2). The water surface diameter of orbiting particles that governs the wave height, however, *exponentially* decreases with depth, and is halved for every one-tenth of the wavelength reached. The orbital diameters thus are reduced to only 4% at the *wave base*, the theoretical limit at which water depth equals one-half of the wavelength. At this depth, water particles cease to oscillate in the vertical.

Each wave in the *wave train* moves forward with a velocity corresponding to its length. However, the advancing train of waves travels only at half the speed of its individual waves (Bascom, 1959). This is because the waves at the front of a group lose energy to those behind, and gradually disappear. They are replaced by waves at the rear. Hence, waves arriving at the shore are remote descendents of waves originally formed. The velocity of a single wave within a group of waves is *wave phase velocity* or celerity, and that of the group of waves is *group phase velocity*.

Swells travel thousands of kilometers across oceans and reach distant shores without losing much energy in the open sea. This energy is divided equally between components of potential energy due to the movement of water in the crest above the still-water level and kinetic energy due to the velocity of water particles within the

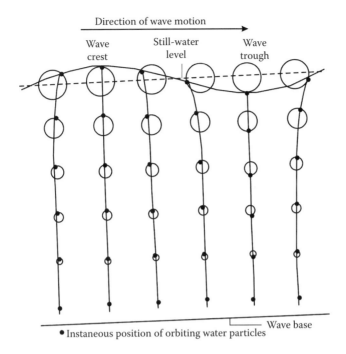

FIGURE 11.2 Orbital path, orbital diameter of water particles, and orbital position of water particles at any given point in time through to the depth of the wave base of a swell.

waveform. Hence, total swell energy (E) is only half the velocity of a waveform. It is related directly to the wave height (King, 1972) as

$$E \propto H^2$$

suggesting that a small change in wave height produces a large change in the wave energy.

MODIFICATION OF SWELL IN SHALLOW WATERS

Deep water waves approaching the shore enter coastal waters shallower than half their wavelength and decelerate and deform. The velocity of shallow water waves, called *waves of translation*, is governed by the water depth (d) and is given as C = √gd. The decelerated waves decrease in length but sharply increase in height without affecting the wave period. The orbital motion of water particles at the *wave base* also changes to that of tilted ellipses with axes horizontal (Figure 11.3), by which sufficient water is not available for the advancing wave to complete its form. The relationship between wave height and wavelength is such that wave height, expressed as H/L, cannot exceed 1/7 (Bascom, 1959; King, 1972). The waves become unstable at this critical limit, causing the wave crest to collapse forward into the trough. This is the phenomenon of *breaking waves* that results in *surf*. The waves also break for

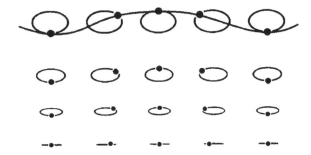

FIGURE 11.3 Water particle orbits in a shallow water. The velocity of these elliptical movements does not decrease with depth as in deep waters. (Source: Figure 3.1 in Pethick, J., *An Introduction to Coastal Geomorphology* [London: Edward Arnold, 1984]. With permission from Edward Arnold Ltd.)

internal angles of less than 120°, at which the velocity of water particles at the crest exceeds the wave speed (Bascom, 1959; Davis, 1991). The breaking waves nevertheless retain some energy and regroup to form a wave train of lesser height that similarly breaks again. The breaking waves generate *strong turbulence*, which places a sufficient quantity of seabed sediments in *suspension* and *saltation* modes for shore normal transport (Wang et al., 1998a).

Other Wave Modifications

Waves refract, reflect, and diffract in shallow coastal waters. Deep water waves travel at an angle to the coast but bend in shallow waters offshore in conformity with the submarine configuration in such a way as to parallel the shoreline. The bending of waves in coastal waters is called *wave refraction*. Wave refraction is governed by variations in the water depth that distinctly segment the passing wave into slower- and faster-moving trains. The slower waves, like over a submarine ridge, increase in height and converge onto the *headland* from all directions (Figure 11.4 left), while faster waves, like over a submarine valley, are longer with divergence outward to the coast (Figure 11.4 right). This difference of energy level in coastal waters generates *longshore currents*, which transport sediments along the littoral tract for building coasts. Wave refraction also redistributes the wave energy in such a way as to *straighten coastlines* (Bascom, 1959; Pethick, 1984; Davis, 1991).

Deep water waves do not break but reflect against small-sized islands. The reflected waves interact with primary waves, forming high and low nodes of standing waves called *clapotis*. In clapotis, the water surface elevation is maximum at antinodes and minimum at nodes (Figure 11.5). Shallow water waves, however, break against natural barriers, seawalls, and other man-made structures. They lose energy in the process and, therefore, reflect poorly offshore (Figure 11.6), and not at all if the wave energy is completely lost in the breaking process. In a general term, clapotis are an important influence on the evolution of *nearshore* configuration of the seabed.

Shallow water waves approaching an obstacle or moving past a barrier at a grazing angle *diffract*, dissipating the wave energy in the form of a parabola with its apex at features creating the diffraction (Davis, 1991). Diffracted waves *erode* in

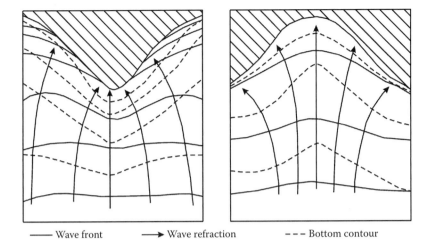

———— Wave front ——→ Wave refraction – – – Bottom contour

FIGURE 11.4   Wave refraction along embayed coasts causes the waves to converge at a submarine ridge (left) and diverge inside a submarine valley (right). The lines of equal energy are perpendicular to the wave crest.

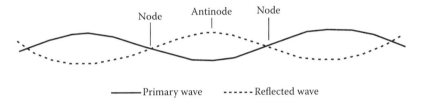

———— Primary wave   - - - - - Reflected wave

FIGURE 11.5   Deep water standing waves or clapotis.

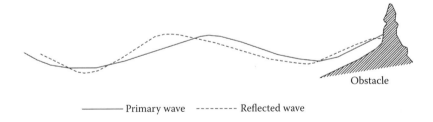

Obstacle

———— Primary wave   - - - - - - - - Reflected wave

FIGURE 11.6   Shallow water clapotis.

the lee of gaps, and those moving into a leeward-positioned open water body induce *longshore currents*.

LUNAR TIDES

Gravitational attraction of the moon and to a lesser extent of the sun on the earth produces a rhythmic rise and fall of the sea surface level called lunar tides. This tide-generating force is highest when the sun and the moon are aligned in conjunction and opposition at new and full moon phases, and lowest when the two heavenly bodies are in quadrature at the first and third quarters of the lunar cycle. The highest and lowest magnitudes of this gravitational force respectively generate *spring* and *neap tides*. The spring and neap tides reoccur twice in a lunar cycle of 28 days or once every 2 weeks. The difference in sea surface level due to tides is called *tidal range*.

Lunar tides result from the tractive force of the moon's gravitational attraction. This force acts parallel to the surface of the earth rotating about its axis, and pulls the water surface toward the moon. The tractive force on the far side of the earth, though, is small, but the water surface is pulled away by the *centrifugal force* of spinning earth, producing another center of high water level in that direction. The two tidal centers, as above, are sustained by the sinking of water at a right angle to the maximum bulge. These centers also migrate westward in resonance with the eastward rotation of the earth, producing high tides at any fixed observation point on the earth 50 min late on successive days. This phenomenon is governed by the earth's rotation period of 24 h about the sun and 24 h 50 min about the moon. Tides generate currents and augment the wave activity over varying heights along the littoral tract. Tidal currents transfer a large volume of water and sediments between the littoral and coastal environments for building coasts.

TIDAL CURRENTS

The alternate rise and fall of the sea surface level due to tides induces a stream-like rhythmic movement of water in and out of estuaries and tidal inlets. The rise of seawater at high tides generates an upstream moving tidal current called *flood*, and the fall of seawater at low tides accompanies a downchannel tidal current known as *ebb*. In certain channels, the upstream moving flood blocks the stream flow and develops a vertical column of water near the mouth of streams. This phenomenon is called *tidal bore*. Tidal bores are commonly a few centimeters high, but those accompanying *spring tides* raise a substantial water column within shallow estuaries. The tidal bore of this type is up to 800 m wide across the head of an estuary in the Bay of Fundy, Canada (Pethick, 1984).

The velocity of tidal currents and manner of sedimentation due to tidal currents depend on the tidal prism and mixing characteristics of saline and fresh waters from the sea and land, respectively (Davis, 1991). *Tidal prism,* which is the total volume of water passing in and out of channels during a tidal cycle, directly governs the velocity of tidal currents. Tidal currents, thus, are strong for a large and weak for a small tidal prism. *Mixing properties* of saline and fresh waters typify estuaries as stratified, partly mixed and mixed (Pethick, 1984; Davis, 1991). *Stratified estuaries* are characteristic of a thick body of upstream thinning *salt-water wedge*, which appreciably reduces the *ebb* velocity. Hence, sea-bound riverine sediments do not travel far into the sea and settle at the mouth of estuaries. *Partly mixed estuaries* carry roughly equal proportions of riverine and marine flows but are relatively stronger in flood than ebb velocity, by which a part of marine sediments moved in flood is retained on land. *Mixed estuaries* are characteristic of strong tidal currents, often exceeding 1 m s^{-1} in speed. The three type of estuaries are also characteristic of a certain *tidal range*: stratified estuaries are typical of a micro-tidal range of 2 m or less, partly mixed estuaries are characteristic of a meso-tidal range between 2 and 4 m, and mixed estuaries are distinguished by a macro-tidal range of 4 m or more.

WAVE-INDUCED CURRENTS

Longshore currents and rip currents are two principal wave-induced current systems of the foreshore and nearshore coastal environments. Longshore currents redistribute sediments along coasts, and rip currents balance the flow of water and sediments between the nearshore and offshore zones.

LONGSHORE CURRENTS

Deep water waves approach coasts at an oblique angle, break in shallow coastal waters, and arrive roughly parallel to the shoreline. The swell energy parallel to the shore and transfer of momentum in breaking waves translate into *longshore currents* (King, 1972; Komar, 1976; Pethick, 1984; Yoo, 1994). Theoretically, the swell energy is highest for waves arriving at 45° to the coast and lowest for waves either reaching parallel to or at a right angle to the shoreline. The wave energy parallel to the shore is given by the angle of wave approach as (King, 1972)

$$E_t = E_0 \sin \alpha_0 \cos \alpha_0 \ (L^*/L_0)$$

in which E_t is the component of wave energy parallel to the shore, α is the angle of wave approach, L^* is the wavelength in shallow waters, and the subscript o denotes deep water values for the angle of wave approach and wavelength.

Shallow water waves breaking in the *nearshore zone* preserve momentum normal to and along the shoreline. The alongshore component of the momentum in breaking waves generates longshore currents. This momentum depends on the angle of wave approach to the shore and orbital velocity of water particles in breaking waves as (Komar, 1975)

$$V_s = 2.7u_m \sin \alpha \cos \alpha$$

where V_s is the longshore current velocity in mid-surf zone, u_m is the orbital velocity in breaking waves, and α is the angle of wave approach. Strong turbulence in breaking waves, however, modifies the above relationship somewhat (Yoo, 1994).

RIP CURRENTS

The movement of water toward the shore traps wave energy and affects the wave setup in parts of the foreshore and backshore zones, producing oscillation of the surface water with crests perpendicular to the shore. This is the phenomenon of *edge waves*, which are much like the standing waves in respect to nodes and antinodes (Bowen and Inman, 1971; Huntley and Bowen, 1975; Komar, 1976; Davis, 1978, 1991). This trapped energy near the shore is released by the return flow of water offshore through

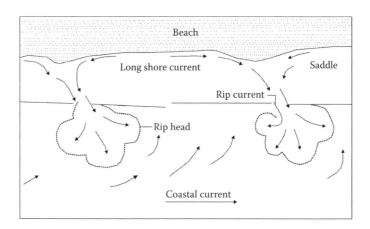

FIGURE 11.7   A rip current system of return flow of water offshore and rip heads. Rip currents are spaced equal to the edge wavelength and spaced three surf zones from shorelines.

an evenly spaced cell circulation system of *rip currents* (Figure 11.7), which escape as undertow at *antinodes* spaced equal to the edge wavelength (Bowen and Inman, 1971) and three-surf zone width from the shoreline (Huntley and Bowen, 1975). Sedimentation from the rip currents evolves a rhythmic submarine topography of shallow sand bars and troughs roughly parallel to the shoreline. These currents reach a velocity of up to 2 m s^{-1} (Sonu, 1972), and return a large volume of sediment to the sea as evidenced in the turbid rip heads (Huntley and Bowen, 1975; Davis, 1978). The rip currents deposit bulk of sediments as *nearshore bars* (Aagard, 1991). However, the association between edge waves and spacing of rip currents is not evident in certain coastal environments (Bowman et al., 1992).

TSUNAMIS

A sudden disturbance of the seabed by volcanic eruptions or submarine earthquakes generates fast-moving waves called tsunamis. Tsunamis generally possess a wavelength in excess of 100 km, an amplitude of about 0.5 m, a wave period of more than 15 min, and a speed in excess of 800 km h^{-1} in the open sea. Hence, they pass unnoticed in oceans but break violently in shallow coastal waters with devastating effect along the impacted coast.

A large number of tsunamis originate along the circum-Pacific belt of *plate subduction*. The Asian tsunami of December 26, 2004, accompanied a major seismic dislocation of the *Indo-Australian Plate* in the zone of subduction off the northwestern coast of Sumatra, Indonesia, causing extensive loss of life and property as far west as the horn of Africa. The exact number of causalities may never be known, but some 200,000 lives were lost by the impact of destructive waves.

SEDIMENT TRANSPORT

Coastal sediments are derived from the materials of the inner continental shelf and shallow waters off coasts. Sediments of the inner continental shelf are nearly 90%

riverine, 3% coastal, and 7% glacial, aeolian, and biogenic in origin (Davis et al., 1992; Wang et al., 1998a). These sediments are entrained and transported in much the same way as fluvial sediments on land, except that the sediment transport is variously affected by dissipation of the wave energy in shallow coastal waters, reversing nature of the flow in waves, hydrodynamic conditions of the surf zone, and the wave climate. Coastal sediments are entrained by *turbulence* in breaking waves, and transported by shore normal and alongshore components of the wave energy in shallow waters offshore. The dynamics of coastal sediment transport is best known in theory (King, 1972; Huntley and Bowen, 1975; Pethick, 1984) and understood in test and field conditions (Kraus, 1987; Wang et al., 1998a; Wang et al., 1998b). These studies suggest that the sediment transport rate depends on hydrodynamic conditions and beach parameters. Of the two, wave height is the most significant variable affecting the sediment transport rate.

SHORE NORMAL SEDIMENT TRANSPORT RATE

Shore normal sediment transport initiates by the shearing of particles in breaking and nonbreaking wave conditions. The theoretical Longuet-Higgins equation predicts that shore normal sediment transport rate, satisfying conditions of wave height, celerity, and water depth as $5/4$ $(H/2d^2)$ C, varies directly with the *wave climate* (King, 1972). Experimental studies suggest that the rate of sediment transport through to the depth of water column in shallow waters increases with wavelength, and that flat waves move sediments onshore and steep waves offshore at a *threshold wave steepness* of 0.012 (King, 1972). Field studies suggest that breaking and nonbreaking waves thoroughly mix the sediment in water column (Kraus et al., 1988; Rosati et al., 1990, 1991). While the turbulence in breaking waves makes a large volume of suspended and bed particles available for shore normal transport, the convective and diffusive processes in nonbreaking waves respectively place coarser and finer fractions in alongshore sediment transport mode (Nielsen, 1983, 1991; Hay and Sheng, 1992; Wang et al., 1998a). Other studies suggest that the rate of energy dissipation in breaking waves (Dean, 1977, 1991), local balance in the transport system (Inman and Bagnold, 1963; Bowen, 1980; Bailard and Inman, 1981), and sediment supply rate (Wang et al., 1998a) affect the shore normal sediment transport rate in many different ways.

LONGSHORE SEDIMENT TRANSPORT RATE

Longshore sediment transport comprises the movement of suspended load and bedload at the *breaker zone* and bedload near the *swash zone* (Kamphuis, 1991). Hence, it is affected by superimposed shore normal and alongshore components of the wave energy. The total longshore sediment transport (TLST) rate theoretically depends on wave power, angle of wave approach to the shore, and cohesion among grains. The *wave power* depends on the wave energy per unit surface area of waves, wave phase velocity, and group phase velocity of progressing waves (Inman and Bagnold, 1963; King, 1972; Pethick, 1984). The sediment transport rate tends to be highest for a 30° wave approach to the shore (King, 1972).

Field and experimental observations suggest that wave power, wave height, wave period, angle of wave approach, beach slope, particle size, and fluid density affect the TLST rate in many different ways (Kamphuis, 1991; Van Rijn et al., 1993). The above variables find expression in several empirical relations predicting the TLST rate (Horikawa, 1988; Fredsoe and Diegaard, 1992). However, the empirical equation based on wave energy per unit area of the waves, given as

$$Q = E \ Cg \ \sin (2\theta_b)$$

is recommended by the *Shore Protection Manual* (CERC, 1984) for the purpose. In the above equation, E is the wave energy, Cg is the group velocity, and θ_b is the angle of incident wave in the breaker zone.

WEATHERING

Nival weathering and salt-layer weathering (Chapter 3) are typical weathering processes respectively of high-latitude coasts of lingering snow covers of sizable extent and tropical coasts of high thermal efficiency environment. *Nival weathering* evolves extensive undulatory lowlands called *strandflats* (Sissons, 1982), such as along the coast of western Scotland, western Norway, Iceland, Svalbard, Baffin Island, and Antarctica. *Salt-layer weathering* is caused by hydration and dehydration of salts on coastal rocks exposed to the action of wave spray. Littoral rocks of tropical coasts to the highest limit of wave spray are particularly susceptible to this form of weathering.

EROSION

Mechanical, chemical, and biological processes variously affect the rocks of littoral tract. *Mechanical erosion* largely manifests in wave erosion and abrasive wear of coastal rocks. *Wave erosion* shatters rocks by the fluid impact of waves, artillery action of clastic fragments in waves breaking against coasts, and wedging of air in rock cavities ahead of the irregular wave front. This form of erosion is characteristic of the middle- and higher-latitude coasts of high-wave energy environment, and is particularly destructive in association with *tsunamis* breaking against coasts. Weathering, erosion, and mass movement along the littoral tract generate clastic fragments, which repeatedly move over the intertidal zone in reversing waves, and cause *abrasive wear* of the surface generally to the depth of the wave base. In general, the efficiency of abrasive wear varies with rock strength, nature and hardness of abrasive tools, and duration of tidal inundation.

Seawater is normally saturated with calcium and carbonate ions, but the nocturnal release of carbon dioxide from *photosynthesis* of algal and marine plants in intertidal pools of *calcareous lithology* enhances the solubility of rock locally. Observations suggest that intertidal pools release carbon dioxide in direct proportion to the freshwater weight of algae and marine plants per liter of water (Newell, 1979).

Coastal tract is the *habitat* of several marine organisms and plant species. While lithophagic animals of the environment and excreta of sea birds sheltering rock ledges promote *biochemical processes* in the colonized area, certain burrowing animals and salt-tolerant vegetation of the *intertidal zone* trap sediments for building coasts.

FEATURES OF MARINE EROSION

Features of marine erosion portray spatial variations in the geologic structure, strength of rocks and sediments, and wave climate along the littoral tract. *Bays* thus develop in incompetent earth materials and *headlands* in resistant coastal rocks. Headlands attacked by waves from two directions leave a *neck*, which when detached from the mainland is called an island. *Caves* at the foot of sea cliffs develop in exploitable rock fissures, while those expanding from two sides of headlands form *arches* and *residual pillars*. Sea cliffs and shore platform are major landforms of marine erosion.

SEA CLIFFS

Sea cliffs are high-angled promontories along 80% of the shoreline (Emery and Kuhn, 1982). They are commonly associated with resistant rocks of the littoral tract and high-wave energy environment of the mid-latitude coasts. Cliffs *retreat* by the destructive impact of waves against coastal rocks of variable strength. The destructive force, which defies quantification, may at best be viewed as a function of wave climate and strength of the material of coastal rocks (Sunamura, 1977, 1982, 1983). In test conditions, the rate of cliff recession varies directly with the *wave climate* (Figure 11.8(a)) and indirectly with the *compressive strength* of cliff-forming material (Figure 11.8(b)) as

$$Dx/dt = k \ (C + \ln \ (\rho \ g \ H/S_c)$$

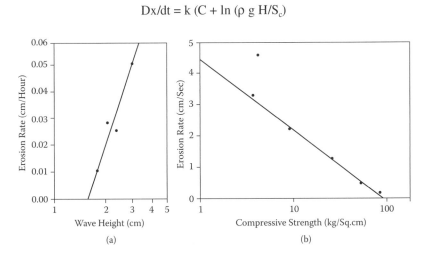

(a) (b)

FIGURE 11.8 The rate of cliff erosion as a function of (a) wave height and (b) material strength. (Source: Figures 5 and 7 in Sunamura, T., "A Relationship between Wave-Induced Erosion and Erosive Forces of Waves," *J. Geol.* 85 [1977]: 613–18. With permission from University of Chicago Press.)

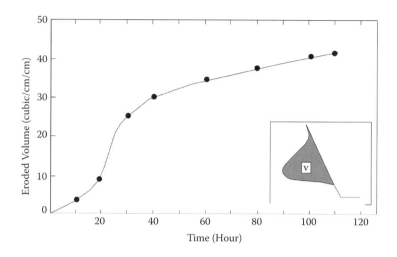

FIGURE 11.9   Cliff recession as a function of time as established from a wave tank experiment. (Source: Figure 2 in Sunamura, T., "A Relationship between Wave-Induced Erosion and Erosive Forces of Waves," *J. Geol.* 85 [1977]: 613–18. With permission from University of Chicago Press.)

in which dx/dt is the rate of cliff recession, k is a constant, C is a function of the wave dimension given as LT^{-1}, H is the wave height, S_c is the compressive strength of cliff-forming material, and ρ is the density of water. The above relation suggests that cliffs retreat only at a certain minimum critical wave height, such that frequent waves of smaller height are inconsequential to cliff erosion. Experimental data further suggest that cliffs initially retreat at a faster rate, but the rate of retreat slows down exponentially with time (Figure 11.9), suggesting dissipation of the wave energy against detrital grains accumulating at the foot of retreating cliffs.

SHORE PLATFORMS

Shore platforms are erosional ramps at the foot of receding sea cliffs. They are commonly intertidal, high-tide, and low-tide morphologic forms of mechanical, chemical, and biochemical activities in marine environments (Figure 11.10). *Intertidal*

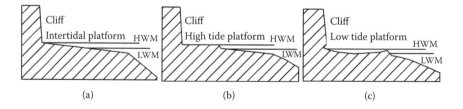

FIGURE 11.10   Types of shore platforms: (a) intertidal platform, (b) high-tide platform, (c) low-tide platform. (Source: Figure 13 in Bird, E. C. F. 1969. *Coasts.* Cambridge, MA: MIT Press. With permission from MIT Press.)

platforms extend between the high- and low-tide level as gently sloping surfaces of wave abrasion. *High-tide platforms* establish above the high-tide level as nearly horizontal surfaces of salt-layer weathering and subaerial denudation. *Low-tide platforms,* with a distinctive wave-cut notch at the foot of backing cliff, evolve at the low-tide level as slightly concave-up ramps of biological weathering, wave erosion, and solution.

The morphologic evolution of *intertidal platforms* is governed by the geologic properties of rocks and the duration of time platforms remain submerged in a tidal cycle. The diurnal tide-gauge data suggest that tides reach peak magnitude about the mid-tide level, such that the higher the tidal range, the smaller is the duration of tidal immersion (Trenhaile and Layzell, 1981). The platforms of high-magnitude tides thus experience stronger wave erosion in the tidal cycle, and are wider and steeper in gradient than platforms of low-magnitude tides. The hourly tide record at the gauge stations, however, suggests two peaks, one each at the rising and falling stages of tides (Carr and Graff, 1982). It, therefore, follows that intertidal platforms may alternatively evolve by *wave erosion* at the rising, and *wave abrasion* at low and mid-tide levels of the tidal cycle.

FEATURES OF MARINE DEPOSITION

Shallow coastal waters are a repository of clastic sediments of littoral origin and coastal sediments of marine origin moved shoreward in breaking and nonbreaking wave conditions. These shallow water sediments are transported and deposited in many different ways by waves, tides and tide-induced currents into features of marine deposition. Beaches are a common form of wave-related deposits, spits and associated forms are the effect of wave refraction, barrier features and chenier plains evolve from the activity of waves and tides, and mudflats and tidal marshes are features of tidal currents.

BEACHES

Beaches are a feature of coastal sedimentation between the low-tide level and the upper limit of wave action (Komar, 1976; Davis, 1978, 1991). They are sand and nonsand in material composition, and "normal" and "storm" by the form of longitudinal profile. *Sand beaches* are common to the low-wave energy environment of the tropical coasts. *Nonsand beaches* comprise mud, shingle, cobble, and biogenic remains of the marine fauna. Mud beaches are confined to sheltered marine environments, shingle and cobble beaches dominate middle- and higher-latitude coasts of moderate- to high-wave energy environment, and beaches of organogenic composition evolve along warm tropical coasts. Normal and storm beaches are symptomatic of the wave climate.

Beach Processes

Sand beaches begin as ephemeral sand bars in a part of the nearshore zone, and extend landward through to the foreshore and backshore coastal zones (Figure 11.11). These bars are an equilibrium form of wave climate and shore normal sediment

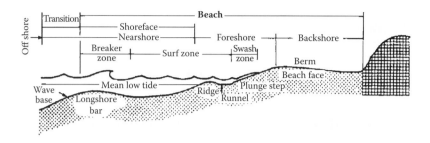

FIGURE 11.11 General profile diagram of a beach and the nearshore zone. (Source: Figure 4 in Davis, R. A., Jr., "Beach and Nearshore Zone," in *Coastal Sedimentary Environments*, ed. R. A. Davis, Jr. [New York: Springer-Verlag, 1978], 237–85. With permission from Springer-Verlag.)

transport initiated at the *breaker zone*. Short-term weather changes (Aubrey and Ross, 1985) and storm conditions (Davis, 1978, 1991; Larson and Kraus, 1994; Boczar-Karakiewicz et al., 1995), however, temporarily affect the sediment transport rate in the *nearshore zone* and disturb the established position of bars. Observations suggest that the nearshore beach gradient varies inversely with water depth, and thus with the dissipation of energy in breaking waves (Dean, 1977, 1983; Schwartz, 1982; Larson, 1991; Bodge, 1992; Wang and Davis, 1998). A part of the beach in the *foreshore zone* comprises ridge and runnel, plung steps, and beach face elements that are in equilibrium with the wave climate and swash gradient (King, 1972). The beach segment in the *backshore zone* is a smooth surface of swash and backwash activity (Strahler, 1969). Here, the beach gradient depends on sediment size, beach's exposure to wave conditions, energy dissipation in shallow waters, and water depth in the nearshore zone. These aspects offer several empirical explanations for the backshore beach gradient (Krumbein, 1944–1947 in Newell, 1979). The beach gradient varies exponentially with particle size (Figure 11.12 top) as

$$D_m = ae^{bS}$$

where S is the beach slope, D_m is the mean particle size, and b is a constant. The beach slope also varies linearly with its exposure to the wave energy (Figure 11.12 middle) as

$$S = kE + K$$

where k and K are constants and E is a measure of relative wave energy. The wave energy is proportional to the square of wave height. Therefore, a small increase in wave height sharply increases the beach slope. The above relationships suggest that the wave energy affects particle size and beach gradient (Figure 11.12(c)) as

$$D_m = ae^{b(kE+K)}$$

Thus, beaches of exposed coasts are built of coarser sediments and are steeper than beaches of protected coasts. The Chesil Beach, England, with an average slope

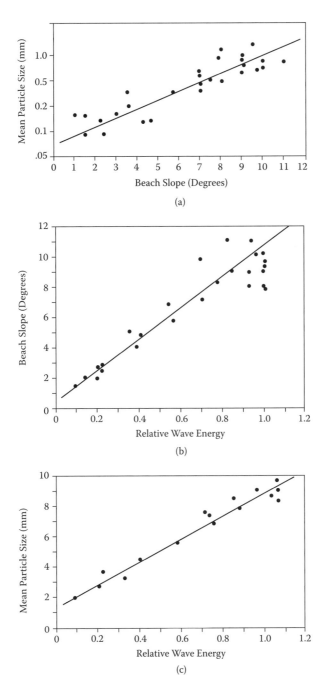

FIGURE 11.12 Relationships for (a) foreshore beach slope and mean particle size, (b) relative wave energy and foreshore beach slope, and (c) and relative wave and mean particle size for Halfmoon Bay, California. (Source: Figures 1.12–1.14 in Newell, R. C., *Biology of Intertidal Animals* [Kent: Marine Ecological Survey, 1979]. With permission.)

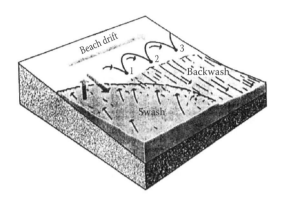

FIGURE 11.13 Beach drifting occurs when waves approach a shoreline obliquely. The swash and backwash move sand particles along the beach in a series of parabolic paths. (Source: Figure 30.5 in Strahler, A. N., *Physical Geography* [New York: John Wiley & Sons, 1969]. With permission.)

of 26° (King, 1972), and the Halfmoon Bay Beach in California, with ½ to 11° of slope (Newell, 1979), illustrate the above effect quite well.

Backshore beach processes are largely governed by the wave height and type of breaking wave in coastal waters. The *wave height* regulates swash and backwash processes. The *swash* arrives at an angle to the beach and, depending upon infiltration losses, reaches a certain limit of uprush before returning down the beach slope as *backwash*. The swash and backwash activity moves beach sediments in a series of parabolic trajectory called *beach drifting* (Figure 11.13). As a rule, beaches of coarser sediments and higher infiltration losses are steeper than beaches of finer sediments and lower infiltration losses.

Breaking waves dissipate wave energy in many different ways, affecting the volume of sediment for building the backshore segment of beaches (Huntley and Bowen, 1975; Dean 1977, 1991). Depending upon the water depth and underwater slope in the nearshore zone, waves break by spilling, plunging, and surging before advancing into the backshore coastal zone (Figure 11.14). The *spilling waves* of typical foaming crests, like at Waikiki in Hawaii and Saunton Sands at Devon, England, are low in velocity and energy attributes. They, therefore, evolve sandy beaches of gentle gradient. The *plunging waves* are forced up the beach by an abrupt underwater slope. They, therefore, curl at crest and release high-wave energy ahead of advancing waves. Steep shingle beaches are commonly associated with plunging waves. The Slapton Sands beach near Devon, England, is one such beach. The *surging waves* break at the wave base, generate turbulence, and place a large volume of coarser sediments for transport up the beach face.

Beach Profile

The shape of a longitudinal beach profile is an adjustment of sediment size to the wave climate. Flat waves move sand onshore and storm waves offshore, evolving normal (nonbarred) and storm (barred) beach profiles, respectively (Figure 11.15). *Normal beaches* develop by aggradation of the berm and beach face elements, and

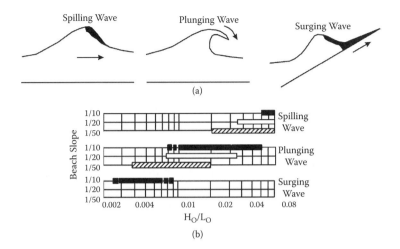

FIGURE 11.14 Types of breaking waves (a) and relation to beach slope and deep water wave steepness (b). (Source: Figure 4.1 in Huntley, D. A., and Bowen, A. J., "Comparison of the Hydrodynamics of Steep and Shallow Beaches," in *Nearshore Sediment Dynamics and Sedimentation*, ed. J. Hails and A. Carr [London: John Wiley & Sons, 1975], 69–110. With permission from John Wiley & Sons.)

are steeper in gradient. *Storm beaches* evolve by erosion of the beach face and deposition of this sediment as a break-point bar, developing a somewhat gentler barred profile. Experimental data suggest that a *threshold wave steepness* of 0.012 distinguishes the two beach forms adequately well (King, 1972). Other models based on wave characteristics and sediment transport rate in variable wave environment conditions have also been developed to distinguish normal and storm beach profiles. Dean (1973) observed that the expression H_o/wT, in which H_o is the deep water wave height, w is the sand fall velocity, and T is the wave period, distinguishes normal and storm beach profiles at the threshold value of 0.85. A sediment transport parameter for deep water waves (π), given as $115^2 \pi/4$, is <10,400 for normal and >10,400 for storm beach profiles (Dalrymple, 1992). The energy dissipation in waves breaking on sandy beaches suggests that a threshold limit of 2/3 for shallow water depth given as $d = Ax^m$, in which d is the water depth, m is the coefficient for beach slope, k is the distance from shore, and A is the sediment size, distinguishes normal and storm beach profiles adequately well (Dean, 1977).

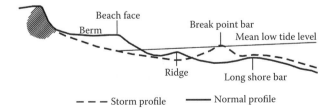

FIGURE 11.15 Normal (barred) and storm (nonbarred) beach profiles. Storm beach profile is gentler in gradient than normal beach profile.

FEATURES OF INDENTED COASTS

Spit forms and tombolos are typical features of indented coasts. *Spits* develop by the effects of longshore currents and wave refraction in seawater across the mouth of tidal inlets and estuaries as simple, curved, and complex forms of sand or pebble accretion, which, while being attached to the coast at one end, prograde into the adjacent body of shallow water as cusps of free limb. Some of the best-known spits exist along the coasts of inland lakes and landlocked seas (Bird, 1969). Spits connecting bays are *baymouth bars*. Spits along coasts of strong wave refraction, however, curl shoreward, and are called *cuspate spits*. The cuspates of Krishna (Figure 11.16) and Godavari rivers in the Bay of Bengal, India, have rapidly developed since the 1960s as a result of large influx of sediment load into the sea from the cultivated land (Vaidyanadhan, 1987). Beach ridges adjusted to wave refraction in the *wave shadow* of offshore islands and along coasts of *convergent waves* develop into triangular cuspates with a sea-facing apex called *cuspate forelands*. The cuspate forelands of Dungeness, England, and Cape Kennedy, Florida, are similarly the result of material adjusting to the *wave environment* (Bird, 1969). The cuspate foreland of Dungeness (Figure 11.17) has particularly evolved from the addition of curved shingle beach ridges across the mouth of a wide shallow channel. The growth pattern of this cuspate suggests shifting of the stream draining into the sea, sea level oscillations of +0.5 to –2.0 m in the last 2500 years, and temporal changes in the pattern of wave erosion and deposition along the shoreline (Strahler, 1969; Bird, 1969; King, 1972).

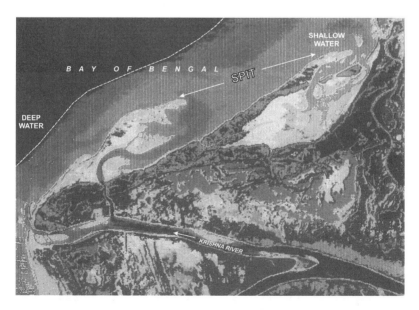

FIGURE 11.16   Spits and cuspate spits at the mouth of Krishna River in the Bay of Bengal, India. They have rapidly evolved as a result of land use changes in the recent decades. (Source: Based on the IRS-LISS-II digitally processed data, National Remote Sensing Agency, Hydrabad, India.)

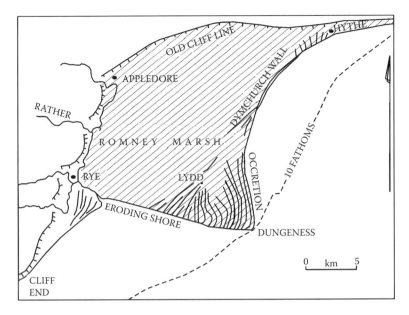

FIGURE 11.17 Dungeness cuspate foreland on the Dover Straits, southeastern England. (Source: Figure 38 in Bird, E. C. F. 1969. *Coasts*. Cambridge, MA: MIT Press. With permission from MIT Press.)

FIGURE 11.18 Two tombolos connecting an offshore island with the mainland. (Source: Figure 30.15 in Strahler, A. N., *Physical Geography* [New York: John Wiley & Sons, 1969]. With permission.)

Tombolos are beach ridges adjusted to wave refraction in the *wave shadow* of islands (Figure 11.18). They are normally built of sand, which had drifted from the sides of islands and moved mainland in the wave shadow of islands.

FEATURES OF FLAT COASTS

Barrier bars, barrier beaches, barrier islands, and chenier plains are characteristic features of flat coasts. *Barrier bars* evolve from alongshore and shore normal migration of sediments (Figure 11.19). The barrier bars of *alongshore* sediment transport are

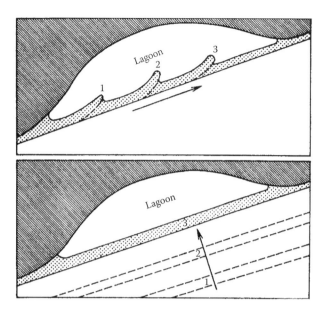

FIGURE 11.19   Evolution of barrier beaches by extension of a spit alongshore (above) and by shore normal migration of a barrier that originated offshore. (Source: Figure 42 in Bird, E. C. F. 1969. *Coasts*. Cambridge, MA: MIT Press. With permission from MIT Press.)

diverse in origin. They are identified with baymouth bars (Bird, 1969), spits drowned by the Holocene rise of sea stand (Oldale, 1985), and ridges built of wave-eroded sand beaches in the middle- and high-latitude coasts (Short, 1975). The barrier bars of *shore normal* sediment transport are shore-migrating ridges in equilibrium with the rising rate of Holocene sea stand (Bird, 1969). *Barrier beaches* are partly emergent ridge-like sand accumulation along low-latitude coasts and coasts of low to moderate tidal range. They are generally absent from rocky coasts and coasts of limited supply of sediments (King, 1972). *Barrier islands* are diverse in origin. They are identified with spits detached at tidal inlets (Fisher, 1968), partly submerged coastal dunes (Hoyt, 1967), segments of shore normal drifted longshore bars (Pierce and Colquhoun, 1970), and aggrading barrier bars of stable sea stands (Otvos, 1970). *Chenier plains* are comprised of lagoons and beach ridges of sand and shell debris perched on mudplains (Figure 11.20). They are an equilibrium form in adjustment with the stream flow and fluvial sediments reaching the delta front, such that relative excess of terrestrial sediments progrades deltas over the continental shelf and that of stream flow erodes the delta front. The wave activity winnows the released sediments and segregates mud. This mud evolves mudplains on which beach ridges with shallow lagoons between them subsequently evolve by the alongshore migration of sediments (Hoyt, 1969; Kraft, 1978). Modern cheniers extend to the limit of *tidal flats*.

LAGOONS

Lagoons are elongated shallow water bodies between the shoreline and open sea. They are estuarine and lake type by access to the sea, and hypersaline and polyhaline

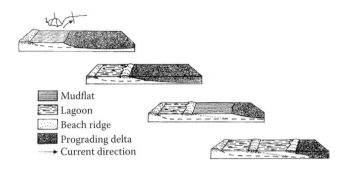

Mudflat
Lagoon
Beach ridge
Prograding delta
→ Current direction

FIGURE 11.20   A schematic representation of the development of a chenier plain. (Source: Figure 13 in Kraft, J. C., "Coastal Stratigraphic Sequence," in *Coastal Sedimentary Environments*, ed. R. A. Davis, Jr. [New York: Springer-Verlag, 1978], 361–83. With permission from Springer-Verlag.)

by the salinity level. The *estuarine-type lagoons* are connected to the sea by tidal inlets, and *lake-type lagoons* are isolated from the sea. *Hypersaline lagoons*, with salinity in excess of 100‰, occur along high-thermal-efficiency coasts and coasts of limited tidal activity. The salinity of *polyhaline lagoons*, however, varies markedly with the seasonal climatic rhythm: being below the normal salinity of ≈35% in wet season and above the normal salinity in dry season.

MUDFLATS

Mudflats are an *intertidal feature* of macro-tidal estuaries. *Macro-tidal estuaries* are characteristic of a tidal range of 4 m and above, a strong *flood*, and a relatively weak *ebb* current in the system. Mudflats, in general, are most extensive along the northern coast of the Netherlands, fairly widespread off the coast of Surinam and southeastern United States, and of limited extent along the coast of India. They are built of finer-sized marine and fluvial sediments, in which the relative contribution of two sources varies from coast to coast. The mudflat of Wash, eastern England, is comprised largely of marine sediments (Evans and Collins, 1975), and the mudflat of the Gulf of Kachchh (Kutch), western India, is built largely of fluvial sediments (Glennie and Evans, 1976).

Mudflats comprise an upper flat surface between the mid- and high-tide levels, and a lower-sloping surface about the mid- and low-tide level (Figure 11.21). The upper segment of the feature is a surface of clay and silt deposits, and the lower segment is built largely of sand-sized fractions. These variations in the plan view and sedimentation aspects are governed by a high concentration of suspended particles in shallow waters offshore and rhythmic changes in the direction of tidal currents passing through estuaries every 6 h. In the saline seawater the suspended clay easily flocculates, forming mud of a lower-density composition and of open structure with up to 90% water content. A large part of the offshore floating mud is carried upstream in the *flood current*, where much of it falls out of suspension and settles along estuaries by progressive weakening of the upchannel tidal current. Thus, only a part of the mud moved landward returns to the sea in the weak *ebb current*.

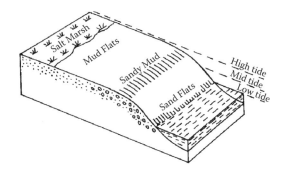

FIGURE 11.21 Relationship between the morphology of mudflats and sediment variation in the Wash, eastern England. (Source: Figure 8.2 in Pethick, J., *An Introduction to Coastal Geomorphology* [London: Edward Arnold, 1984]. With permission from Edward Arnold Ltd.)

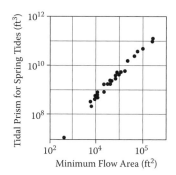

FIGURE 11.22 Relationship of the volume of tidal prism for spring range to the minimum cross-sectional area of the entrance for some natural inlets in the United States with the Wash also plotted. (Source: Figure 11.10 in Evans, G., and Collins, M. B., "The Transportation and Deposition of Suspended Sediment over the Intertidal Flats of the Wash," in *Nearshore Sediment Dynamics and Sedimentation*, ed. J. Hails and A. Carr [London: John Wiley & Sons, 1975], 273–306. With permission from John Wiley & Sons.)

By comparison, the lower segment of mudflats is a surface of high tidal current velocity and low rate of sedimentation. The size of mudflats, therefore, depends on the *tidal prism* that governs the relative strength of flood and ebb currents. In most cases, the mudflat size is positively related to the tidal prism for *spring tides* (Figure 11.22).

The mudflats of the Ranns of Kachchh (Kutch) and Gulf of Khambhat (Cambay), India, are *supratidal*. The sediments of the Ranns of Kachchh were derived from the mouth of Indus draining into the Arabian Sea, and moved alongshore in historic times (Glennie and Evans, 1976). The Indus flow and the peninsular drainage into the Arabian Sea, however, do not contribute as much sediment now for the building of mudflats.

TIDAL MARSHES

Tidal marshes are *saline wetlands* of temperate and tropical coasts. They extend between the neap tide and storm wave limits as low and high marsh, in which part of

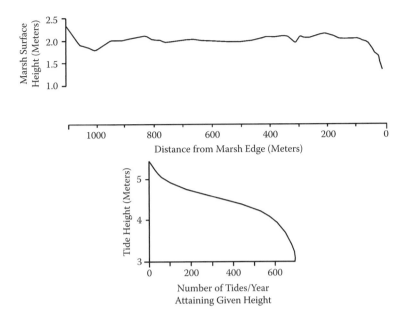

FIGURE 11.23 Profiles across tidal marshes in the Tamar Estuary, Cornwall. The bottom figure shows the distribution of high-tide levels for that estuary. (Source: Figure 8.10 in Pethick, J., *An Introduction to Coastal Geomorphology* [London: Edward Arnold, 1984]. With permission from Edward Arnold Ltd.)

the surface between the neap and spring tide level is *low marsh* and that between the spring tide and storm wave limit is *high marsh*. The relative size of two marsh areas is commonly used as an index of the maturity of marsh, such that a *mature marsh* carries a higher proportion of area in high marsh and salt-tolerant vegetation (Davis, 1991). The decay of this vegetation produces *peat*.

Tidal marshes are convex up facing the sea and concave up toward the land (Figure 11.23). This profile is adjusted to the frequency and volume of suspended load in tidal currents (Pethick, 1984). In general, the sea-facing convex-up surface receives the highest frequency of fairly strong tidal currents and an equally large volume of suspended sediments, which the salt-tolerant vegetation efficiently traps. The strength of tidal current and the volume of suspended load, however, progressively decrease upchannel, evolving a concave-up segment of the tidal marshes. Tropical coasts of a meso-tidal range, low coastal relief, a large supply of sediments, and a low-wave energy environment, however, develop tidally submerged extensive *mangrove swamps*, such as along the southwestern coasts of Florida, the Pacific coast of Colombia and Ecuador, and the coast of Borneo, the Philippines, and Malaysia. Mangrove swamps in the Ganga-Brahmaputra delta region of India and Bangladesh are called *sunderbans*.

ORGANIC REEFS

Several marine life forms precipitate carbonate in skeletal and nonskeletal forms, evolving the structure of organogenic eminencies called reefs. Organic reefs are

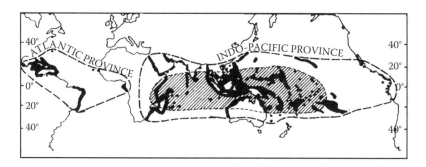

FIGURE 11.24 World distribution of coral reefs. Areas shown in solid black represent the most prolific reef development. The hatched region includes areas of oceanic atolls. (Source: Figure 47 in Davies, J. L., *Geographical Variation in Coastal Development* [Edinburgh: Oliver & Boyd, 1980]. With permission from Addison Wesley Longman.)

noncoral and coral in type. *Noncoral reefs* are the remains of oyster and shellfish adapted to estuarine and mangrove environments. Modern *coral reefs* belong to a group of sedentary Coelenterate invertebrates of the class Anthazoa, which dwell to the depth of sunlight penetration in shallow waters offshore. The coral polyps respire shell-forming *calcium carbonate*, which they extract from the seawater, and provide for the composition of the hard and resistant structure of organic reefs.

Scleratina or stone-like colonial polyps are common contemporary reef-forming marine organisms. They are adapted to the mean annual seawater temperature of more than 25°C and the mean water temperature of not less than 18°C for the coldest month, a firm sea bottom not exceeding 20 m depth for photosynthesis, clear and oxygenated water, and normal salinity of 33 to 37%. These tolerance limits are better satisfied in tropical waters of the Indo-Pacific province than at comparable latitudes in the Atlantic province (Figure 11.24). Even within the *Indo-Pacific province*, corals are absent near the mouth of streams draining a large volume of discharge and suspended load into the sea. The freshwater discharge dilutes salinity and suspended load interferes with the depth of sunlight penetration for *photosynthesis* in muddy waters. In the *Atlantic province*, the coral habitat appears to have been altered by cooler temperatures of the seawater during glacial phases of the *Pleistocene climate* (Davies, 1980).

Types of Coral Reef

Coral reef is a structure of organogenic calcium carbonate, which rises from the sea bottom to the tide limit. Charles Darwin first suggested that atoll reefs evolve through a sequence of fringing and barrier reefs (Figure 11.25). This classification of 1838 has since been expanded many times to incorporate the great diversity of reef forms present in nature (Table 11.1).

Fringing reefs are narrow eminences of reef limestone some distance from the shoreline and offshore islands. They present an uneven rough surface at about the low water level. Fringing reefs are absent in turbid waters and off the mouth of streams discharging freshwater into the sea.

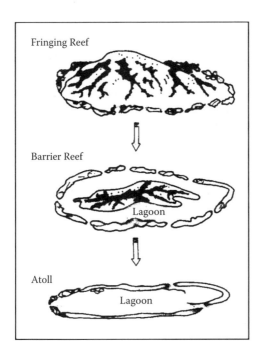

FIGURE 11.25 Stages in the evolution of atolls, from fringing through to a barrier reef. (Source: Adapted from Figure B15.1 in Davis, R. A., Jr., *Oceanography* [Dubuque: Wm. C. Brown Publishers, 1991].)

TABLE 11.1
Descriptive Classification of Organic Reef Forms

Characteristics/Association	Type
1. Adjacent to coast or separated by shallow channel	Fringing reef
2. Separated from coast by deep channel	Barrier reef
3. Forming islands without central lagoon:	
Large	Platform reef
Small	Patch reef
Very small	Coral pinnacle
4. Forming islands with central lagoon:	
Deep lagoon	Oceanic atoll
Shallow lagoon	
Large	Shelf atoll
Small	Faro

Source: Davies, J. L., *Geographical Variation in Coastal Development* (Edinburgh: Oliver & Boyd, 1980). With permission from Addison Wesley Longman.

Barrier reefs are a massive ridge-like structure of coral on continental shelves that now lie 1500 m or more down the limit of sunlight penetration for *photosynthesis* in the seawater. This form of the reef is related to progressive slow subsidence of the seabed and rise of sea level, enabling the coral polyps to grow toward the sea surface level at the rate of net increase in the seawater depth. For reasons not understood, the barrier reef over the outer continental shelf in the Gulf of Sunda, Indonesia, however, could not maintain a regular growth during the Holocene rise of sea level. The *Great Barrier Reef* off the coast of northeastern Australia is 1930 km long, and the barrier reefs off Borneo, Palau Island, New Caledonia, and Fiji in the Pacific Ocean are noteworthy for their size and splendor.

Platform and *patch reefs* are some 20 to 40 m deep submerged tabular eminences on the continental shelf, in which the platform reefs distinguish from the patch reefs by being more than 1.5 km long. The two reef forms establish around *headlands* and grow toward the sea surface level in harmony with gradually rising Holocene sea stand. *Coral pinnacles* are up to 10 m high isolated stumps of wave erosion on the sea-facing surface of barrier reefs.

Atolls and *faros* form islands around central lagoons. Atolls are oceanic and shelf type by the nature of foundation and evolutionary process. *Oceanic atolls* are some 2000 m thick ring-shaped deposits of modern and pre-Quaternary coral skeletons based on submarine volcanoes called seamounts. *Shelf atolls* represent the organogenic structure of calcium carbonate on continental shelves and preexisting platform reefs that are generally 400 to 500 m below the sea surface level. Faros evolve at the rim of major atolls, suggesting cratonic subsidence of the oceanic crust and glacioeustatic rise of the sea level. Lakshadweep Islands, India, and Maldives, with nearly two thousand islands in the Indian Ocean, are the faro form of reefs.

SEA LEVEL CHANGE

Sea level is the water surface elevation at tide gauge stations corrected for the effects of waves and tides, upwelling from deep sea currents, variation of pressure, temperature, salinity, and stream discharge, and movement of continental blocks. Global climate change and instability of the earth, however, have resulted in long-term sea level fluctuations, called sea level changes, which can be reconstructed with confidence to the beginning of the Quaternary period. The pattern and magnitude of sea level changes is central to the explanation of coastal landforms and interpretation of the chronology of coastal development.

CLIMATIC CAUSES

Climate-driven growth and melting of massive continental ice 600 to 700 Ma in the Precambrian, 350 to 450 Ma in the Silurian–Devonian, 200 to 300 Ma in the Triassic–Jurassic, and 2 Ma to 10,000 years BP in the Quaternary period affected the global sea stands on a regular basis. Of these, the signature only of the Pleistocene climate change can be inferred with confidence from attributes of the contemporary landscape and evidence present in the environment-sensitive indices of the sea level change. The evidence of climate change for the remote past is known to be completely lost in crustal deformations of the earth.

The multiple Pleistocene *glacial events* (Chapter 7) required massive transfer of the seawater for building successive bodies of thick continental ice, which lowered the sea stand with certain regularity. Melting of the ice in *interglacial periods*, however, returned the locked water in the ice to its source and raised the sea surface level periodically. The evidence for glacial-interglacial sea level changes is recorded in the form of a temperature-sensitive *oxygen isotope ratio* ($^{18}O/^{16}O$) in the calcium carbonate skeletal remains of marine *foraminifera* (Bowen, 1990). A difference of 0.1% in the ratio of $^{18}O/^{16}O$ theoretically equals 10 m change in the sea level, the total difference of which in the remains of foraminifera provides for a net 100 m change in the Quaternary sea stand (Shackleton and Opdyke, 1976), and another 75 m inferred rise in the sea stand if all the existing global ice were to melt (Fairbridge, 1960). The Pleistocene sea level changes also correlate well with the theoretical temporal pattern of solar radiation and temperature of the seawater (Figure 11.26). The oxygen isotope data further suggest that the sea level rapidly increased between 17,000 and 8000 years BP at the rate of 1 m/century and stabilized thereafter at about 6000 BP (Figure 11.27). This period of near-stable sea stand is called *Holocene marine transgression*. Raised coastal beaches and organic reefs, however, do not correlate with the above pattern of sea level change, suggesting that the sea stand has also been affected by other causes.

OTHER CAUSES

A record of the environmental change in rocks and landforms reveals that a major tectonic instability struck the earth at the beginning of the Tertiary and continued into the Quaternary period, causing rise of the landmass and subsidence of the sea floor on a grand scale. The relative movement of land and sea caused an *apparent sea level change*, which provides for the explanation of raised beaches, thalassostatic terraces, and organogenic eminences of atoll and faro worldwide. The drowned mouth of the Hudson River and relatively shallow inlets like the Chesapeake Bay along the east coast of North America, and the opening of the Straight of Dover separating Briton from continental Europe, are illustrations of the effect of postglacial relative movements of the land and sea. In this sense, the coasts worldwide are relatively young.

The *loading* of continental crust by more than 3 km thick Pleistocene ice at places and its *unloading* due to thaw of the ice has caused *isostatic rebound* of the continental crust. The loading depressed the crust by an amount roughly equal to one-third of the maximum thickness of ice, producing a compensatory downwarping, subsidence, and tilting of continental blocks adjacent to the glaciated areas (Dillon and Oldale, 1978). The isostatic rebound and crustal deformation have affected the sea stand in many different ways.

CLASSIFICATION OF COASTS

Coastal morphology has evolved in many different ways by the combined effects of geologic structure, tectonic activity, present and past littoral and marine processes,

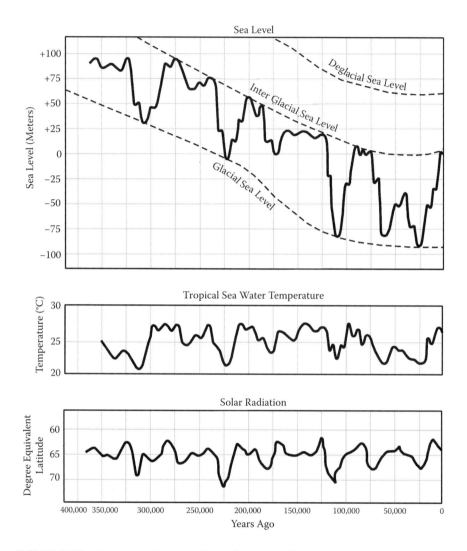

FIGURE 11.26 Long-term changes in the sea level caused by glacial cycles in the Pleistocene (above), seawater temperature (middle), and solar radiation patterns (below). (Source: Figure on p. 77 in Fairbridge, R. W., "The Changing Level of the Sea," *Sci. Am.* 202 [1960]: 70–79. With permission from Scientific American © 1960.)

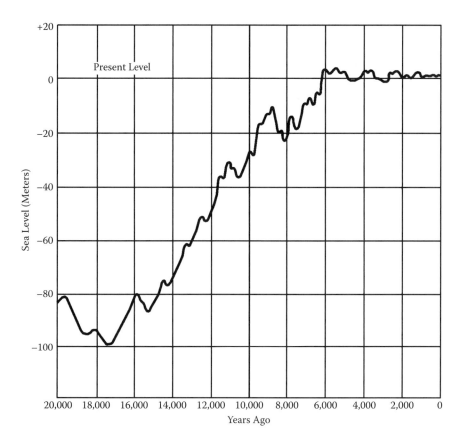

FIGURE 11.27 Sea level change showing rapid upsurge between 17,000 and 6,000 years ago. The sea level has remained relatively stable for the past 6,000 years. (Source: Figure on p. 76 in Fairbridge, R. W., "The Changing Level of the Sea," *Sci. Am.* 202 [1960]: 70–79. With permission from Scientific American © 1960.)

and variations in the Quaternary sea stand. Hence, several descriptive classifications emphasizing one or another's control of coastal morphogenesis have been suggested in the literature. In 1919, Johnson proposed that a sea level change in relation to adjacent land produces submergent, emergent, neutral, and compound coasts (King, 1972). *Submergent coasts* identify with ria and fjord coasts, *emergent coasts* with beach and barrier features, *neutral coasts* with deltas, alluvial plains, volcanic rocks, and organogenic eminences, and *compound coasts* with the assemblage of features from the above types. Valentin's 1952 classification recognized advanced and retreated coasts by apparent stability of the sea stand (King, 1972). *Advanced coasts* suggest emergence and progradation, and *retreated coasts* submergence and loss of landmass by wave erosion. Bloom (1965) introduced the element of time in coastal morphogenesis, and suggested that coasts either *advance* by deposition and emergence or *retreat* by subsidence and wave erosion. Cotton's 1952 structural classification recognizes coasts of stable and unstable regions by sea level oscillations of the Quaternary and tectonic instability of landmass (King, 1972). The coasts of

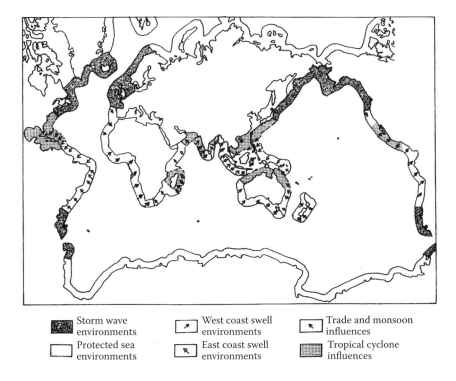

Storm wave environments

Protected sea environments

West coast swell environments

East coast swell environments

Trade and monsoon influences

Tropical cyclone influences

FIGURE 11.28 Major global wave environments for the classification of coasts. (Source: Figure 27 in Davies, J. L., *Geographical Variation in Coastal Development* [Edinburgh: Oliver & Boyd, 1980]. With permission from Addison Wesley Longman.)

stable regions present the affect of Quaternary sea level oscillations and of *unstable regions* the affect of Quaternary earth movements on emergence and submergence of the coasts. Coastal landscapes, however, do not differentiate the climatic signature from the tectonic signal. Shepard's 1963 classification distinguishes primary and secondary coasts, such that *primary coasts* evolve by nonmarine and *secondary coasts* by marine processes (King, 1972).

Inman and Nordstrom (1971) proposed a geophysical classification of coasts based on the concept of *plate tectonics*, suggesting that the nature of interaction between continental and oceanic plates at plate boundaries evolves morphologically distinct collision, trailing edge, and marginal coasts, and their subtypes. *Collision coasts* occur along the subduction zone of plate interaction, *trailing edge coasts* align with the trailing edge of continental plates, and *marginal coasts* occur along oceanic plates facing island arcs.

Davies (1980) recognized that latitude-dependent storm, swell, and protected wave environments strongly affect the contemporary morphology of coasts (Figure 11.28). The *storm wave environment* of mid-latitude coasts is typically associated with shingle beaches, wide shore platforms, and sea cliffs. The east and west coast *swell environment* at tropical latitudes develops sandy beaches, and coasts commonly devoid of sea cliffs and shore platforms. The *protected wave environment* at high latitudes, and along bays, estuaries, and enclosed seas, is commonly reflected in the dominance of mud beaches and the near absence of shore platforms along coasts.

COAST OF INDIA

Nearly 5,700 km of coast of India, including that of the Andaman and Nicobar Islands in the Bay of Bengal and Lakshadweep Islands in the Arabian Sea, is diverse in morphology, evolution, and chronology (Vaidyanadhan, 1987). The east coast, in general, has prograded into the Bay of Bengal over a subsiding continental shelf. Parts of the west coast have evolved by faulting and subsidence of the littoral tract in the remote geologic past and neotectonic subsidence of the continental shelf in the Arabian Sea (Rao et al., 1996). The northern segment of the coast, which had been an island in the Tertiary, has evolved by shore normal and alongshore accumulation of sediments in the Quaternary period of rising sea stand over a graben structure (Gupta, 1972; Glennie and Evans, 1976). A large segment of the west coast in the south is receding by wave erosion. Shell and reef debris are common to the group of islands in the Bay of Bengal and Arabian Sea.

Using morphologic data from topographic maps and one or another *inferred* aspect of Quaternary sea level oscillations, apparent stability of the sea stand, and relative significance of terrestrial and marine processes, Ahmad (1972) classified the coast of India into seven empirical types (Figure 11.29). The *Kathiawar Coast* over an active graben structure is extensively built of supratidal mudflats, saltpans, noncoral reefs, raised beaches, and tidal flats. The *Gujarat Coast* is similar to the Kathiawar Coast in structural and tectonic aspects. It largely comprises mudflats and tidal marshes around the Gulf of Khambhat. The *Konkan Coast* owes its morphology to faulting and subsidence of the littoral tract. The *Malabar Coast* is complex in morphology, suggesting emergence, submergence, and recession along segments of the shoreline. The *Coromandal Coast* is comprised of raised beaches, coastal dunes, bars, and spits that are so well characteristic of the flat coasts. The *Deltaic Coast* is identified with prograding deltas of the Ganga, Mahanadi, Godavari, Krishna, and Cauvery rivers draining into the Bay of Bengal. The *Andaman and Nicobar Coast* is rugged, moderately indented along the northern group of islands, and built of volcanic ejecta, modern coral, and coastal swamps. The coastal tract of Lakshadweep Islands in the Arabian Sea comprises nearly 3,000-year-old coral debris.

SUMMARY

Waves, tides, and currents are dominant processes of the coastal environment. Waves, called swell, are generated by the transfer of wind energy on to the surface of oceans. These deep water waves are sinusoidal in form, and regular in length, height, and period attributes. They travel at an angle to the shore, enter shallow waters offshore, modify in dimensional properties except the period, and break, dissipating the wave energy by refraction, diffraction, and reflection in the system. The breaking of waves makes coastal sediments available for transport and building the coast. Lunar tides result from the gravitational attraction of the moon and the sun on the earth's surface, which sets in motion a rhythmic rise and fall of the sea surface level. Tidal

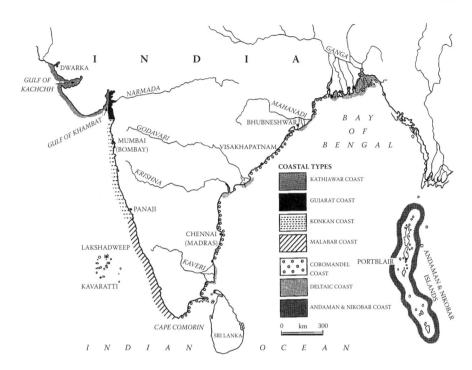

FIGURE 11.29 A descriptive classification of the coast of India. (Source: Figure 34 in Ahmad, E., *Coastal Geomorphology of India* [New Delhi: Orient Longman, 1972]. With permission from Orient Longman.)

effects generate flood and ebb currents in estuaries and tidal inlets. These currents completely reverse direction of flow once or twice daily, and transfer a large volume of sediments between the coastal and littoral environments. Shallow water waves induce longshore currents and rip currents offshore. Longshore currents are caused by the alongshore component of momentum in breaking waves, and have an important effect on long-term building of coasts. The movement of water toward the shore traps water and wave energy that affects the wave setup in shallow coastal waters. This excess water returns as undertow to the sea through a cell circulation system of rip currents. Rip currents are essentially a balancing mechanism for the flow of water and sediments between the nearshore and foreshore coastal zones.

Coastal sediment transport, which is a mix of suspended and bedload particles, is initiated by the turbulence in breaking waves and dissipation of wave energy shore normal and alongshore at the breaker zone. Theoretical and experimental sediment transport models suggest that shore normal sediment transport rate varies directly with the wave climate, water depth, and energy dissipation in breaking waves, and that longshore sediment transport rate varies with the wave energy, angle of wave approach to the shore, and cohesion among grains.

Climate-dependent nival and salt-layer weathering activities, and mechanical, chemical, and biochemical processes of erosion, variously affect the development of coasts between limits of the wave spray and low-tide level. The destructive impact of waves evolves retreating sea cliffs against more resistant rocks of the littoral tract. In test conditions, the rate of cliff recession varies directly with the wave climate and indirectly with the strength of rock-forming material. The receding cliffs leave intertidal, low-tide, and high-tide shore platforms at their foot. The morphology of intertidal platforms depends on the duration of wave abrasion in tidal cycles, of low-tide platforms on wave abrasion and chemical and biochemical erosion, and of high-tide platforms on salt-layer weathering to the highest limit of wave spray against sea cliffs.

Beaches are a morphologic form of shingle, sand, or clay accretion between a part of the nearshore through to the foreshore and backshore coastal zones, in which the beach material size varies directly with the wave environment. Beaches are normal (or nonbarred) and storm (or barred) at certain thresholds of shallow and deep water wave environments.

Wave refraction along indented coasts evolves spits, cuspate spits, cuspate forelands, and tombolos. The spits-forming sediments are moved in longshore currents and shaped by wave refraction in coastal waters. Spits connecting bays are baymouth bars. Spits along coasts of strong wave refraction, however, strongly curve landward to become cuspate spits. Triangle-shaped cuspates with a sea-facing apex evolve in the wave shadow of offshore islands and along coasts of convergent waves. Tombolos are beach ridges adjusted to wave refraction in the wave shadow of offshore islands.

Barrier bars, barrier beaches, barrier islands, and chenier plains are typical features of flat coasts. Barrier bars are built of the sediments moved alongshore and shore normal at the breaker zone. The barrier bars of alongshore sediment transport are diverse in origin, and of shore normal sediment transport evolve barrier ridges and barrier islands. Barrier islands are identified with spits detached at tidal inlets, partly submerged coastal dunes, and segments of longshore bars. Chenier plains are mudplains carrying lagoons and beach ridges. They evolve in the intervening period of coastal retreat and delta progradation.

Mudflats extend between high- and low-tide levels as a feature of the macro-tidal environment. Their longitudinal profile is adjusted to the concentration of suspended load in the flood current moving up tidal channels. Tidal marshes extend between the spring tide and storm wave limit as saline wetlands of temperate and tropical coasts of meso-tidal rage.

Coral reefs are a hard structure of calcium carbonate at and near the sea level. This structure is built of sedentary colonies of coral polyps that thrive in well-oxygenated and clear tropical waters to the depth of photosynthesis in the seawater. Coral reefs are fringing reef, platform and patch reef, barrier reefs, atoll, and faro by morphology and association with coastal and submarine features. Barrier reefs, atoll, and faro forms rise from great depths in the seawater, suggesting subsidence of the sea floor and glacioeustatic rise of the sea level.

Quaternary sea level oscillations due to the climate change, instability of the landmass, subsidence of the sea floor, and isostatic rebound have affected the morphology, evolution, and chronology of coasts in many different ways. A record of the environmental change in oxygen-bearing carbonate skeletal remains of marine organisms called *foraminifera* suggests a 100 m change in the sea level between glacial-interglacial phases of the Pleistocene climate.

Geologic structure, littoral and marine processes of the present and past, and apparent sea level change have evolved coastal morphology in many different ways. Hence, no less than sixteen major descriptive classifications have been proposed since 1912 for the morphodynamics and chronology of coastal landscapes. These classic hypotheses form the basis of empirical classification of coasts. The more recent classifications, however, explain coastal morphology by the concept of plate tectonics and the global wave environment.

REFERENCES

Aagard, T. 1991. Multiple bar morphodynamics and its relation to low frequency edge waves. *J. Coastal Res.* 7:801–13.

Ahmad, E. 1972. *Coastal geomorphology of India.* New Delhi: Orient Longman.

Aubrey, D. G., and Ross, R. M. 1975. The quantitative description of beach cycles. *Marine Geol.* 69:155–71.

Bailard, J. A., and Inman, D. L. 1981. An energetic bedload model for a plane sloping beach: Local transport. *J. Geophys. Res.* 86:2035–43.

Basocm, W. 1959. Ocean waves. *Sci. Am.* 74–84.

Bird, E. C. F. 1969. *Coasts.* Cambridge, MA: MIT Press.

Bloom, A. L. 1965. The explanatory description of coasts. *Z. Geomorph.* 9:422–36.

Boczar-Karakiewicz, B., Forbes, D. L., and Drapean, G. 1995. Nearshore bar development in southern Gulf of St. Lawrence. *J. Wtrwy. Harb. Coast. Eng.* 121:49–60.

Bodge, K. R. 1992. Representing equilibrium beach profiles with an exponential expression. *J. Coastal Res.* 8:47–55.

Bowen, A. J. 1980. Simple models of nearshore sedimentation: Beach profiles and longshore bars. In *The coastline of Canada: Littoral processes and nearshore bars*, ed. S. B. McCann, 1–11. Ottawa: Geological Survey of Canada.

Bowen, A. J., and Inman D. L. 1971. Edge waves and crescentic bars. *J. Geophys. Res.* 76:8662–71.

Bowen, R. 1990. *Isotope and climate.* London: Elsevier Applied Sciences.

Bowman, D., Birkenfeld, H., and Rosen, D. S. 1992. The longshore flow component in low energy rip channels; the Mediterranean, Israel. *Marine Geol.* 108:259–74.

Carr, A. P., and Graff, J. 1982. The tidal immersion factor and shore platform development: Discussion. *Trans. Inst. Br. Geogr.* NS7:240–45.

CERC. 1984. *Shore protection manual.* Washington, DC: U.S. Army Corps of Engineers, Coastal Engineering Research Center.

Dalrymple, R. A. 1992. Prediction of storm/normal beach profiles. *J. Wtrwy. Harb. Coast. Eng.* 118:193–200.

Davies, J. L. 1980. *Geographical variation in coastal development.* Edinburgh: Oliver & Boyd.

Davis, R. A., Jr. 1978. Beach and nearshore zone. In *Coastal sedimentary environments*, ed. R. A. Davis, Jr., 237–85. New York: Springer-Verlag.

Davis, R. A., Jr. 1991. *Oceanography.* Dubuque: Wm. C. Brown Publishers.

Davis, R. A., Jr., Hine, A. C., and Shinn, E. A. 1992. Holocene coastal development of the Florida Peninsula. In *Quaternary coasts of the United States: Marine and Lacustrine systems*, ed. C. H. Fletcher III and J. F. Wehmiller, 193–213. SEPM Special Publication 48. SEPM.

Dean, R. G. 1973. Heuristic models of sand transport in the surf zone. In *Proceedings of the Conference on Engineering Dynamics in the Surf Zone,* Institute of Engineers Sydney, Australia, pp. 208–14.

Dean, R. G. 1977. *Equilibrium beach profiles: U.S. Atlantic and Gulf coasts.* Ocean Engineering Report 12. Newark, DE: Department of Civil Engineering, University of Delaware.

Dean, R. G. 1983. Principles of beach nourishment. In *Handbook of coastal processes and erosion,* ed. P. Komar, 217–31. Boca Raton, FL: CRC Press.

Dean, R. G. 1991. Equilibrium beach profiles: Characteristics and applications. *J. Coastal Res.* 7:53–84.

Dillon, W. P., and Oldale, R. N. 1978. Late Quaternary sea-level curve: Reinterpretation based on glaciotectonic influence. *Geology* 6:56–60.

Emergy, K. O., and Kuhn, G. G. 1982. Sea cliffs: Processes, profiles, classification. *Bull. Geol. Soc. Am.* 93:644–53.

Evans, G., and Collins, M. B. 1975. The transportation and deposition of suspended sediment over the intertidal flats of the Wash. In *Nearshore sediment dynamics and sedimentation,* ed. J. Hails and A. Carr, 273–306. London: John Wiley & Sons.

Fairbridge, R. W. 1960. The changing level of the sea. *Sci. Am.* 202:70–79.

Fisher, J. J. 1968. Barrier island formation: A discussion. *Bull. Geol. Soc. Am.* 76:1421–26.

Glennie, K. W., and Evans, G. 1976. A reconnaissance of the recent sediments of the Ranns of Kutch, India. *Sedimentology* 23:625–47.

Gupta, S. K. 1972. Chronology of raised beaches and Indian coral reefs of the Sourashtra coast. *J. Geol.* 80:357–61.

Hay, A. E., and Sheng, J. 1992. Vertical profiles of suspended sand concentration and size from multifrequency acoustics backscatter. *J. Geophys. Res.* 97:15661–77.

Hoyt, J. H. 1967. Barrier island formation. *Bull. Geol. Soc. Am.* 78:1125–36.

Hoyt, J. H. 1969. Chenier versus barrier, genetic and stratigraphic distinction. *Bull. Am. Assoc. Petrol. Geol.* 53:299–306.

Huntley, D. A., and Bowen, A. J. 1975. Comparison of the hydrodynamics of steep and shallow beaches. In *Nearshore sediment dynamics and sedimentation,* ed. J. Hails and A. Carr, 69–110. London: John Wiley & Sons.

Inman, D. L., and Bagnold, R. A. 1963. Littoral processes. In *The sea,* ed. M. N. Hill, 529–33. New York: Wiley Interscience.

Inman, D. L., and Nordstrom, C. E. 1971. On the tectonic and morphologic classification of coasts. *J. Geol.* 79:1–21.

Kamphuis, J. W. 1991. Alongshore sediment transport rate. *J. Wtrwy. Port. Coast. Ocean Eng.* 117:624–40.

Kamphuis, J. W., Davies, M. H., Nairn, R. B., and Sayao, O. J. 1986. Calculation of littoral sand transport rate. *Coastal Eng.* 10:1–21.

Kidson, C. 1968. Coastal geomorphology.In *The encyclopedia of geomorphology,* ed. R. W. Fairbridge, 134–39. New York: Reinhold Book Corp.

King, C. A. M. 1972. *Beaches and coasts.* London: Edward Arnold.

Komar, P. D. 1976. *Beach processes and sedimentation.* Englewood Cliffs, NJ: Prentice Hall.

Kraft, J. C. 1978. Coastal stratigraphic sequence. In *Coastal sedimentary environments,* ed. R. A. Davis, Jr., 361–83. New York: Springer-Verlag.

Kraus, N. C. 1987. Application of portable traps for obtaining point measurement of sediment transport rates in the surf zone. *J. Coastal Res.* 2:139–52.

Kraus, N. C., Gingerich, K. J., and Rosati, J. D., 1988. Towards an improved empirical formula for longshore sand transport. *Proceedings 21st Coastal Engineering Conference,* ASCE, 1183–96.

Larson, M. 1991. Equilibrium profile of a beach with varying grain size. In *Proceedings of Coastal Sediments '91,* ASCE, New York. pp. 861–74.

Larson, M., and Kraus, N. C. 1994. Temporal and spatial scales of beach profile change, DUCK, North Carolina. *Marine Geol.* 117:75–94.

Newell, R. C. 1979. *Biology of intertidal animals.* Kent: Marine Ecological Survey.

Nielsen, P. 1983. Entertainment and distribution of different sand sizes under water waves. *J. Sediment. Petrol.* 53:423–28.

Nielsen, P. 1991. Combined convection and diffusion: A new framework for suspended sediment modeling. In *Proceedings Coastal Sediments '91*, pp. 419–31.

Oldale, R. N. 1985. A drowned Holocene barrier spit off Cape Ann, Massachusetts. *Geology* 13:375–77.

Otvos, E. G. 1970. Development and migration of barrier islands, northern Gulf of Mexico. *Bull. Geol. Soc. Am.* 81:241–46.

Pethick, J. 1984. *An introduction to coastal geomorphology*. London: Edward Arnold.

Pierce, J. W., and Colquhoun, D. J. 1970. Holocene evolution of a portion of North Carolina coast. *Bull. Geol. Soc. Am.* 81:3697–714.

Rao, V. P., Veerayya, M., Thamban, M., and Wagle, B. G. 1996. Evidence of Late Quaternary neotectonic activity and sea-level changes along the western continental margin of India. *Curr. Sci.* 71:213–19.

Rosati, J. D., Gingerich, K. J., and Kraus, N. C. 1990. Superduc surf zone sand transport experiment. *Tech. Rep.* CERC-90-10, U.S. Army Engineer Waterways Experiment Station, CERC, Vicksburg, MS.

Rosati, J. D., Gingerich, K. J., and Kraus, N. C. 1991. East pass and Ludington sand transport data collect project: data report. *Tech. Rep.* CERC-91-3, U.S. Army Engineering Waterway Experiment Station, CERC, Vicksburg, MS.

Shackleton, N. J., and Opdyke, N. D. 1976. Oxygen-isotope and palaeo-magnetic stratigraphy of Pacific core V28-239, Late Pliocene to latest Pleistocene. *Geol. Soc. Am. Mem.* 145:449–64.

Short, A. D. 1975. Offshore bars along the Alaskan arctic coast. *J. Geol.* 83:209–21.

Sissons, J. B. 1982. The so-called interglacial rock shoreline of western Scotland. *Trans. Inst. Br. Geogr.* NS7:205–16.

Sonu, C. J. 1972. Field observation of nearshore circulation and meandering currents. *J. Geophys. Res.* 77:3232–47.

Strahler, A. N. 1969. *Physical geography*. New York: John Wiley & Sons.

Sunamura, T. 1977. A relationship between wave-induced erosion and erosive forces of waves. *J. Geol.* 85:613–18.

Sunamura, T. 1982. A predictive model for wave-induced erosion, with application to Pacific coast of Japan. *J. Geol.* 90:167–78.

Sunamura, T. 1983. Processes of sea cliff and platform erosion. In *CRC handbook of coastal processes and erosion*, ed. P. Komar, 233–65. Boca Raton, FL: CRC Press.

Trenhaile, A. S., and Layzell, M. G. S. 1981. Shore platform morphology and the tidal duration factor. *Trans. Inst. Br. Geogr.* NS6:82–102.

Vaidyanadhan, R. 1987. Coastal geomorphology of India. *J. Geol. Soc. Ind.* 29:373–78.

Van Rijn, L. C., Nieuwjaar, M. W. C., Van der Kaay, T., Napp, E., and Van Kampen, A. 1993. Transport of fine sands by currents and waves. *J. Wtrwy. Port Coast Ocean Eng.* 119:123–43.

Wang, P. 1998. Longshore sediment flux in a water column and cross surf zone. *J. Wtrwy. Harb. Coast. Eng.* 124:108–17.

Wang, P., and Davis, R. A. 1998. A beach profile for a barred coast—Case study from Sand Kay, west-central Florida. *J. Coastal Res.* 14:981–91.

Wang, P., Davis, R. A., and Kraus, N. C. 1998a. Cross-shore distribution of sediment texture under breaking waves along low-wave-energy coasts. *J. Sediment. Res.* 68:497–506.

Wang, P., Kraus, N. C., and Davis, R. A. 1998b. Total longshore sediment transport rate in the surf zone: Field measurements and empirical predictions. *J. Coastal Res.* 14:269–82.

Yoo, D. 1994. Wave-induced longshore currents in surf zone. *J. Wtrwy. Harb. Coast. Eng.* 120:557–75.

12 Applied Geomorphology

Geomorphology is the study of landforms and their processes of evolution. Human activities, however, accelerate *process rates* and thereby affect the stability of geomorphic systems and the environment in many different ways. Applied geomorphology analyzes the rates of earth surface processes stressed by human activities and interprets their effect on short- and long-term stability of geomorphic systems and quality of the environment (Craig and Crafts, 1982). Applied geomorphology also generates spatial data on optimum utilization and management of the land resource and predicts environmental hazards due to human impacts (Hails, 1977; Brunsden, 1988; Cowell, 1997). Applied geomorphology is *environmental geology* when geologic information is additionally translated for resolving environmental problems.

Early civilizations living off the land had maintained a symbiotic relationship with the earth's resources and, therefore, did not alter the environment much. In more recent decades, the introduction of newer technology for exploitation and transformation of natural resources into a usable form, however, has remarkably stressed the *man-environment relationship* by altering the frequency and magnitude of geomorphic processes manifold. The accelerated process rates disturb the nature of exchange between energy and materials and feedback mechanisms, locally affecting the stability of natural systems that otherwise are simple in function (Tuan, 1971).

Successive stages of *developmental activity* introduce cumulative effects on geomorphic systems. A sequence of frequent forest burning to pave the way for first-ever artificial use of land for agriculture through to dwelling sites, generation of waste and resulting pollution, and emergence of a fuel-based economy has remarkably disturbed the ecological balance and long-term stability of geomorphic systems. Hence, *land use changes* invariably produce an *irreversible change* in the attributes of geomorphic systems and properties of the environment.

The impact of certain human actions on the environment, however, is *reversible*. For example, planners had the flood flow of *inland* Ghaggar River diverted to sixteen interconnected dunal depressions in a part of the eastern Thar Desert, India, in the hope that the sand would absorb surface flow and save the villages of the Suratgarh-Baropal-Kanaur area from the fury of seasonal floods (Figure 12.1). This action offered welcome respite to the target villages, but the problem surfaced a few years later in areas beyond the flow confinement. The waterlogging was traced to the presence of an impervious *gypsiferous hardpan* at a shallow depth of 2 to 4 m from the sandy surface, subsurface drainage of the routed flow through *sinkhole openings* in thin sections of the gypsum column, unpredictable lateral flow of the subsurface water through porous substratum, and accumulation as a subsurface body of water, generating strong *hydrostatic pressure* to repeatedly burst through to the surface (Kar, 1992). This practice of relieving the seasonal *flood hazard* has been discontinued, but it will take a long time to reclaim the affected land.

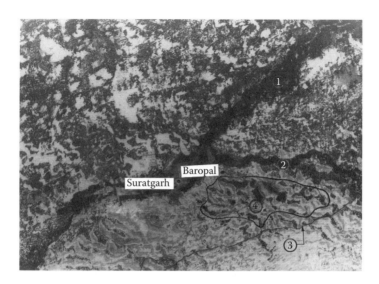

FIGURE 12.1   Landsat TM band 3 image (February 23, 1988) of the Suratgarh-Baropal area (1) and its tributary, (2) and the Indira Ghandi Canal (3) are the interdunal depressions where flood flow was released through a diversion canal from the north. Sixteen such depressions (4) were linked through the diversion canal that led to waterlogging and consequent spread of salinity-alkalinity in the nearby areas. In the west, the Suratgarh area in the dry valley is also threatened by waterlogging. (Source: Imagery and description courtesy Dr. Amal Kar, Central Arid Zone Research Institute, Jodhpur, India.)

APPLICATIONS OF GEOMORPHOLOGY

Applications of geomorphology encompass a wide spectrum of human activities, which variously utilize the finite resources of the earth and affect the environment in many different ways. Applied geomorphology integrates the physical and cultural process rates for planning and management of the earth's resources and takes advantage of the earth science information and principles of geomorphology for analyzing the impact of human activities on components of the geomorphic landscape and quality of the environment.

LAND RESOURCE ATTRIBUTES

Morphology and material composition of land are basic attributes that find application in site-specific and need-specific evaluation of this finite resource. Morphological maps, geomorphic maps, land systems maps, and terrain classification maps are well-known end products of the categorization of land resource attributes.

Morphological Maps

Morphological maps depict the surface configuration in terms of slope steepness, slope form, break of slope, and direction of slope change by suitable symbols (Figure 12.2). These resource attributes provide a range of limiting conditions for

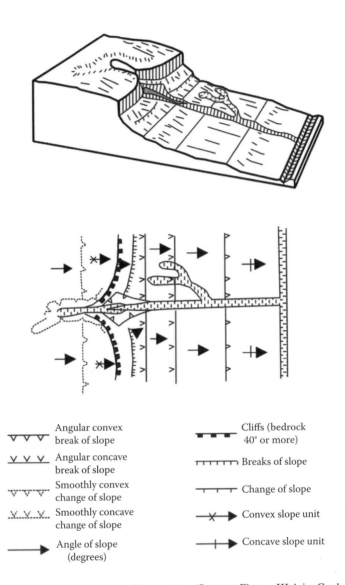

FIGURE 12.2 A morphologic mapping system. (Source: Figure III.4 in Cooke, R. U., Brunsden, D., Doornkamp, J. C., and Jones, D. K. C., *Urban Geomorphology in Drylands* [New York: Oxford University Press, 1982]. With permission from Oxford University Press.)

site planning and alignment of surface transport routes on 1:5000 and 1:50,000 map scales (Cooke et al., 1982). The information on morphological maps is also useful for monitoring the direction of urban growth and land cover change.

Geomorphic Maps

Geomorphic maps depict the morphology (appearance), morphometry (dimension), morphogeny (origin), and morphochronology (age) of each landform or group of landforms by color and symbol schemes for projecting the *morphodynamic controls* of the landscape (St.-Onge, 1968). Although representation of diverse data on a single map makes the document complex, landforms and their processes can be compared for areas differing in geologic controls and climatic regime. Besides, the wealth of information on geomorphic maps enables identification of areas favorable for a proposed activity, areas problematic to engineering construction, and areas susceptible to slope failure.

Land Systems Maps

The land systems approach of the CSIRO Division of Land Research, Australia, evaluates the inherent potential and limitations of terrain for its intended use. Land systems are a mosaic of recurring land units, in which *land unit* is a distinct entity in terms of material composition, process of evolution, past and present climates, and associated topographic, soil, and vegetation attributes (Christian, 1957). Therefore, the land systems maps are similar to *geomorphic maps*, except that they provide additional data on the *rate of landscape change* (Christian and Stewart, 1952). The boundary of land systems, as in Figure 12.3, changes with a change in the nature of land units.

The size of land units is particularly a sensitive index of the regional potential of an area and strategies for area development. Hence, a land systems approach provides a framework for classification, planning, and management of the land resource. Sharma (1991) similarly classified an area to the southwest of Delhi, India, into dryland, inland lake evaporite, and structural hill land systems (Figure 12.4), and evaluated their inherent agricultural productivity potential in relation to the existing cropping pattern.

Terrain Classification Maps

The Military Engineering Experimental Establishment (MEXE), UK, rates the *physical attributes* of terrain for the movement of military vehicles of various types, and generates predictive data for modeling the terrain attributes of unknown areas. Studies of the MEXE group are largely available in mimeo form.

The MEXE scheme recognizes a *facet* to be the fundamental unit of terrain, in which each facet is distinct in morphology, material composition, and water regime. The facets of a given geomorphic environment exhibit similar spatial association within the terrain and, thereby, provide for the *recurrent landscape pattern*. The *suitability status* of facets, and of terrain for military purposes, is governed by slope steepness and soil strength for dry and wet conditions. Besides, additional data on locally available drinking water, construction material, and visibility and camouflage characteristics of terrain are always useful for the mobility of men and equipment in

1 Piedmont and Alluvial Plain Land System
2 Hill Land System
3 Structural Valley Land System
4 Mountain Land System

FIGURE 12.3 Four types of land systems depicted on IRS-1A LISS 2 imagery. The Himalayan Mountain land system covers the larger part of the image. The Siwalik Hills land system in the southeast of the image may be differentiated from the former by terrain texture. A structural valley land system is visible in the area of Dehra Dun town. A piedmont and alluvial plain land system is prominent to the south and southwest of Haridwar. (Source: Imagery courtesy Indian Space Research Organization, Department of Space, Bangalore.)

traditional warfare scenarios. The *predictability* of physical attributes of terrain in unknown regions is based on the premise that the recurring landscape pattern repeats in areas of similar climatic, geologic, and structural controls. Hence, similar aspects of terrain at a known site provide an identical landscape pattern at unknown sites.

LAND RESOURCE PLANNING

A parcel of land holds certain potential and a range of limitations for its existing and proposed use. The land resource potential and limitations vis-à-vis economic use of the land are inherent in the physical attributes and process controls of the terrain. Land resource planning evaluates these attributes and their controls for judicious utilization of the land for *sustainable development.* The land resource plans are short and long term in perspective, and site specific and regional in context.

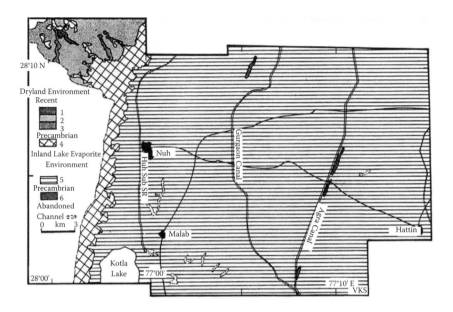

FIGURE 12.4   Land units in a part of dryland, inland lake evaporite, and structural hill land systems, southeasten Haryana, India. The dryland land system is comprised of (1) alluvium, (2) sandy plain, (3) broken land, and (4) the Precambrian remnantal hills. The inland lake evaporite land system consists of a recent lake flat (5) and the Precambrian remnantal hills (6). The drainage from the Precambrian Aravalli Hills to the west of the area accumulated in a lake basin, forming the flat surface of an evaporite lake environment. (Source: Figure 4.1 in Sharma, V. K., *Remote Sensing for Land Resource Planning* [New Delhi: Concept Publishing Co., 1991]. With permission from Concept Publishing Company.)

LAND USE PLANNING

Land use planning is the process of evaluating a proposed use of the land resource at local (site) and regional scales. The *site planning* on scales of 1:2500 to 1:10,000 requires detailed quantitative data on slope attributes, intensity of erosion, and locally available construction material. The *regional planning* on scales ranging between 1:25,000 and 1:100,000 integrates data as diverse as physical and geotechnical aspects of terrain, flood hazards, land use characteristics, soil conservation, and others. Planners, though, do not always have access to comprehensive and reliable information on these prerequisites to base the documented plan, and defend their short- and long-term land use decisions. Hence, some of the constraints that are not obvious initially for want of data or other reasons become irritants in the late stage of the planning process, requiring expensive remedial measures for avoidable decisions (Brunsden, 1988).

The land use planning process is comprised of five conceptually distinct and continuously interacting phases of *problem identification* (Figure 12.5): aims and objectives of the land use plan, collection and interpretation of data for the proposed use, formulation of the plan, review and adoption of the plan, and implementation of the plan (Spangle and Associates et al., 1976).

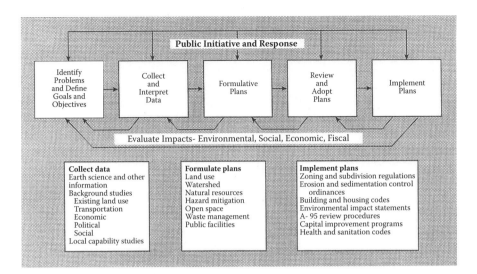

FIGURE 12.5 Diagram of the land use planning process. (Source: Figure 2 in Spangle, W., and Associates, F. Beach Leiguton and Associates, and Baxter, McDonald and Company, *Earth-Science in Land-Use Planning—Guidelines for Earth Scientists and Planners*, U.S. Geological Survey Circular 721 [U.S. Geological Survey, 1976]. With permission from U.S. Geological Survey.)

Land use planning integrates diverse data on topography, geology, soils, and hydrology with the envisaged use of the land resource. *Topographic data* provide interpretative information on slope steepness and slope instability, *geologic data* are useful in the identification of geomorphic hazards, *soil data* are invaluable for rating the capability of land for agriculture and other economic uses of the resource, and *hydrologic data* enable evaluation of the quantity and quality of available water, severity of flooding, and related potential problems bearing on the proposed use of the land resource. The planning document also considers the effect of human activities on natural process rates and their likely impact on the stability of landscape. Hence, *impact assessment* of short- and long-term consequences of land use decisions is an integral component of the planning process. Identification of *hazards* and evaluation of *likely risk* from hazards are similarly built into the planning process (Spangle and Associates et al., 1976). It is obvious, therefore, that planners require a variety of thematic maps of relevance to the objectives of the envisaged plan. Landscape stability, land sensitivity, land capability, and land suitability maps are a few documents that planners frequently consult for defending land use decisions.

Landscape Stability Maps

Landscape stability maps project the graded stability of landscape in terms of natural or cultural process rates. The stability of landscape due to *natural processes*, like in Kim-me-ni-oli basin, northwestern New Mexico, varies with the infiltration rate and surface runoff, giving stable, metastable, and unstable status to parts of the desert basin (Connors and Gardner, 1987). *Stable areas* are identified with sand-mantled uplands of high infiltration rate and of adequate vegetation cover that protects the surface from deflation; *metastable areas* represent valley fills, coppice dunes, and exposed saltpans of low deflation rate; and *unstable areas* comprise poorly vegetated sandy tracts of high deflation rate and sand-bedded channels migrating in flood spasms. *Cultural processes* also affect the stability of landscape in many different ways. Singh et al. (1978) assessed the impact of pastoral, cultivation, and irrigation practices on long-term stability of the geomorphic landscape in a part of the Thar Desert, India, and identified areas likely to be affected by deflation, aeolian sedimentation, and soil salts, if the current land use practices were to continue at the same rate in the future.

Documentation of *geomorphic hazards* due to natural or cultural processes, or both, aims to provide suitable protection to the proposed engineering structures, and to the population and economy of planning areas at likely risk in the event of hazards. The spatial pattern of the type and magnitude of active and potential zones of land instability due to slope failure and flood hazards in a part of the Front Range near Colorado (Dow et al., 1981), and susceptibility of unconsolidated sediments in Santa Cruz Mountains, California, to slope failure when wet and in earthquake events (Spangle and Associates et al., 1976) provide planners vital feedback on areas to be excluded from the preview of certain developmental activities. The French ZERMOS scheme (*zones exposées aux risques liés aux mouvement du sol et du sous sol*), however, depicts the spatial pattern of *risk magnitude* due to both natural and cultural processes in the planning document (Brunsden, 1988). The frequency and magnitude of slope instability and likely risk to the population and economy provide a measure of the *gross stability* of landscape (Owen et al., 1995). The geomorphic hazards map of a part of Bhagirathi valley in the northwestern Himalayas, India, identifies such areas that are susceptible in different degrees to rainfall and earthquake-induced mass movement activities (Figure 12.6).

Land Sensitivity Maps

The inherent stability of the components of landscape is governed by externally driven process rates (Carthew and Drysdale, 2003) and internal process adjustments in periods of infrequent extreme events (Calver and Anderson, 2004). Therefore, sensitivity of landscape to change may also be rated by the contemporary and past processes of landscape evolution and judged by the indices of landform change (Table 12.1). Obviously, the land sensitivity cannot be rated for the data-sparse regions.

Land Capability Maps

The utilization of land broadly varies with the morphologic characteristics of terrain. Farming practices thus require a nearly level land of good fertility status, heavy industries seek a vast flat surface with good foundation, and recreational use of land for golf needs an undulating land of variable surface and subsurface conditions (Keller, 1979). However, the ability of resources to accommodate a variety of land uses, called land capability, depends on the inherent potential and limitations of terrain attributes.

Land capability is evaluated by the *analytic hierarchy procedure* (AHP), which grades the relative significance of each terrain attribute to the envisaged land use.

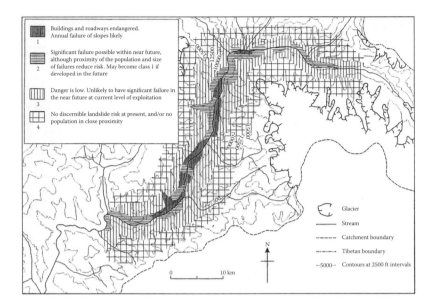

Buildings and roadways endangered.
Annual failure of slopes likely
1

Significant failure possible within near future,
although proximity of the population and size
of failures reduce risk. May become class 1 if
developed in the future
2

Danger is low. Unlikely to have significant failure in
the near future at current level of exploitation
3

No discernible landslide risk at present, and/or no
population in close proximity
4

⊂ Glacier

───── Stream

─ ─ ─ ─ Catchment boundary

─ ·· ─ ·· Tibetan boundary

─5000─ Contours at 2500 ft intervals

0   10 km

N

FIGURE 12.6 Landslide hazard compiled from the mapping at a scale of 1:50,000 for the whole of Bhagirathi valley between Gangotri and Uttarkashi. (Source: Figure 5.9 in Owen, L. A., Sharma, M. C., and Bigwood, R., "Mass Movement Hazard in the Garhwal Himalaya: The Effect of the 20 October 1991 Garhwal Earthquake and the July–August 1992 Monsoon Season," in *Geomorphology and Land Management in a Changing Environment*, ed. D. F M. McGregor and D. A. Thompson [New York: John Wiley & Sons, 1995], 69–88. With permission from John Wiley & Sons.)

Livingston and Blayney et al. (1971; Spangle and Associates et al., 1976) proposed a methodological framework, listed below, for rating the capability status of terrain.

1. *Identify* the type of land use for which land capability is to be evaluated.
2. *Determine* the natural factors that have a significant effect on the capability of land to accommodate each use.
3. *Develop* a scale of values for rating each natural factor in relation to its effect on land capability. Such ratings quantify the inherent potential and limiting conditions of each factor.
4. *Assign* a weight to each natural factor by its relative significance to the land capability. This analytical hierarchical procedure judges the relative bearing of each factor on the capability of land.
5. *Divide* the area into units by the diversity of physical conditions, requirements of proposed land use, and envisaged land use policy.
6. *Rate* each unit by its factor, calculate the weighted ratings for each factor, and aggregate the weighted ratings for each unit to give it a capability status.

Livingston and Blayney et al. (1976) evaluated land capability for Palo Alto, California, by the rated weightage of limiting conditions of the physical factors of terrain (Table 12.2). Here, each of the eight physical factors considered significant

TABLE 12.1

Methodology for Preparing Landscape Sensitivity Maps for Planning Purpose

1. *Prepare* factual base maps on topography, geology, geomorphology, hydrology, and soils.
2. *Tabulate* large-scale cycles, epicycles, and episodes of landscape evolution.
3. *Collate* evidence on dates, frequency, magnitude, and duration of climatic, seismic, flooding, landslides, and other events.
4. *Determine* probabilities and thresholds for each process.
5. *Classify* vulnerability of landscape.
6. *Devise* a scheme of proxy variables for assigning probability values of recurrence intervals to the classified landscape areas.
7. *Determine* the diagnostic large events that cause landscape change using the past events as a guide to future risk.
8. *Map* the direction of aggregate erosional stress (e.g., erosion propagation).
9. *Map* the variable spatial sensitivity or resistance to change (e.g., flat slopes, distance from erosion axis, low drainage density, resistant rock).
10. *Isolate* vulnerable rock types, swelling clays, shattered rocks, metastable soils, and the like.
11. *Predict* the worst state (e.g., saturated ground).
12. *Map* any known threshold relationships (e.g., slope angle vs. slope failure, translational sliding vs. regolith, soil thickness vs. slope).
13. *Utilize* postaudit surveys and learn from mistakes by examining what happened in particular events or situations in the past.

Source: Table 7.3 in Brunsden, D., "Slope Instability, Planning and Geomorphology in the United Kingdom," in *Geomorphology in Environmental Planning*, ed. J. M. Hooke (Chichester: John Wiley & Sons, 1988), 105–19. With permission from John Wiley & Sons.

(column 1) were categorized and assigned a weight, such that the higher the rating, the greater its status to the capability of land (column 2). The overall significance of each physical factor to the land capability is judged and assigned an aggregate weight (column 3). Finally, the weighted ratings for individual categories of physical factors were obtained as products of rated values and aggregate weight (column 4). The weighted ratings were compiled for the county divided into suitable-sized grids, providing the spatial pattern of land capability status for the region.

Several other land capability classifications have also been proposed. Notable among these are the Land Capability Classification of the U.S. Department of Agriculture, Land Inventory of Canada, Land Use Capability of Britain, and Land Evaluation of the Food and Agriculture Organization (McRae and Burnham, 1981).

Land Suitability Maps

Ever-increasing competing demands on the finite land resource, and pressing social, economic, and political compulsions often require the allocation of a parcel of land to a use to which it is not ideally suited. This use of land for a predetermined activity is called land suitability, wherein the capability and suitability status of land are not expected to match. Experience suggests that a wide disparity in the land status of two types results in abandoning of the regulated use or costly revision of the

TABLE 12.2
Rating System for Land Capability Factors

Geologic, Topographic, and Soil Factors (1)	Rating (2)	Aggregate Weight (3)	Weighted Rating (4)
Average Slope			
Over 50%	1		10
31–50%	2	10	20
16–30%	4		40
0–15%	5		50
San Andreas Fault Zone			
Within the zone	1	7	7
Outside the zone	5		35
Landslides			
Within the slide area	1	6	6
Outside the slide area	5		30
Natural Slope Stability			
Poor	1		5
Fair	3		15
Good	5		25
Cut-Slope Stability			
Poor	1		4
Fair	3	4	12
Good	5		20
Soil Suitability as Fill			
Poor	1		2
Fair	3	2	6
Good	5		10
Soil Erosion			
Severe	1	6	6
Moderate	5		30
Soil Expansion			
High	1		3
Moderate	3	3	9

Source: Table 4 in Spangle, W., and Associates, F. Beach Leiguton and Associates, and Baxter, McDonald and Company, *Earth-Science in Land-Use Planning—Guidelines for Earth Scientists and Planners*, U.S. Geological Survey Circular 721 (U.S. Geological Survey, 1976). With permission from U.S. Geological Survey.

implemented plan at a later stage of plan execution (Keller, 1979). In this context, the selection of a part of Irwell Valley in northwest England for developing a housing complex could well have been avoided only if the presence of shallow *subsurface faults* that pass through the area was known at the time of plan formulation. The severe damage to houses soon after they were built and occupied was traced to the reactivation of these faults due to natural causes or to the stress of construction activity (Douglas, 1988).

Selection of sites for dams, tunnels, tall buildings, airports, and highways is also based on the suitability status of land. Solid municipal waste disposal sites and the layout of urban housing, among others, are similarly decided on the suitability status of land, and social, cultural, economic, and political aspects of area development.

SOLID MUNICIPAL WASTE DISPOSAL SITES

All societies generate a vast amount of industrial, agricultural, and municipal liquid and solid waste each year. The technology permits recycling only of a part of the waste, making the management of a large volume of practically indestructible waste a challenge. Even the affluent nations find difficulty in managing their industrial waste. They have, at times, disposed of a part of the hazardous waste in territorial waters of less privileged nations and liberally sent the scrap to *soft states* for recycling at the cost of human health and quality of the environment.

The solid municipal waste, comprising the biodegradable and nondegradable components, is disposed of in part by feeding to pigs, composting, and incineration. *Sanitary landfills*, the other conventional procedure for handling solid municipal waste, are sites for the *decomposition* of biodegradable waste, where the waste is reduced to a minimum volume and covered with a layer of compact soil at the end of each operation to deny insects, rodents, and animals access to the waste. Ideal sanitary landfill sites, however, are nonexistent, and even suitable sites for solid waste disposal are scarce due to the constraints of physical space, public opposition to the facility, and the environmental impacts (Ward and Li, 1993; Gray 1997).

The decomposition of solid waste releases leachates of heavy and trace metals and generates toxic organic compounds, obnoxious gases, and pathogens from landfill sites, which enter the environment in six different ways (Figure 12.7). The *metal contaminants* stay in soils and their ionic species mobilize into the groundwater system. Landfill sites also release carbon dioxide and methane into the atmosphere and generate pathogens that move into shallow subsurface waters through overland flow and infiltration processes. *Pathogens* are the primary cause of *waterborne diseases.*

Determinants of Landfill Sites

Aspects of terrain, existing land use, aesthetic value, government regulations, and public opposition to the proposed site are major determinants of landfill sites. The physical factors of terrain, however, exert an overriding influence on the suitability of land for waste disposal, ensuring a reasonable decomposition of the solid waste without affecting the quality of environment much. In humid areas, landfills are avoided on shallow soils less than 10 m deep and above a limestone terrain in order

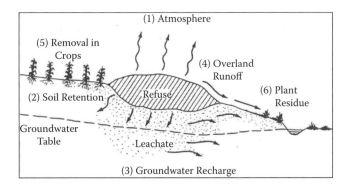

FIGURE 12.7   Idealized diagram showing several ways by which hazardous waste pollutants from a solid-waste disposal site may enter the environment. (Source: Figure 10.9 in Keller, E. A., *Environmental Geology*, 2nd ed. [Columbus, Ohio: Bell & Howell, 1979]. With permission from Macmillan Press Ltd.)

to deny leachates access to the groundwater system (Keller, 1979). The cultural aspects are nevertheless also significant in many different ways in the evaluation of landfill sites. The selected physical and cultural aspects are assessed for the suitability status of terrain and graded into least to most suitable sites by the analytic hierarchy procedure (APH) and computer-based geographical information systems (GIS) to the spatial data.

The suitability of terrain for landfill sites is evaluated by the exclusionary and inclusionary criteria. The mandatory *exclusionary criteria* as in Oklahoma lay down that landfill sites not be within 0.8 km of a stream course, permanent body of water, and wetlands, and below the level of the 100-year flood surface. These sites should also not be within 1.6 km upslope of a public water supply source and at sites from where the groundwater travel time to a public water supply well is less than 1 year. Further, the landfills are to be situated at least 1.52 and 3.05 km from airport runways used by piston engine and turbojet aircrafts, respectively, and be at least 1.6 km from areas frequented by endangered species. Finally, the proposed landfill sites should have a life span of at least 20 years. By these criteria, Siddiqui et al. (1996) excluded the *negative areas* from further consideration and evaluated the remaining land for landfill sites in Cleveland County, Oklahoma, by the attributes of physical and cultural landscape at three levels of spatial organization (Table 12.3). Level 1 decision variables refer to hydrogeology/geology, land use characteristics, and settlement pattern of the area. Except for land use, the level 1 variables are graded by attributes of relevance to the landfill site (level 2), and assessed further by the suitability of geology, topography, subsoil water depth, soils and land use, and proximity to settlements (level 3). In this case, the suitability of terrain for landfill sites is graded by greater weightage to the physical than cultural factors.

Other decision hierarchies have also been proposed. Ramasamy et al. (2002) graded the suitability of geologic structure, depth to weathering profile, and hydrogeologic conditions of terrain for ranking waste disposal sites in a part of the hills of South India. This classification, however, altogether ignores the environmental and cultural aspects of land suitability for the purpose.

TABLE 12.3
Decision Hierarchy Framework for Landfill Suitability Ranking

Decision Factors (Level 1)	Building Blocks (Level 2)	Limiting Attributes (Level 3)
Hydrogeology/geology	Depth to water table	Limiting
		Not limiting
	Depth to bedrock	Limiting
		Not limiting
	Permeability	Limiting
		Not limiting
	Slope	Limiting
		Not limiting
	Texture	Limiting
		Not limiting
Land use	—	Agricultural land
		Forestland
Proximity[a]	Moore proximity	0–10 km zone
		10–20 km zone
		>20 km zone
	Norman proximity	0–10 km zone
		10–20 km zone
		>20 km zone
	Noble proximity	0–10 km zone
		10–20 km zone
		>20 km zone

Source: Modified from Figure 1 in Siddiqui, M. Z., Everett, J. M., and Vieux, B. E., "Landfill Siting Using Geographical Information System: A Demonstration," *J. Environ. Eng.* 122 (1996): 515–23.

[a] Refers to sociocultural and economic aspects of settlements.

SITE SUITABILITY FOR HOUSING PLAN

Aspects of topography, geology, climate, soils, hydrology, vegetation, and ecology are key components in the ranking of *physical space* for layout and growth direction of urban settlements. In mountainous regions and high-latitude locations, however, the site suitability depends more on slope aspect or the amount of solar energy received at the surface than on other aspects of the landscape. These variations in *slope aspect* offer five suitability classes for the residential sector in Vladivostok, Russia (Govorushko, 1996). Prior experience from the northerly latitude of 43° 06', at which Vladivostok is located, further suggests that the south-facing building walls receive almost 3.6 times less solar energy in January than in July, and that the north-facing walls lose 30 to 70% more heat than the south-facing ones. Hence, in view of the slope aspect, the buildings oriented north–south or northwest–southeast would be thermally more comfortable to live in.

ENVIRONMENTAL IMPACTS

Humans exploit, transform, and utilize the resources of the earth, and return the end product as an out-of-place resource that cannot be conveniently disposed of in the environment. This is the environmental impact, which appears in the final stage of the resource process (Goudie, 1994). The environmental impact also refers to the environmental consequences of the utilization of the earth's resources. The environmental impacts are direct and indirect (Goudie, 1986). *Direct impacts* are planned, and *indirect impacts* are unplanned or accidental. The consequences of direct impacts are commonly predictable, and of indirect impacts are relatively unpredictable, severe, and long lasting in the environment.

Human activities alter the environment in many different ways, but the manner of the environment's response to the stress of human activities is far from understood. Of these, a *change in land use* produces a cascading effect on the process rates, stability of geomorphic landscape, quality of water resources, and biotic and the abiotic associations in natural systems (Adger and Subak, 1996).

LAND USE CHANGE

Deforestation and intensive mechanized agriculture are the major causes of soil and nutrient loss from watersheds. Goudie (1986) observed that the rate of soil erosion in the United States nearly doubles for every 20% loss of the forest cover. The rate of soil erosion even from poorly managed subsistence agriculture with forest clearing, as in a part of the Himalayas of north India, is fifty to one hundred times higher than the natural rate of soil erosion from undisturbed areas of comparable size (Rawat and Rawat, 1994). Land use changes, therefore, alter not only the sediment load but also the hydrologic response of streams to floods. However, the biophysical mechanisms releasing the anthropogenic load into hydrologic systems are inadequately understood (Johnson, 1995) and poorly quantified as well (Rawat and Rawat, 1994).

A predictable relationship exists between the rate of soil erosion and land use change (Wolman, 1967). The rate of soil loss in conditions unaltered by human activities is remarkably low, but sharply increases in areas of primitive to mechanized agriculture without conservation practices, and dwindles thereafter to a low value again for the land returning to its natural state (Figure 12.8). The rate of soil loss also varies with the land use practice. For a dust storm event, the rate of soil loss in a desert terrain is fourteen times higher from deeply plowed fields than from fallow lands with 2 to 4% vegetation cover, and seven times higher from degraded pastures with 8 to 10% vegetation cover (Dhir et al., 1989). Urban use of the land, which increases the area in pavement, however, suppresses the rate of soil loss but increases the frequency and magnitude of *flood events*. Several studies on *urban watersheds* in the United States suggest that the magnitude of mean annual flood progressively increases with the increasing state of urbanization, such that small- and intermediate-sized

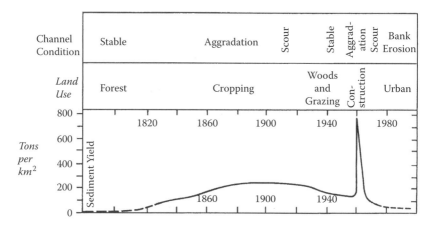

FIGURE 12.8   Sediment yield as a function of land use. (Source: Figure on p. 263 in Goudie, A., Simmons, I. G., Atkinson, B. W., and Gregory, K. J., *The Encyclopaedic Dictionary of Physical Geography*, 2nd ed. [Oxford: Blackwell, 1994]. With permission from Blackwell Publishers.)

urban watersheds generate three times more floods than is expected of the preurban environment (Leopold, 1968). Observations from a small-sized East Meadow Brook urban watershed in New York similarly suggest that direct runoff from individual storms is 1.1 to 4.6 times the preurban runoff (Figure 12.9).

Deforestation and Soil Loss

Archaeological evidence from Mohanjodaro, Pakistan, suggests that the Indus Valley civilization that had existed before 2500 BC and thrived for some 1000 years, and possibly decayed as a result of progressive deforestation throughout the Indus Basin (Wheeler, 1959). The loss of forest cover accelerated the rate of soil erosion and silted streams, which increased the frequency and magnitude of floods in the region. The recurring floods of higher magnitude destroyed the canal irrigation support system of the economy and caused the collapse of this prehistoric civilization. A shift in the overall land use pattern affects local and regional climates (O'Brien, 1998), but the deforestation forcing of climate change is not known for the Indus Valley in certain terms. It is a rare coincidence, though, that a phase of intense *climatic aridity* that had already set in around 1800 BC in northwestern India (Singh, 1971) may have caused the initial gradual decline and subsequent decay of the Indus civilization due to alien invasions.

A relationship between the density of vegetation cover and soil loss is well documented at local and regional scales. At a local scale, an open grazing land with diverse vegetation cover generates a meager runoff of about 2% of the precipitation and scant soil loss of the order of 12 t km^{-2} (Noble, 1965; Goudie, 1994). The volume of runoff, however, reaches 73% of the precipitation and soil loss to 1349 t km^{-2} as vegetation cover progressively loses to area development. In a regional context, the effect of deforestation on soil loss is dampened by the sedimentation of a major part of eroded sediments within catchments. The gross rate of soil loss from the Indian

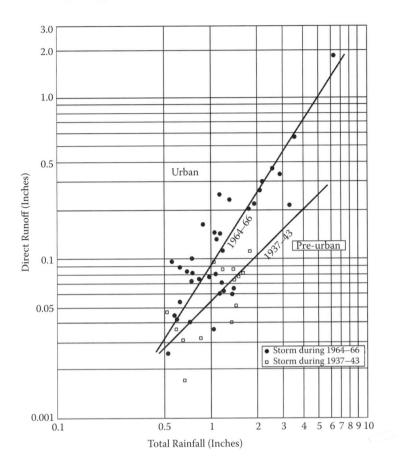

FIGURE 12.9 Comparison of the rainfall-runoff relationships for preurban and urban conditions. The data are for individual storms in one subarea of Nassau County, New York. (Source: Seaburn, G. E., *Effects of Urban Development on Direct Runoff to East Meadow Brooks, Nassau County, Long Island, New York*, U.S. Geological Survey Professional Paper 627B [U.S. Geological Survey, 1969]. With permission from U.S. Geological Survey.)

Ocean island of Madagascar, which had its forest wealth depleted to the extent of 80% by 1975, thus varies only between 12 and 50 t year^{-1} (Helfret and Wood, 1986). The rate of soil loss is specific to a given geomorphic environment and to the magnitude of anthropogenic impacts in the biosphere. Hence, it is not comparable between and within regions.

Land Use and Slope Instability

In mountainous regions, areas along roads, under settlements, and in agricultural use are particularly susceptible to various forms of mass movement activity (Eisbacher and Clague, 1981; Saczuk, 2001). The frequency of slope failure in a small-sized Himalayan catchment of north India is highest (5.57 km^{-2}) along roads and lowest (0.55 km^{-2}) within forested lands (Valdiya and Bartarya, 1989). In general, the high

TABLE 12.4

Land Use and Nutrient Discharge from Mamlay Watershed, Sikkim Himalayas, India

Land Use	Overland Flow as Percentage of Rainfall	Soil Loss (kg ha⁻¹)	Nutrient Loss (kg ha⁻¹)		
			Nitrogen	Organic Carbon	Phosphorous
Forest					
Temperate	2.56	6	0.068	3.08	0.007
Subtropical	4.55	137	0.537	3.887	0.089
Agriculture					
Cardamom agroforestry	2.17	30	0.415	1.865	0.020
Mandrin agroforestry	4.76	145	1.072	9.793	0.099
Cropped area	9.55	477	2.345	11.863	0.677
Wasteland					
Fallow land	3.37	43	0.398	2.324	0.103

Source: Table 3 in Rai, S. C., and Sharma, E., "Land-Use Change and Resource Degradation in Sikkim Himalaya: A Case Study from Mamlay Watershed," in *Proceedings of the Third International Symposium on Headwater Control: Sustainable Reconstruction of Highland and Headwater Regions*, New Delhi, October 6–8, 1995, pp. 265–78. With permission from Oxford & IBH Publishing Company.

incidence of slope instability along roads is related to cut excavation and removal of vegetation ahead of a construction site, and reactivation of *ancient landslides* in some cases (Schuster and Hübl, 1995).

Soil Erosion and Nutrient Losses

Intensive mechanized agriculture is the primary cause of soil and nutrient losses from cultivated lands. This observation in Table 12.4 is in general agreement with other local and regional studies, although soil and nutrient losses vary widely within and between regions. The nutrients reaching surface waters of the earth affect the *biological productivity* of aquatic systems.

WATER QUALITY

Water quality depends on the concentration of suspended and dissolved load in the water column, and is specific to the intended use of water. The particulate matter and dissolved load of *surface waters* are derived from basin geology and land use practices, and of *groundwater systems* are based in the mineral and chemical composition of aquifers. In addition, *bacterial contamination* of surface and subsurface waters by human activities also contributes to the deterioration of the quality of the earth's water resources.

Surface Water Quality

The *suspended load* of water bodies largely arrives from agricultural, deforested, and overgrazed lands, and the land disturbed by surface transport networks, mining, and other similar activities. However, it is not feasible to isolate the contribution of each economic activity to the quantity of suspended load in the water column. Water bodies also carry a variety of *dissolved load*, much of which is derived from agricultural lands, municipal sewers, and industrial waste. Of these, the agricultural sector and untreated liquid waste disposal practices are key sources that directly affect the chemical composition of streams and freshwater lakes and indirectly contribute to oxygen depletion and accumulation of toxic gases in the system.

Agricultural lands are the principal source of phosphorous and nitrogen content in overland flows, and the cause of *nutrient loading* of water bodies (Burt and Johnes, 1997). These nutrients enhance the primary production rate, magnify biomass, and affect *eutrophication* of the aquatic systems. The eutrophication produces *algal bloom*, which consumes the dissolved oxygen in water more rapidly than can be recouped by natural processes. This deoxygenation increases the *biological oxygen demand* (BOD) on the system, leading to a high mortality rate among aquatic organisms.

Algal blooms have similarly caused oxygen depletion and fish kills in Lake Pepin on the border of Minnesota and Wisconsin (Lung and Larson, 1995). This eutrophication was initially thought to result from a *point source* of phosphorous discharge of an effluent treatment plant upstream of the lake. Analysis of the chemical composition of affluent flow, however, suggested that the treatment plant does not contribute much to the nutrient loading and biomass production of the lake (Figure 12.10). Nutrient loading was instead traced to the *area source* in agricultural lands upstream of Lake Pepin. The chemical quality of aquatic systems can, however, be improved by changing the cropping pattern and restricting the quantity of fertilizer use for each type of crop (Burt and Johnes, 1997).

Sewer flows are inherently rich in organic matter and nutrient load. Therefore, some of the storm sewers that overflow into urban streams adversely affect the chemical quality of water bodies (Crabtree, 1988). Observations suggest that the *reduction of organic matter* at and downstream of overflow confluence depletes the dissolved oxygen quite rapidly in water bodies. Of the total *oxygen depletion*, 39% occurs in the vicinity of overflow confluence, 4% by the decomposition of organic matter and adsorption in the water column, 35% by the adsorption and decomposition of organic matter at the bed of streams, and the remaining 22% by the mixing of two waters downstream (Figure 12.11). The *nutrient-loaded* streams draining directly into the freshwater lakes and coastal waters are also known to cause *eutrophication*, and long-term accumulation of partly decomposed organic matter and toxic waste in aquatic systems.

Stream-borne nutrients eventually pass into shallow coastal waters, causing algal growth and oxygen depletion in the water column. Rapid proliferation of *phytoplankton* and *hypoxic conditions* (dissolved oxygen < 2 mg l^{-1}) in shallow coastal waters off the mouth of the Mississippi River are similarly related to heavy nitrogen loading of the river (Turner and Rabalais, 1994). The water quality of the system is unlikely

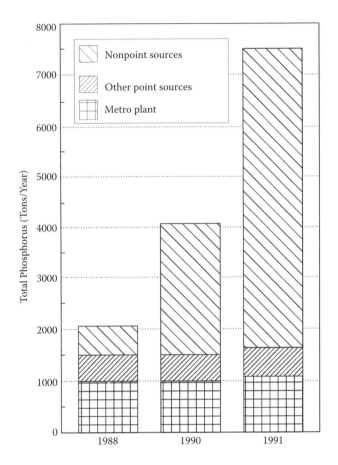

FIGURE 12.10   Annual loading of total phosphorous from Mississippi River watershed to Lake Pepin. Source: Figure 9 in Lung, W.-S., and Larson, C. E., "Water Quality Modeling of Upper Mississippi River and Lake Pepin," *J. Environ. Eng.* 121 [1995]: 691–99. With permission from American Society of Civil Engineers.)

to change, though, as upstream agricultural systems have come to depend entirely on heavy doses of fertilizer use.

Groundwater Quality

The quality of subsurface water varies with the presence of metallic and nonmetallic species in aquifers, redox-controlled movement of the mineral front, and bacterial contamination of the groundwater discharge. The *oxidation* of *pyrite* in groundwater systems of Bangladesh and adjoining India releases *arsenic* in high concentrations of up to 5 mg l^{-1}, compared to the permissible limit of 1 mg l^{-1} for potable water (Mandal et al., 1996). Similar hydrogeologic conditions and arsenic-related diseases are also common to parts of Thailand, Taiwan, Chile, Argentina, and Mexico.

Intensive farming based on irrigation and nitrogenous fertilizers contributes significantly to the pollution of groundwater systems. The *return flow* from the irrigated

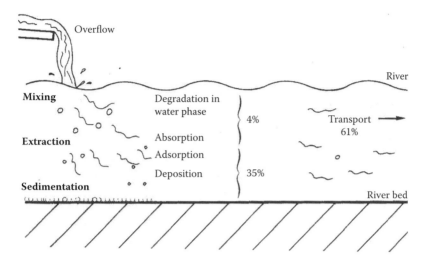

FIGURE 12.11   A schematic removal process for organic pollutants discharged to a river in the vicinity of an overflow. (Source: Figure 10.5 in Crabtree, B., "Urban River Pollution in the UK: The WRc River Basin Management Programme," in *Geomorphology in Environmental Planning*, ed. J. M. Hooke [Chichester: John Wiley & Sons, 1988], 169–85. With permission from John Wiley & Sons.)

alluvial plain of northern India had exceeded the concentration of nitrates in shallow sources of groundwater discharge to 100 mg l⁻¹, compared to the permissible limit of 45 mg l⁻¹ for human consumption. The excess of nitrates in the groundwater discharge had been the cause of *methemoglobinemia*, or blue baby disease, in the late sixties (Handa, 1994).

Unscientific disposal of liquid and solid municipal waste produces metallic and nonmetallic ions from the refuge and generates pathogenic and nonpathogenic bacteria. These bacteria reach shallow subsurface waters through infiltration and seepage processes. *Pathogenic bacteria* cause waterborne diseases and *nonpathogenic* bacteria produce foul-smelling hydrogen sulfide from the *reduction* of iron and sulfate in groundwater systems.

HEAVY METAL CONTAMINATION OF SOILS

Heavy metal contaminants exist in the biosphere as *trace elements*. They are *toxic* above a certain limit for human consumption and a *health hazard* in the active state. The toxicity of trace elements, however, depends on the geochemical environment. Selenium is nontoxic at a high concentration of 2000 ppm in acidic soils and toxic even at a low concentration of 4 ppm in alkali soils (Keller, 1979).

General regional relationships between the heavy metal contamination of soils and disease are discussed in several texts on environmental geology. However, local studies on heavy metal contamination of soils and human health are scant for the developed and nonexistent for the developing countries. Aspinall et al. (1988) observed a relationship between the industrial activity, soil contamination, and health hazard in Tyneside, UK, where the soils are contaminated with inactive cadmium and inactive and active lead and zinc elements. Inactive cadmium naturally exits in the parent material of soils and is enriched there by the industrial activity to the south of the Tyneside River. The presence of inactive lead and zinc in soils correlates with the distribution pattern of shipbuilding, engineering, chemical, and other industries. In addition, the existence of a strong linear pattern of inactive lead along a surface transport network correlates well with the emission from vehicular traffic. Active lead and zinc, however, occur in plant-available form. Their concentration level in fruit and vegetable produce of the kitchen gardens in Tyneside's residential sector suggests that 58 and 97% of the population of the area is at the threshold respectively of lead- and zinc-related health problems.

SALT HAZARDS

A variety of salts present in the environment of human habitat mobilize in the *ionic state* and interact with the building material, producing stress in man-made structures. In the 1970s, Fookes and associates discussed the dynamics of salt damage to asphalt roads, the key aspects of which findings are summarized in several texts on the subject. In recent years, the dynamics of salt damage to concrete structures, though, is being increasingly emphasized.

SOURCE OF SALTS

A variety of salts, such as sulfates, nitrites, nitrates, chlorides, and carbonates, have a *geologic source* in the earth materials from which bricks and cement are prepared. These, and other salts, mobilize in *capillary suction*, migrate in aqueous phase, and precipitate at or near the surface of the earth. They react with the minerals in bricks, cement, and reinforced steel, forming *alteration products* that are inherently weak in strength attributes.

Processes of Salt Damage to Asphalt Roads

Asphalt-surfaced roads are damaged from the effects of axle load, poor quality of subgrade, poor workmanship, and a variety of natural processes, including the migration of near-surface salts in capillary suction (Cooke et al., 1982). The migrating salts *hydrate and dehydrate* at or in the vicinity of subgrade, generating strong hydration expansion and dehydration contraction stress to cause blisters, potholes, and surface cracks in asphalt roads.

Concrete Disruption

Salt solution migrates slowly through the concrete, reacts with the aggregate, produces alteration products, and generates stress in the structure. The 1994 failure of the Tapo Canyon tailings dam in California similarly followed the weakened strength of saturated tailings slime and retention embankment that could not withstand the stress of loading and unloading cycles of the Northridge earthquake (Harder and Stewart, 1996). The mortar of the Saunders Power Station on St. Lawrence River in Ontario, Canada, likewise shows signs of cracking either to moisture or to variations in the reactivity of aggregate in wet areas of the facility (Grattan-Bellow, 1995).

Solution of carbon dioxide, chlorides, nitrites, and sulfates is disruptive to concrete. Carbonic acid from the diffusion of atmospheric carbon dioxide into the wet portion of concrete reacts with the calcium content of the binder, forming inherently weak alteration products. A solution of chlorides corrodes the reinforced steel and increases the volume of steel bars by three to seven times, exerting an enormous pressure on the concrete structure (Nilsson, 1996). A reaction between sulfate solution and hydrated aluminates of concrete produces *ettringite* or *condolt salt* that causes volumetric expansion, upheaving, and cracking of the concrete structures. The concrete also upheaves and cracks by reacting with the solution of sodium and magnesium sulfates (Hope and Thompson, 1995). Reaction of sodium and magnesium sulfates with the concrete forms the porous product of *gypsum*, which indirectly accelerates the process of carbonation in concrete structures (Al-Amoudi et al., 1995).

Hydrocarbons are other stress-generating compounds, which break down to alcohol and fatty acids at high temperatures, such as in an aircraft maintenance facility (McVay et al., 1993). The *fatty acids* react with calcium hydroxide ($Ca(OH)_2$) in the concrete, and form a highly porous product of calcium silicate hydrate. This process, called *saponification*, flakes the concrete aprons.

CHANNELIZATION

Channelization is channel structuring for flood control, drainage and navigation improvement, and stabilization of banks. Channel modification initiates certain *feedbacks* in fluvial systems, which variously affect the fluvial dynamics, channel morphology, and biology of streams some distance up- and downchannel of the structured reach (Brookes and Gregory, 1988; Downs and Thorne, 1996; Thorne et al., 1996). The feedback mechanisms, however, are unique to each fluvial system.

CHANNELIZATION AND HYDRAULIC VARIABLES

Hydraulic variables up- and downstream of the structured reaches adjust in many different ways. Therefore, only the general nature of channel adjustment is understood for engineered fluvial systems. In general, the straightening of channels and construction of reservoirs across streams alter the channel width and bed slope characteristics of each system in several ways. Experience suggests that the straightened segments of meandering streams evolve steeper gradient, increasing the *channel competence*. This effect had been most devastating for the Lang Lang River in Australia, where the straightening of meander loops between 1920 and 1930 had caused severe bed and bank erosion, and loss of seven bridges across the stream (Brookes and Gregory, 1988).

Construction of reservoirs affects the discharge and sediment load equilibrium in channel segments up- and downstream of the impounded water. The disturbed systems adjust to a new state of equilibrium by channel silting to the limit of the backwater-stream profile intersection, and bed and bank scour immediately downstream

TABLE 12.5
Effect of Sedimentation on Channel Gradient Upstream of Selected Reservoirs in India

Reservoir/State	Year Completed	Channel Slope (m/km) Original	Channel Slope (m/km) Present	Upstream Reach Affected (km)	Depth of Deposit (m)
Nizamsagar, Andhra Pradesh	1931	Deep	Silted	—	—
Bhakra, Punjab	1958	1.9	0.38	24.0	40.0
Panchet Hill, Bihar	1959	0.95	0.475	17.6	3.7–4.3
Maithon, Bihar	1959	1.9	0.95	—	—

Source: Abstracted from Murthy, B. N., "River Bed Variations—Aggradation and Degradation," in *Hydraulics of Alluvial Streams*, Status Report 3 (New Delhi: Central Board of Irrigation and Power, 1974), 59–73. With permission from Central Board of Irrigation and Power.

of the silt-free discharge from reservoirs. Observations in Table 12.5 suggest that channels flatten for a considerable length upstream of reservoirs. The silt-free regulated flow released from reservoirs, however, scours channels and erodes banks up to a point where prevailing bed slope and discharge do not permit effective transport of sediments. Hence, channels thereafter silt farther downstream. The silt-free regulated flow in the Gavin Point Dam on the Missouri River had similarly scoured the channel and eroded banks, destroying 890 ha of land between 1970 and 1973 (Rahn, 1979). The silt-free water released since 1975 from the Farakha Barrage on Hugli River, India, has similarly scoured the bed, eroded banks, and straightened the channel (Basu, 1998).

Observations suggest that the morphologic *recovery time* for structured streams in India is between 4 and 5 years (Murthy, 1974), and for streams elsewhere is between 10 and 100 years (Brookes and Gregory, 1988). Further, the recovery time is less than 5 years for streams disturbed by urbanization and more than 100 years for streams disturbed by extreme floods (ASCE, 1998). It may be remembered, however, that channels widened in floods do not recover the width change even over long periods of recovery time.

Manipulation of the *hydraulic variables* is often advantageous to the management of fluvial systems for a given purpose. Certain reaches of the Huangpu River in eastern China, for example, were known to frequently silt due to poor ebb and required frequent dredging of the bed for year-round navigation of the channel. This problem was resolved by manipulating the width-depth relationship for problematic channel segments in such a way that enhanced the *ebb velocity* for transport of bed sediments to the sea, maintaining sufficiently deep water in the channel for navigation at all stages of flow (Zhu, 1994).

Implications of the regulated flow on economy downstream of reservoirs require evaluation. The *impact assessment* of Bakolori Dam in northwest Nigeria suggests a shift in the cropping pattern toward less-water-demanding crops, decline in the dry season cultivated area, and near collapse of the fishing industry from a decline of *flood frequency* downstream of the reservoir (Adams, 1985). A comparison of pre- and postreservoir period flow in the Damodar River system, India, also suggests a sharp decline in the magnitude of flow from 8000 m^3s^{-1} to 3800 m^3s^{-1}, and high variability of the regulated flow (Sharma, 1976).

ENVIRONMENTAL ASPECTS

Channelization promotes greater uniformity of channel morphology, reduces biological diversity of the engineered stream segment, and alters the physical and biochemical environment of the structured system. Observations suggest that temporal variations in discharge and sediment load of streams affect the spawning habitat of aquatic species and, thereby, regulate the biological diversity of engineered streams (Moon et al., 1997). The physical and chemical aspects of structured streams, such as thermal stratification of the water column, sedimentation of reservoirs, and changes in the chemical composition of water due to nutrient loading from upstream sources variously affect the chemical environment of the system and biologic diversity of aquatic life in many ways. The *nutrient loading* increases the BOD on the system, and *decomposition* of the submerged biomass changes the pH of water column with depth. These aspects respectively affect the aquatic habitat and survival of a few faunal species. Preliminary studies on the environmental aspects of Bhakra Reservoir and Pong Dam in northwestern India suggest that the content of dissolved oxygen decreases with depth at a variable rate, which possibly reflects differences in the nutrient loading and decay of biomass within the two systems (Government of Himachal Pradesh, 1996).

SUMMARY

Human activities accelerate process rates, which affect the stability of geomorphic landscape and the environment in many different ways. Applied geomorphology synthesizes the earth science information for general and purpose-oriented classification of land, providing a rationale for planning, optimum utilization, impact assessment, and management of this finite resource. The methodology and content of major land resource classifications are discussed, and their application to the existing and proposed use of the land resource is highlighted in this chapter.

Cultural processes rapidly alter the land use pattern over short periods of time. Land use changes accelerate the rate of soil and nutrient losses from the land manifold and enhance the concentration of suspended and dissolved load in water bodies, affecting the hydrologic response of streams to floods and the quality of surface and shallow coastal waters offshore. Economic activities also pollute groundwater discharge and contaminate soils with toxic trace elements.

Earth materials and shallow subsoil waters release a variety of salts in the environment. They mobilize in the ionic state, interact with the man-made facilities, and cause their deterioration. The migration of capillary salts in the aqueous phase and the stress of mineral hydration expansion and dehydration contraction damage the asphalt roads. The reaction of carbon dioxide, chlorides, nitrites, and sulfates with concrete produces alteration products that are inherently weak in strength. The reinforced steel corrodes and expands by reacting with a solution of chlorides, generating additional stress in the concrete structures. Hydrocarbons cause flaking of the concrete at very high temperature conditions.

Channel structuring for flood control, drainage and navigation improvement, and stabilization of banks affect fluvial dynamics, channel conditions, biological diversity, and physical and chemical properties of the impounded water in ways that are unique to each system. The above aspects are highlighted in this chapter with selected studies.

REFERENCES

Adams, W. M. 1985. The downstream impacts of dam construction: A case study from Nigeria. *Trans. Inst. Br. Geogr.* 10:292–302.

Adger, W. N., and Subak, S. 1996. Estimating above-ground carbon fluxes from UK agricultural lands. *Geogr. J.* 162:191–204.

Al-Amoudi, O. S. B., Maslehuddin, M., and Saadi, M. M. 1995. Effect of magnesium and sodium sulfate on the durability performance of plain and blended cements. *ASI Mater. J.* 92:15–24.

ASCE, Task Committee on Hydraulics, Bank Mechanics, and Modeling of River Width Adjustment. 1998. River width adjustment. I. Processes and mechanisms. *J. Hydraulic Eng.* 124:881–902.

Aspinall, R., Macklin, M., and Openshaw, S. 1988. Heavy metal contamination in soils of Tyneside: A geographically-based assessment of environmental quality in an urban area. In *Geomorphology in environmental planning*, ed. J. M. Hooke, 87–102. Chichester: John Wiley & Sons.

Basu, S. R. 1998. Bank erosion of the river Bhagirathi-Hugli. In *Ninth Conference of Indian Institute of Geomorphologists (IIG)*, Delhi, January 30–February 1, 1998, p. 144.

Brookes, A., and Gregory, K. 1988. Channelization, river engineering and geomorphology. In *Geomorphology in environmental planning*, ed. J. M. Hooke, 145–68. Chichester: John Wiley & Sons.

Brunsden, D. 1988. Slope instability, planning and geomorphology in the United Kingdom. In *Geomorphology in environmental planning*, ed. J. M. Hooke, 105–19. Chichester: John Wiley & Sons.

Burt, T. P., and Johnes, P. J. 1997. Managing water quality in agricultural catchments. *Trans. Inst. Br. Geogr.* 22:61–68.

Calver, A., and Anderson, M. G. 2004. Conceptual framework for persistence of flood-initiated geomorphological features. *Trans. Inst. Br. Geogr.* 29:129–37.

Carthew, K. D., and Drysdale, R. N. 2003. Late Holocene fluvial change in a tufa-depositing stream: Davys Creek, New South Wales, Australia. *Austr. Geogr.* 34:123–39.

Christian, C. S. 1957. The concept of land units and land systems. *Proc. Ninth Pacific Congr.* 20:74–81.

Christian, C. S., and Stewart, G. A. 1952. *Survey of Katherine—Darwin Region, 1946.* Land Research Series 1. CSIRO.

Connors, K. F., and Gardner, T. W. 1987. Classification of geomorphic features and landscape stability in northwestern New Mexico using simulated SPOT imagery. *Remote Sensing Environ.* 22:187–207.

Cooke, R. U., Brunsden, D., Doornkamp, J. C., and Jones, D. K. C. 1982. *Urban geomorphology in drylands.* New York: Oxford University Press.

Cowell, R. 1997. Stretching the limits: Environmental compensation, habitat creation and sustainable development. *Trans. Inst. Br. Geogr.* 22:292–306.

Crabtree, B. 1988. Urban river pollution in the UK: The WRc river basin management programme. In *Geomorphology in environmental planning*, ed. J. M. Hooke, 169–85. Chichester: John Wiley & Sons.

Craig, R., and Crafts, J. L. 1982. Preface. In *Applied geomorphology*, ed. R. G. Craig and J. L. Crafts, v–vi. London: George Allen & Unwin.

Dhir, R. P. 1989. Wind erosion in relation to landuse and management in Indian arid zone. In *International Symposium on Managing Sandy Soils*, Jodhpur, February 6–10, 1989.

Douglas, I., 1988. Urban planning policies for physical constraints and environmental change. In *Geomorphology in environmental planning*, ed. J. M. Hooke, 63–86. Chichester: John Wiley & Sons.

Dow, V., Kienholz, H., Palm, M., and Ives, J. D. 1981. Mountain hazard mapping: The development of a prototype combined hazards map, Monarch Lake quadrangle, Colorado, USA. *Mt. Res. Dev.* 1:55–64.

Downs, P., and Thorne, C. R. 1996. A geomorphological justification of river channel reconnaissance surveys. *Trans. Inst. Br. Geogr.* 21:455–68.

Eisbacher, G. H., and Clague, J. J. 1981. Urban landslides in the vicinity of Vancouver, British Columbia, with reference to the December 1979 rainstorm. *Can. Geotech. J.* 18:205–16.

Goudie, A. 1986. *The human impact on the natural environment.* Oxford: Basil Blackwell.

Goudie, A., Simmons, I. G., Atkins, B. W., and Gregory, K. J., 1994. *The encyclopaedic dictionary of physical geography.* 2nd ed. Oxford: Blackwell.

Government of Himachal Pradesh. 1996. *Impact of hydropower projects on water quality.* Shimla: State Pollution Control Board.

Govorushko, S. M. 1996. Environmental assessment of a site for civil construction. *J. Urban Planning Dev.* 122:18–31.

Grattan-Bellow, P. E. 1995. Laboratory evaluation of alkali-silica reaction in concrete from Saunders Generating Station. *ACI Mater. J.* 92:126–34.

Gray, J. M. 1997. Environment, policy and municipal waste management in the UK. *Trans. Inst. Br. Geogr.* 22:61–68.

Hails, J. R. 1977. Introduction. In *Applied geomorphology*, ed. J. R. Hails, 1–8. Amsterdam: Elsevier Science Publishing Co.

Handa, B. K. 1994. Groundwater contamination in India. In *Regional Workshop on Environmental Aspects of Groundwater Development*, Kurukshetra, October 17–19, pp. I-1–33.

Harder Jr., L. F., and Stewart, J. P. 1996. Failure of Topo Canyon tailing dam. *J. Perform. Constructed Facilities* 10:109–14.

Helfret, M. R. and Wood, C. A. 1986. Shuttle photos show Madagascar erosion. *Geotimes*, March 4–5.

Hope, B. B., and Thompson, S. V. 1995. Damage to concrete induced by calcium nitrite. *ACI Mater. J.* 95:529–31.

Johnson, R. C. 1995. Framework of a methodology for classifying sediments in Himalayan rivers. In *Proceedings of the Third International Symposium on Headwater Control: Sustainable Reconstruction of Highland and Headwater Regions*, New Delhi, October 6–8, 1995, pp. 289–98.

Keller, E. A. 1979. *Environmental geology.* 2nd ed. Columbus, Ohio: Bell & Howell.

Leopold, L. B. 1968. *Hydrology for urban land planning.* U.S. Geological Survey Circular 559. U.S. Geological Survey.

Lung, W.-S., and Larson, C. E. 1995. Water quality modeling of upper Mississippi River and Lake Pepin. *J. Environ. Eng.* 121:691–99.

Mandal, B. K., Chowdhury, T. R., Samanta, G., Basu, G. K., Chowdhury, P. P., Chanda, C. R. Lodh, D., Karan, N. K., Dhar, R. K., Tamili, D. P., Das, D., Saha, K. C., and Chakraborti, D. 1996. Arsenic in groundwater in seven districts of West Bengal, India—The biggest arsenic calamity in the world. *Curr. Sci.* 70:976–86.

McRae, S. G., and Burnham, C. P. 1981. *Land evaluation.* Oxford: Clarendon Press.

McVay, M. C., Smithson, L. D., and Manzione, C. 1993. Chemical damage to airfield aprons from heat and oils. *ACI Mater. J.* 90:253–58.

Moon, B. P., van Niekerk, A. W., Heritage, G. L., Rogers, H. H., and James, C. S. 1997. A geographical approach to the ecological management of rivers in the Kruger National Park: The case study of the Sabie River. *Trans. Inst. Br. Geogr.* 22:31–48.

Murthy, B. N. 1974. River bed variations—Aggradation and degradation. In *Hydraulics of alluvial streams*, 59–73. Status Report 3. New Delhi: Central Board of Irrigation and Power.

Nilsson, L.-O. 1996. Interaction between microclimate and concrete—A prerequisite for deterioration. *Const. Bldg. Mater.* 10:301–8.

Noble, E. L. 1965. Sediment reduction through watershed rehabilitation. In *Proceedings of the Federal Interagency Sedimentation Conference*, U.S. Department of Agriculture Miscellaneous Publication 970, pp. 114–23.

O'Brien, K. L. 1998. Tropical deforestation and climate change: What does the record reveal? *Prof. Geogr.* 50:140–53.

Owen, L. A., Sharma, M. C., and Bigwood, R. 1995. Mass movement hazard in the Garhwal Himalaya: The effect of the 20 October 1991 Garhwal earthquake and the July–August 1992 monsoon season. In *Geomorphology and land management in a changing environment*, ed. D. F M. McGregor and D. A. Thompson, 69–88. New York: John Wiley & Sons.

Rahn, P. H. 1979. Remote sensing of bank erosion along the Missouri River, South Dakota. In *Satellite hydrology*, ed. M. Deutsch, D. R. Wisenet, and A. Rango, 697–700. American Water Resources Association.

Rai, S. C., and Sharma, E. 1995. Land-use change and resource degradation in Sikkim Himalaya: A case study from Mamlay watershed. In *Proceedings of the Third International Symposium on Headwater Control: Sustainable Reconstruction of Highland and Headwater Regions*, New Delhi, October 6–8, 1995, pp. 265–78.

Ramasamy, S. M., Kumanan, C. J., and Palanivel, K. 2002. GIS based solutions for waste disposals. *GIS Dev.* 6:40–42.

Rawat, J. S., and Rawat, M. S. 1994. Accelerated erosion and denudation in the Nana Kosi watershed, Central Himalaya, India. Part I. Sediment load. *Mt. Res. Dev.* 14:25–38.

Saczuk, E. A. 2001. Has the Kullu district experienced an increase in natural hazard activity? *GIS Dev.* 5:34–41, 44.

Schuster, M., and Hübl, J. 1995. Impact of road construction on the Pokhara-Baglung highway, Kaski district, Nepal. In *Proceedings of the Third International Symposium on Headwater Control: Sustainable Reconstruction of Highland and Headwater Regions*, New Delhi, October 6–8, 1995, pp. 175–82.

Seaburn, G. E. 1969. *Effects of urban development on direct runoff to East Meadow Brooks, Nassau County, Long Island, New York.* U.S. Geological Survey Professional Paper 627B. U.S. Geological Survey.

Sharma, V. K. 1976. Some hydrologic characteristics of Damodar River. *Geogr. Rev. India* 38:330–43.

Sharma, V. K. 1991. *Remote sensing for land resource planning.* New Delhi: Concept Publishing Co.

Siddiqui, M. Z., Everett, J. M., and Vieux, B. E. 1996. Landfill siting using geographical information system: A demonstration. *J. Environ. Eng.* 122:515–23.

Singh, G., 1971. The Indus Valley culture. *Archaeol. Phys. Anthropol. Oceania* 6:177–89.

Singh, S., Ghosh, B., and Kar, A. 1978. Geomorphic changes as evidence of palaeoclimate and desertification in Rajasthan Desert, India (Luni Development Block: A case study). *Man Environ.* II:1.13.

Spangle, W., and Associates, F. Beach Leiguton and Associates, and Baxter, McDonald and Company. 1976. *Earth-science in land-use planning—Guidelines for earth scientists and planners.* U.S. Geological Survey Circular 721. U.S. Geological Survey.

St.-Onge, D. A. 1968. Geomorphic maps. In *The encyclopedia of geomorphology*, ed. R. W. Fairbridge, 388–403. New York: Reinhold Book Corp.

Thorne, C. R., Allen, R. G., and Simon, A. 1996. Geomorphological river channel reconnaissance for river analysis, engineering and management. *Trans. Inst. Br. Geogr.* 21:469–83.

Tuan, Y.-F. 1971. *Man and nature.* Resource Paper 10. Washington, DC: Commission on College Geography.

Turner, R. E., and Rabalais, N. N. 1994. Coastal eutrophication near the Mississippi river delta. *Nature* 368:619–21.

Valdiya, K. S., and Bartarya, S. K. 1989. Problem of mass movement in a part of Kumaun Himalaya. *Curr. Sci.* 58:486–91.

Ward, R. M., and Li, J. 1993. Solid-waste disposal in Shanghai. *Geogr. Rev.* 83:29–42.

Wheeler, R. E. M. 1959. *The Indus civilization.* New York: Frederick A. Praeger.

Wolman, M. G. 1967. A cycle of sedimentation and erosion in urban river channels. *Geogr. Annlr.* 49A:385–95.

Zhu, G. 1994. Recent improvement of Gao Qiao new channel on Huangpu River. *J. Wtrwy. Port Coastal Ocean Eng.* 120:368–81.

Abbreviations

α **(alpha):** Positively charged radioactive particles; angle of the wave approach to coasts

α_b: Longshore component of the wave power

α_o: Angle of wave propagation in deep waters

β **(beta):** Negatively charged radioactive particles

γ **(gamma):** Unit weight of sediment; internal radius of the capillary tube; specific weight of water; radioactive particles of neutral charge

γ_b: Buoyant weight of moist soils

γ_t: Unit weight of a soil mass

γ_s: Bulk density

γ_w: Unit weight of water

δ_w **(delta):** Energy dissipation per unit volume

ΔT **(delta):** Activation energy

ε **(epsilon):** Eddy viscosity

ε·: Strain rate

θ **(theta):** Slope inclination; surface slope in percent; angle of the meniscus with capillary tube; surface slope of the glacier ice

θ_a: Half-angle of the subglacial asperity tip of rock fragments

θ_b: Angle of the incident wave in breaker zone

κ **(kappa):** Constant in von Karman equation

λ **(lambda):** Wavelength; displacement distance; linear concentration of particles in the granular flow

μ **(mu):** Dynamic viscosity; coefficient of viscosity; coefficient of friction; micron (particle size)

ν **(nu):** Kinematic viscosity

π **(pi):** Diameter of the capillary tube; a dimensionless sediment transport parameter

ρ **(rho):** Density; density of water; density of ice

ρ_a: Density of air

ρ_s: Particle density

σ **(sigma):** Normal stress

σ'_c: Initial confining pressure

σ_e: Effective normal stress

τ **(tau):** Shear stress

τ_b: Basal shear stress; driving stress; buoyant unit weight of soils, basal shear stress of the glacier ice

τ_c: Critical shear stress for grain movement

τ_{max}: Shear stress at failure

τ_o: Average shear stress for bed particles

τ_y: Yield stress

τ_t: Unit weight of a soil

φ **(phi):** Angle of internal friction; angle of intergranular cohesion; rate of ice discharge

φ': Apparent angle of internal friction

ω **(omega):** Available power due to the bed drag

A: Drainage basin area; surface area; constant for soil types; sediment size

A_b: Rate of subglacial abrasion by rock fragments

A_d: Drainage basin area

A_m: Wave amplitude

A_p: Surface area of normal stress application

B: Boltzman constant for temperature-dependent hardness of ice; rate of chemical weathering; a constant for beach profile expression

C: Wave phase velocity; Chézy coefficient; constant for factors contributing to flow resistance

D: Particle size diameter

D_h: Vertical displacement in liquefied sediments

D_m: Mean particle size

D_r: Relative density of sediments

E: Wave energy

E_n: Resisting force acting horizontally in rotational failures

F: Froude number; inertial force; tensile force due to the water film in sediments

F_N: Normal force

F_S: Ratio of shear strength to shear stress for slope stability; shear stress not available for resisting the weight of mass in rotational failure

G_g: Group velocity of waves

H: Wave height

H': Dimensionless bed form height

H'_c: Cliff height above a rupture plane

H_o: Deep water wave height

K: Kilo

ka: Thousand years

L: Wavelength of sea waves; suspended load of streams by weight per unit time

L_I: Liquidity index

L_o: Angle of wave propagation in deep waters

L*: Wavelength in shallow coastal waters

M: Rank of an extreme event in series of observation

Ma: Million years

N: Newton (force); number of observations

P: Normal pressure

PCO_2: Partial pressure of carbon dioxide

P_a: Pascal (Newton per square meter)

P_I: Plasticity index

P_w: Pore pressure

Q: Discharge; sediment transport rate; total longshore sediment transport rate

Q_b: Bankful discharge

Q_m: Mean annual discharge

R: Hydraulic radius; radius of bedrock curvature

R_e: Reynolds number

S: Channel slope; water surface slope; beach slope

S_a: Susceptibility to abrasion

S_c: Compressive strength of cliff-forming materials

T: Absolute temperature; half-life period of radioactive elements; surface temperature; recurrence interval of an event; period of measurement; surface tension; thickness of liquefied layer; mean water hardness; wave period (celerity)

U_p: Relative velocity of ice; relative velocity of subglacial rock fragments

U_s: Sliding velocity of ice; mass transport velocity

V: Mean velocity of fluid flow

V_p: Particle velocity of saltating grains

V_s: Longshore current velocity in the mid-surf zone

W: Rate of subglacial abrasion by clean ice; watt; weight of a unit slice of failed mass

W_b: Effective weight of the failed mass

W_l: Plastic limit

W_p: Liquid limit

W^*: Volume of the clastic load in basal ice

X_n: Driving force acting vertically through a rotational failure

X_t: Mean dissolution rate of the karst terrain

Z: Critical joint depth

a: Ionic activity of calcium and carbonate in aqueous solution; angle of arc from the center of curvature; constant for hydraulic relations in the hydraulic geometry of streams

b: Exponent for relationships in the hydraulic geometry of streams

c: Wave phase velocity; constant for surface conditions in aeolian environments; constant for hydraulic relations in the hydraulic geometry of streams

c′: Effective stress, intergranular cohesion; apparent cohesion; dimensionless bed form celerity

d: Water depth; constant for hydraulic relations in the hydraulic geometry of streams; particle size diameter

f: Darcy–Weisbach resistance coefficient; constant for hydraulic relations in the hydraulic geometry of streams

f*: Average applied force in chemical reactions

f(v): Sliding velocity of the failed mass

h: Capillary head; thickness of the glacier ice

i: Mass transport

k: Constant for an index of channel erosion; an index for cliff recession; curvature of sand beach; constant for an index of beach slope; constant for hydraulic relations in the hydraulic geometry of streams

ka: Thousand years

l: Length of arc in rotational failure

ln: Natural logarithm

m: Beach slope; coefficient for beach shape; weight of clastic sediments; exponent for hydraulic relations in the hydraulic geometry of streams

n: Bed roughness; channel roughness factor in the Manning equation; proportion of the karst area in a drainage basin; a constant for expressing group phase velocity; an exponent for hydraulic relations in the hydraulic geometry of streams

r: Radius; radius of suspended particles; density of karst

$\mathbf{r_m}$**:** Radius of channel curvature

s: Shear strength of unconsolidated sediments; sliding velocity; channel steepness

u: Dynamic viscosity; pore pressure; wind velocity

$\mathbf{u_m}$**:** Orbital velocity

$\mathbf{u_t}$**:** Dynamic threshold velocity

$\mathbf{u_*}$**:** Wind shear velocity

$\mathbf{u_*t}$**:** Fluid threshold velocity

v: Mean velocity of the fluid flow

v': Proportion of flow units displaced per unit time

w: Water surface width in stream channels; sand fall velocity in coastal waters

z: Wind height above the surface; an exponent for the variation of channel slope with mean annual discharge

μ/ρ**:** Kinematic viscosity

τ/v^2**:** Resistance to channel flow

$\delta\varphi/\delta\mathbf{x}$**:** Rate of change of ice accumulation over a certain distance

du/dy: Velocity gradient in a deforming mass

dv/dy: Rate of change of velocity with depth of a fluid mass

dx/dt: Rate of cliff recession

Glossary

Abiotic: Nonliving components of the ecosystem. They mostly include atmospheric gases, water, mineral soils, and inorganic salts.

Ablation zone: The lower part of a glacier where annual melting of snow and ice exceeds the annual accumulation of snow.

Abrasion: A physical process of rubbing, scouring, or scraping by friction.

Absorption: Passing or diffusion through the surface into the interior of substances.

Accumulation zone: The upper part of a glacier where annual accumulation of snow exceeds evaporation and melting.

Acidic rock: An igneous rock of 66% or more silica content or comprised predominantly of silicate minerals.

Activation energy: The amount of energy a particle or group of particles requires to escape from one energy state to the other. The term is also applied to a phase change in chemical reactions and movement of particles.

Active layer: Mineral soil above the permafrost table that freezes each winter and thaws in summer.

Activity index: Ratio of the plastic index of soils and percentage weight of soil particles less than 0.002 mm in size. It is a useful measure of the swelling potential of soils.

Adiabatic: Change of temperature or pressure when a gas or other fluid is compressed or expanded without exchanging heat with the system.

Adsorption: Adhesion of molecules, such as gases and water, to the surface of solid bodies they are in contact with.

Advective flow: The movement of fluid due entirely to the velocity of fluid in the medium.

Aeolianite: Cemented sand dune in which carbonate is the cementing agent.

Aeration zone: A subsoil zone in sediments through which vadose water moves downward under the influence of gravity.

Aerial photograph: A photograph of a part of the earth taken with a fixed camera on an aeroplane flying along a predetermined path and altitude.

Aerobic: Active or living in the presence of air or oxygen. It is also the utilization of molecular oxygen in chemical reactions.

Aerodynamic form: A form of a length-to-width ratio of 3 or 4:1 that offers least resistance to the activity of process (glacier, wind) forming it.

A-horizon: The uppermost layer of soils containing organic matter and leached minerals.

Air-freezing index: The duration and magnitude of winter season temperature above and below the freezing point.

Albedo: A measure of the capacity of a body or surface to reflected incident radiation.

Algae: A group of simple nonvascular plants from minute plankton to enormous seaweeds capable of photosynthesis.

Algal bloom: Proliferation of living algae by the nutrient loading of water bodies.

Alkali: A substance that produces an excess of hydroxyl ion in solution.

Alkaline rock: An igneous rock rich in the content of sodium and potassium.

Allochthonous: Rocks that have not formed *in situ*.

Alluvial channel: A channel in noncohesive bed material. It deforms by the channel flow.

Alluvial fan: A fan-shaped deposit of the detritus load of highland basins onto an adjacent surface of gentle gradient.

Alluvial fill: Alluvium of a single stratigraphic unit of an uninterrupted period of aggradation.

Alluvium: Clastic fractions, chemical sediments, and organic matter transported and deposited by the running water.

Altiplanation: A cold region process of the leveling down of relief by mass wasting.

Amino acid: Building block of protein having amino group(s) and carboxylic group(s) in the same molecule.

Anabranch: Channels interspersed by flood plains of sizable extent.

Anaerobic: Living or active. Also pertains to chemical reactions induced by organisms that live in the absence of free oxygen.

Anastomosing: A process of rejoining of major channel branches with the main channel.

Andesite: A volcanic rock essentially of andesine and one or more mafic minerals.

Angle of internal friction: The angle at which sliding particles settle to a more stable position about intergranular spaces.

Angle of repose: The angle at which granular materials come to rest.

Anion: An ion with negative charge.

Anisotropic: A term for physical properties that vary in different directions.

Anorthite: A common rock-forming mineral comprised of plagioclase minerals of a general composition $(Na, Ca) Al (Si, Al) Si_2O_8$.

Antidune: Bed form of symmetric sand wave, which moves downstream or remains stationary while being in phase with the water surface level.

Aplite: A granite-like rock consisting generally of quartz and orthoclase.

Aquifer: A permeable geologic formation that can hold and transmit water in sufficient quantity.

Aragonite: Calcium carbonate mineral, which is a mirror reflection of calcite.

Arenaceous: Sedimentary rock composed mainly of sand particles.

Arête: An irregular, sharp-crested mountain ridge shaped by cirques on either side of the hillslope.

Argillaceous: Sedimentary rock composed mainly of clay fractions.

Artesian water: Subsoil water that rises to the surface under hydrostatic head.

Asthenosphere: Probably a weak and molten layer of an unspecified thickness below the lithosphere.

Atm: Force per unit area of the weight of atmosphere at sea level. It is 76.0 cm of mercury column or 1,013.25 millibars (1 bar = 1,000,000 dynes cm^{-2}).

Atoll: An enormously thick ring or horseshoe-shaped reef attached to seamounts and guyots. It forms islands around central lagoons.

Atom: The smallest charged particle of an element.

Atterberg limits: Approximate soil moisture content separating solid, plastic, and liquid behavior of fine-grained soils.

Attrition: The wear of clastic sediments by collision in transport.

Aureole: Area surrounding the intrusion of magma in rock bodies. It is affected by contact metamorphism.

Authogenic: A term applied to minerals formed on the spot by chemical and biochemical processes.

Autochthonous: A term for features formed *in situ*. It is often applied to evaporites that develop in place and to geologic structures folded and faulted without displacement.

Autotrophs: Microorganisms using inorganic materials as a source of nutrients and carbon dioxide as source of carbon.

Avalanche: Sudden and catastrophic movement of a vast mass of ice, snow, and earth materials.

Back-arc basin: An ocean depression on the concave side of an island arc.

Backshore: Portion of a beach above the high-tide level, which can only be reached by storm waves.

Backswamp: A low-lying marsh or swamp area on a flood plain.

Backwash: Return flow of water down the beach gradient.

Bacteria: Microscopic (2 to 5 μ long and 0.5 to 1 μ wide) organisms devoid of a defined nucleus and other membrane-bound organelles.

Badland: A term for intensely dissected fluvial topography on unconsolidated sediments of scant vegetation cover.

Bajada: A continuous apron of alluvial fans and pediments at the foot of mountains and hillslopes surrounding desert basins.

Bar, coastal: Submarine sand bars at the break point of steep waves in shallow coastal waters.

Barchan: A crescent-shaped sand dune with horns pointing downwind.

Barophilic: Existing under high-pressure conditions.

Barred (storm) beach: A sand beach of gentler gradient formed by the erosion of berm and beach face.

Barrier island: An elongated sandy feature that parallels the coast but remains separated from the shore by a lagoon or tidal inlet.

Barrier reef: A coral reef separated from the coast by a lagoon or water body.

Barrier spit: A part of the barrier island attached to an eroding headland. It generally parallels and separates the coast by a lagoon.

Basalt: A term for fine-grained, dark-colored igneous rock.

Base level: The level below which streams cannot erode. Sea level is the ultimate base level, but lakes and resistant rock formations also provide regional and local controls, respectively, on the base level of erosion.

Basic rock: An igneous rock rich in mafic minerals of iron and magnesium.

Batholith: A stock or shield-shaped mass of intrusive igneous rock with 100 km^2 or more of exposed surface area.

Bauxite: Principal ore of aluminum comprised aluminum oxides and hydroxides.

Bay: A wide and shallow inundation of the land by the sea.

Baymouth bar: A coastal bar that extends across bays.

Beach: A feature of coastal sediments that extend between the low-tide level and first major topographic change on land.

Bed configuration: Irregularity of the bed of alluvial channels produced by the movement of water and sediments.

Bedding plane: A division plane separating individual layers, beds, or strata of stratified sedimentary rocks.

Bedload: The coarser fraction of bed material load that moves as a contact and saltation load at or near the bed of streams.

Bed material: Material comprising the bed of streams.

Bed shear: The fluid force, fluid stress, or tractive force required to entrain bed material of a given size.

Benioff zone: Seismically active zone at the bottom of deep sea trenches where the oceanic crust subducts rapidly into the mantle material.

Benthos: The living plant or animal organisms at the floor of oceans.

Bergschrund: A deep crevasse at the head of a few cirques and valley glaciers.

B-horizon: A soil layer comprising the accumulation of clay and oxide minerals beneath the A-horizon.

Bioameleorant: Organisms that convert toxic compounds in the environment to non-toxic products.

Biochemical oxygen demand (BOD): The amount of dissolved oxygen in water required for decomposition of the organic matter by aerobic bacteria. The BOD is also an index of water contamination.

Biodegradable: Organic matter decomposed by microorganisms.

Biomass: Organic matter derived from plants, animals, and other organisms. It is usually expressed by the oven-dry weight per unit area.

Biosorbent: Biotic organisms that absorb toxic compounds and recycle them in the environment.

Biosphere: Interface of the earth's crust, ocean, and atmosphere with living organisms. This interface environment extends to about 10 m in soils and 100 m in oceans.

Blockfield: *In situ* accumulation of large-sized frost-shattered angular blocks of bedrock in areas of continuous permafrost.

Blowout: A deflation hollow or breach in a vegetated sand dune.

Bog: Waterlogged area carrying thick accumulation of dead or partly decomposed plants.

Bog hummock: Small-sized rounded knobs in the bogs of continuous permafrost.

Bond strength: The linkage between two atoms in a molecule due to the sharing of electron pairs. Also the amount of energy required for breaking one mole of the bond.

Bornhardt: A domed-shaped residual hill in granite and gneiss terrain.

Boundary layer: A zone in the fluid flow nearest to the solid surface where velocity gradient develops by the frictional contact of the surface.

BP: Before present.

Brackish: Water of a salt-content intermediate between freshwater and seawater. The salt content is described as salinity.

Braided channel: A stream carrying the flow and sediment load through a number of interlaced branches that divide and rejoin across channel bars and islands.

Breaking antidune: A curved-shaped symmetric wave on the water surface and channel bottom, which lunges forward by building and breaking downstream.

Brine: A solution of dissolved salts in the seawater.

Brittle: A property of rocks that despite being hard causes them to break at a critical stress.

Butte: A steep-sided and flat-topped hill formed by the erosion of flat-lying strata.

Cainozoic: The era, which includes the Tertiary and Quaternary periods.

Calcite: A mineral of calcium carbonate. It is the principal constituent of limestone.

Caldera: A large basin-shaped crater bound by steep slopes. It is often formed by the collapse of the top of an active volcano and sometimes occupied by a lake.

Cambrian: The oldest geological period, which lasted from 500 to 600 million years ago.

Capillarity: A phenomenon of undersaturated flow in porous media.

Carbohydrate: A compound in which hydrogen and oxygen atoms bind to carbon in a 2:1 ratio.

Carbon-14 (^{14}C): Radioactive isotope of carbon.

Carbonation: A process of chemical weathering that replaces minerals in the earth materials by carbonates.

Carbonic acid: A weak acid formed by the diffusion of carbon dioxide in water.

Carcinogenic: Any physical or chemical agent leading to the development of cancer.

Catalyst: A substance or molecule that activates a reaction without being consumed or altered in the reaction.

Cation: An ion with positive charge.

Cave: A large-enough solutionally enlarged cavity in limestone beds.

Cave-vein ice: Ground ice formed within the cavities of a frozen ground.

Chalk: Soft limestone formed from the tests of floating microorganisms and bottom-dwelling marine forms.

Channel avulsion: Process of channel migration in which a major segment of the channel shifts in entirety.

Channelization: Structural modification of streams for flood control, navigation, and prevention of erosion.

Channel pattern: A pattern of large-sized bed forms in plan view. Channel patterns are commonly classified as braided, meandering, and straight.

Channel roughness: The roughness due to bed and bank materials, vegetation, sinuosity, obstruction within channels, and irregularity of channel cross sections. Channel roughness is expressed mathematically by the Manning equation.

Chelation: Biochemical removal of metallic and nonmetallic ions from the earth materials.

Chemical element: A substance of the same atomic number. Atmospheric gases like nitrogen and oxygen, and solids like iron and nickel are all individual elements.

Chemical sedimentary rock: A rock formed at or near its place of deposition by chemical precipitation in the sea and large lakes.

Chemical weathering: Decomposition of the earth materials by chemical alteration of the rock-forming minerals to products in equilibrium with the environmental stress conditions.

Chenier: A beach ridge composed of sand and shell debris surrounded by low-lying swamp deposits.

Chézy equation: An expression for the stream velocity as a function of hydraulic radius and slope.

Chlorophyll: The green coloring agents of plants, which are made up of chelates.

Chute and pool: Flow phenomenon and bed configuration at steep slopes and large bed material discharge.

Chute cutoff: Closing of meander loops by lateral and downchannel migration of streams, forming meander necks and oxbows.

Cirque: A deep closed depression at the head of many valley glaciers and on mountain slopes.

Clapotis: A phenomenon of standing waves due to wave reflection.

Clastic: A term for rock and organic fragmentary material that has moved from the place of its origin.

Clay: Hydrous aluminosilicate minerals, less than 0.002 mm in diameter, formed by weathering and hydration of other silicates.

Climatic optimum: A warming trend for the mid-latitude Holocene climate, during which the mean annual temperatures were 2.5°C higher than the present mean.

Climbing dune: An anchored dune that climes up the windward face of low-angled cliffs.

Closed system: A system with closed boundaries across which gain or loss of energy and mass is not permissible.

Coagulation: A process that thickens the suspended sediments into semisolid mass.

Coast: A strip of land that extends from the shoreline to the land where first major change occurs in terrain features.

Coefficient of thermal expansion: Volumetric increase of a material with increasing temperature.

Cohesion: Mutual attraction among particles, particularly in clay-sized fractions. Cohesive soils possess greater shear strength.

Cold-based glacier: See *polar glacier.*

Colloids: Generally amorphous substances and their solution that are without a definite crystalline structure.

Colluvium: A deposit of a wide variety of particle sizes moved downslope by the gravitative influence.

Competent velocity: Flow velocity at which particles of a given size are entrained in the flow.

Complex: Molecular compounds that retain identity even when dissolved in water or other solvents. The property of complexes is completely different from the constituents.

Complex dune: Sand dune of a fundamental type on which dunes of other types are superimposed.

Compound: A substance formed by the combination of cation(s) and anion(s) in a specific ratio, such as H_2O.

Compressive strength: The load per unit area at which the earth materials fail by shearing or splitting.

Conglomerate: A sedimentary rock comprising lightly cemented rounded pebbles and boulders.

Conservation: Rational use of the environment for improving the quality of life.

Constituted ice: Ground ice formed by the freezing of water in frozen ground environments.

Contact metamorphism: Mineralogical and textural changes caused by heat and pressure of magmatic intrusion at or near its contact with rock.

Continental crust: The outermost layer of the lithosphere. It is 2 to 3 billion years old, 30 to 70 km thick, 2.7 in density, brittle in nature, and comprises relatively light-density minerals, predominantly of silica and aluminum (Sial).

Continental glacier: A substantially thick ice sheet of continental dimension, which deforms largely by sliding over its bed.

Continentality: Characteristic of a region determined by low specific heat and poor thermal conductivity of land relative to the sea.

Continental shelf: A gently sloping featureless portion of the continental crust, 200 to 500 m below the sea level.

Continuous permafrost: Frozen ground reaching great depths in areas of –5°C or less mean annual temperature for at least two consecutive summers and the intervening winter.

Coral: Sedentary marine organisms, which grow in colonies to the depth of sunlight penetration in tropical and subtropical waters. They form massive calcareous structures called coral reefs.

Coral reef: An extensive eminence built of coral, coral sand, and coralline algae, which grow from the depth of sunlight penetration in the seawater.

Cordillera: A group of mountain systems in which each system comprises several ranges.

Core: Central part of the earth beginning at a depth of 2900 km from the earth's surface. It probably consists of iron-nickel alloy (Nife) and comprises an inner solid core roughly 1300 km in radius and an outer core that may be in a molten state.

Coulomb equation: A widely used relationship for describing the sediment strength in terms of cohesion and friction among grains.

Craton: A portion of a continent that has remained stable and not affected by major deformation since the Early Palaeozoic time.

Creep: Imperceptibly slow movement of sand, soil, and rock fragments.

Crevasse: A tensional crack at the surface of glacier ice.

Crevasse splay: An overbank bedload deposit laid beyond levees by accidental breach in flood events.

Cryoplanation: A cold region process of the redistribution of sediments from higher to lower ground.

Crystal structure: The manner of packing of chemical elements in a mineral.

Cuspate spit: A compound spit or bar formed by the progradation of barrier bar or island in a hooked form.

Cyanobacteria: A division of bacteria that obtain energy for sustenance from photosynthesis. Traditionally, cyanobacteria have been called blue-green algae.

Cyclic time: A geologic time of thousands to millions of years during which geomorphic systems progressively evolve by losing relief and potential energy.

Cymatogeny: Warping of the earth's crust across tens to hundreds of kilometers without significant rock deformation. The process produces vertical displacement of the affected crust by tens to hundreds of meters in elevation. Cymatogeny is intermediate between orogeny and epeirogeny.

Dead-ice moraine: A hummocky moraine comprising stratified and nonstratified glacial drift of a dead-ice mass.

Deccan trap: Fine-grained basaltic lava covering an area of some 52,000 km^2 in the Deccan of southern India.

Deflation: A process of the removal of loose surface particles by the wind.

Degree-days: The sum of positive or negative air temperatures for a specified period of time.

Delta: A fan-shaped alluvial tract of subaerial and subaqueous deposition at the mouth of rivers.

Dendrochronology: Study of tree ring widths for interpretation of the environmental change.

Denudation: General lowering of the earth's surface by weathering, erosion, and mass wasting.

Desert pavement: A surface of pebble- to gravel-sized stones in aeolian environments.

Desert varnish: Coating of the oxides of iron and manganese on relatively stable surface stones and rock surfaces in aeolian environments.

Desilication: A process of the removal of silica from the earth materials by hydrolysis.

Deviator stress: The algebraic difference between the maximum and minimum principal stresses.

Diagenesis: A process of physical and chemical change in sediments during lithification and compaction.

Dike: A roughly planar body of intrusive igneous rock with a discordant contact with the rock into which it has intruded.

Dip: The angle at which a stratum is inclined from the horizontal.

Discontinuous permafrost: Frozen ground with pockets of thawed areas within. The mean annual temperature of areas in the discontinuous permafrost varies between −5 and −1° C.

Doline: A fundamental unit of solutionally enlarged depression commonly at the karst surface. Also called a sinkhole.

Dolomite: A common rock-forming calcareous mineral of the composition Ca Mg $(CO_3)_2$.

Drag: Force of the fluid on its bed. This force resolves into types of surface and form drag. Surface drag results from the shear across surfaces, and form drag is caused by the difference of pressure at the up- and downfluid sides of particles.

Drainage density: The ratio of total stream length to drainage basin area. Drainage density is commonly the effect of climate, basin lithology, topography, soils, vegetation, and land use characteristics.

Drift potential: Maximum amount of sand moved by the wind of certain velocity. The wind velocity is measured in knots.

Drumlin: An aerodynamically shaped smooth low hill composed generally of till and sometimes of till, glaciofluvial sediments, or bedrock surfaces.

Dry valley: A feature commonly of chalk terrain.

Ductile: A property of materials that deform plastically.

Duricrust: A generic term for indurated surfaces formed at or near the earth's surface by weathering of rocks and soil formation. Duricrusts are common in tropical and arid regions.

Dust: Silt and clay-sized particles carried in the atmosphere by suspension. It is the source of loess.

Dynamic (impact) threshold: A threshold velocity for maintaining the movement of sediment already in motion.

Dynamic equilibrium: Adjustment between contemporary processes and landforms, balance between process rates in an area of specified size, or balance between matter and energy entering and leaving geomorphic systems.

Dynamic viscosity: Ratio of the stress intensity to the accompanying rate of fluid deformation.

Earth hummock: Rounded knobs of silt and clay in frozen ground environments.

Earth's crust: The outer shell of the earth, comprising a continental crust and an oceanic crust.

Ebb: A tidal current that moves seaward through estuaries and inlets at low tides.

Echo dune: A dune upwind of high-angled cliffs. Its shape obtains from compression of the wind flow and development of reverse eddy flow.

Ecological succession: A gradual directional vegetation change at a particular site, producing a sequence of plant communities.

Ecology: A study of the interaction of organisms with one another and their abiotic environment.

Ecosystem: An area of any size within which organisms interact with the abiotic environment by the flow of energy across the system.

Edge wave: Surface waves with crests perpendicular to the shore. They occur where the wave energy is trapped against the coast.

Efflorescence: Accumulation of capillary salts at or near the earth's surface, usually as a powdery encrustation.

Eh: The activity of ionic species of the same element changing to a stable oxidized or reduced state. It is measured in volts.

Elastic deformation: A type of deformation in materials under stress in which affected substances return to the original shape once the stress is relieved.

Electromagnetic radiation: Short-wave radiation from the sun that propagates through space or various solid media in the manner of vibrating electric and magnetic fields.

Electron: An elementary negatively charged particle of mass 9.1×10^{-31} kg and unit charge 1.6×10^{-19} coulomb.

Endangered species: Plant or animal species, which no longer can be relied upon to reproduce themselves in a sufficient number to ensure their survival.

Endokarst: Subsurface features of dissolution and precipitation of karst.

Endolithic lichen: Algae and fungi subsisting on rock materials.

Energy gradient: The water surface gradient in streams, such that streams of steeper gradient are higher in energy level.

Environment: The sum total of all abiotic and biotic conditions that act upon an organism or community to influence its development or existence.

Environmental change: Broad spectrum of changes in climate, continents and sea floors, floral and faunal species, and others, usually since the beginning of the Pleistocene epoch.

Environmental geology: Application of geologic information to the study of environmental problems.

Environmental impacts: Direct and indirect impacts of human activities on the stability of ecosystems.

Environmental resource unit: A distinct division of the land surface with similar geology, soils, and other biotic associations.

Epicycle: Small cycle within a larger cycle.

Erg: A sand sea.

Erosion: Wearing down of the land surface by water, ice, and wind, and transport of resulting products.

Esker: An irregular-crested ridge of glaciofluvial deposition in ice tunnels and supraglacial streams.

Ester: A fruity-smelling reaction product of the carboxylic acid and alcohol.

Estuary: A funnel-shaped mouth of a river experiencing tidal effects due to the mixing of freshwater and seawater.

Eustacy: Global sea level change.

Eutrophication: Addition of nitrogen and phosphorous nutrients to an ecosystem, raising its net primary productivity. It usually causes algal bloom in fresh and offshore water bodies.

Evaporite: Mineral matter precipitated from the evaporation of aqueous solutions.

Evapotranspiration: A combined process of evaporation of water from soils and transpiration by plants.

Exfoliation: The peeling off of concentric layers of bare rock surfaces by the mechanical and chemical stress of the environment.

Exhokarst: Surface features of the dissolution of karst.

Existence domain: The dominance of a certain singular process activity.

Extremophiles: Organisms adapted to extreme environmental stress conditions.

Extrusive lava: Surface deposit of lava escaping volcanic vents and fissures in the earth's crust.

Facies: Characteristics of a sediment or sedimentary rock that indicate its particular subenvironment of deposition.

Fallow land: Plowed or cultivated land allowed to rest for one or more cropping seasons.

Faro: A reef that evolves around an atoll, forming islands around a central lagoon.

Fatty acid: A general term for group of monobasic organic acids derived from hydrocarbons.

Fault: A fracture or break in the rock or rock sequences that causes vertical or lateral movement of the affected block.

Feedback mechanism: A process of interadjustment among the form attributes of geomorphic systems to the effects of changes intrinsic and extrinsic to the system.

Felsic mineral: A light-colored mineral rich in the content of aluminum and silica.

Field capacity: The maximum amount of moisture soils can hold.

Firn: Snow compacted into a granular form.

First-order landforms: Global scale differentiation of landforms, giving continents and oceans as two geologically and geographically distinct landforms of the earth.

Fjord: Glacially eroded long, narrow, and deep gorge terminating into the sea.

Flood (tidal): A tidal current that moves landward through estuaries and inlets at high tides.

Flood basalt: Extrusive lava of a regional dimension.

Flood frequency: Recurrence interval of overbank flows of a given magnitude.

Flood plain: A surface adjacent to streams built by the vertical and lateral accretion of sediments.

Flow failure: Downslope movement of viscous sediments by interparticle deformation.

Flow law of ice: Predicts that the rate of ice deformation is proportional to the thickness and surface slope of ice mass.

Flow regime: The range of flows producing similar bed forms, resistance to flow, and mode of sediment transport.

Fluid: A substance that deforms continuously by the application of stress.

Fluid mechanics: The science of the behavior of liquid, gas, and vapor at rest and in motion. Hydraulics, hydrodynamics, and aerodynamics are three branches of fluid mechanics.

Fluvial terrace: An abandoned flood plain.

Focal depth: The depth point at which earthquakes originate.

Foraminifera: Microscopic planktonian unicellular organisms that contribute siliceous tests to the deep sea sediments.

Foreland: The side of mountain range toward which the overturned folds incline.

Foreset beds: High-angled deposits dipping in the direction of fluid flow.

Foreshore: A seaward-sloping intertidal portion of the beach.

Fossil: The remains or traces of animals or plants.

Fracture zone: A fracture system in the oceanic crust that may or may not be active.

Free radical: Atom or group of atoms in a compound that carry an odd number of electrons in the outer shell configuration. Hence, free radicals are highly unstable and chemically reactive.

Freeze-thaw: A process of cold region environments generating temperature-regulated freeze-and-thaw cycles in the subsoil and rock cavities.

Freezing index: The sum total of below-freezing temperatures in a year. It is expressed in degree-days.

Fringing reef: A basic form of coral reef around headlands of the tropical coasts.

Frost cracking: Thermal cracking of the frozen ground on an annual basis.

Frost creep: An imperceptibly slow downslope movement of particles in frost-susceptible soils.

Frost heave: A process of vertical displacement of soil particles and blocks of rock due to the development of segregated ice in the active layer.

Frost shattering: See *frost wedging.*

Frost-susceptible soil: Silt-sized mineral soils that frost heave many times the original volume by the development of segregated ice.

Frost thrust: A process of horizontal displacement of soil particles and rock fragments in the active layer.

Frost wedging: A process of the freezing of water in rock cavities, causing dilation and disintegration of rocks along the cavity space.

Froude number: A dimensionless ratio of inertial to gravity forces in flowing water, providing threshold values for tranquil and rapid stream flows.

Frozen ground: Below-freezing subsoil temperature in rocks and sediments of cold region environments.

Gelifluction: Redistribution of sediments from higher to lower ground by the movement of saturated active layer.

Geocryology: A study of the processes and landforms of cold region environments.

Geomorphic hazard: Process that makes our life difficult. It causes loss of life, property damage, or both.

Geomorphic system: Objects or elements, their instantaneous state, and relationships between interacting and interrelated components within a defined space. Geomorphic systems are open systems.

Geomorphic threshold: Refers to a shift from one system state to the other. The shift can be initiated by conditions external to geomorphic systems or internal to landform instability.

Geosyncline: A linear depression of regional dimension that subsided deeply throughout a long period of geologic time.

Geothermal heat: Internal heat of the earth produced largely by the decay of radioactive elements. It escapes the earth at the rate of 65 erg cm^{-1} s^{-1} or 38 cal cm^{-2} $year^{-1}$.

Glacial drift: Nonstratified and stratified sediments deposited by glaciers and meltwater flow.

Glacier: A mass of ice that flows or has flowed under the influence of gravity.

Glacier flood: Catastrophic flood of a glacier melt ponded behind an ice-dammed lake.

Glaciofluvial: The activity of meltwater streams.

Glei soil: A clayey soil rich in organic matter. It develops in waterlogged and poorly drained conditions.

Glen's law: A relationship for the deformation rate of ice with shear stress.

Gneiss: A coarse-grained metamorphosed igneous rock.

Gondwanaland: A large ancient continental mass that rifted from a super continent called Pangea. Component blocks of Gondwanaland now exist in parts of Antarctica, Australia, India, Africa, and South America.

Graben: A valley or trough bound by two normal faults of parallel strike but of converging dips.

Gradational process: Processes of aggradation and degradation that bring the land surface to a common level.

Graded time: A time of the order of hundreds to thousands of years during which components of a geomorphic system do not change with time.

Granite: A coarse-grained igneous rock comprised predominantly of quartz and alkali feldspar. Mica and hornblende are its other constituents.

Gravitational fractionation: A process of density-differentiated evolution of the layered structure of the earth.

Gravity waves: See *swell*.

Great Ice Age: See *Pleistocene epoch*.

Ground ice: More or less clear ice in the subsoil of frozen ground environments.

Ground moraine: A nearly smooth vast area of subglacial till deposited from active glaciers.

Grus: Fragmental product of local decomposition of granite.

Guyot: A tabular form of seamount.

Gypsum: A common evaporite mineral of hydrated calcium sulfate ($CaSO_4.2H_2O$). Also known as alabaster.

Habitat: Natural abode of an organism comprising the biotic and abiotic components of the environment.

Haboob: Dust raised by the downdraft of wind from a thunderstorm squall.

Half-life: Time required for 50% of the atoms of a radioactive isotope to decay into the next radioactive daughter products.

Halophilic: Organisms adapted to high sunlight conditions.

Hardpan: See *duricrust*.

Headland: A comparatively high rock promontory with a steep face projecting into the sea.

Heavy metal: A metal usually with a density of more than 5 g cm^{-3}.

Helical flow: Superposed transverse circulation of water on the main channel flow.

Helmholtz wave: A phenomenon of wave-like unstable wind flow a little above the ground.

Hetrotroph: Microorganisms that are not able to use carbon dioxide as a source of carbon.

Holocene: Latter part of the Quaternary period from about 10,000 years ago to the present.

Holocene climatic optimum: A phase of Holocene climate in which the mean annual temperatures and precipitation 4000 to 5000 years ago were an order of magnitude higher than that obtained presently.

Holocene marine transgression: A phase of nearly stable sea stands in the Holocene beginning about 6000 years before the present.

Homolytic fission: Cleavage of a covalent bond of two atoms usually of similar electonegativity, resulting in the formation of electrically neutral atoms called radicals.

Horst: An elevated block of the earth's crust between parallel faults. It evolves either as an uplifted block between faults or by the sinking of land outside faults.

Humic acid: An organic acid formed by the water circulating through humus.

Humus: Partly decomposed organic matter, which accumulates on and within soils.

Hydration: Incorporation of water in crystal structures, forming crystallographically distinct minerals.

Hydration shattering: Shattering of rocks along cavities by the expansive force of minerals' hydration.

Hydraulic conductivity: The ease of liquid flow through porous media. It has the dimension of velocity and is expressed in $m\ s^{-1}$.

Hydraulic geometry: At-a-station and downstream interadjustment of channel variables of width, depth, velocity, and suspended load to discharge.

Hydraulic gradient: The rate of change of hydraulic head with distance in aquifers.

Hydraulic head: Sum of the elevation head and pressure head in the medium.

Hydraulic radius: Ratio of the cross-sectional area of a channel to its wetted perimeter.

Hydraulics: Application of the principles of fluid mechanics for the behavior of water flowing in pipes and open channels of rigid and mobile boundaries, respectively.

Hydrocarbon: A compound entirely of the elements of carbon and hydrogen.

Hydrodynamics: Analytical study of the motion of ideal, nonviscous fluids.

Hydrogen ion: The proton (H^+) that exists as a hydrated form in aqueous solutions.

Hydrology: A study of the circulation and distribution of water at or near the surface of the earth.

Hydrolysis: A process of the dissociation of silica in ionic species.

Hydrophobic: Compounds that repel interaction with water.

Hydrothermal: Processes, rocks, ore deposits, alteration products, and springs produced by hot magmatic emanations rich in water content.

Ice: A mass of roughly intergranular and interlocked crystals formed by the freezing and refreezing of water.

Ice cap: A dome-shaped glacier that flows outward from thick centers of ice accumulation on mountain sites.

Ice-thrust ridge: A ridge of deformed bedrock and glacial drift roughly perpendicular to the push of continental glaciers.

Ice wedge: A vertical mass of pure ice in the frozen ground environment.

Ice-wedge cast: A relict ice wedge.

Igneous rock: A rock formed by the solidification of hot magma.

Illite: A group of clay minerals composed of interlayered mica and montmorillonite.

Inselberg: A large steep-sided residual hill usually above erosional plains. Small residual hills are called tors and larger ones bornhardts.

Intensity domain: A process that dominates other concurrently functioning processes at a site by virtue of its higher functional intensity.

Interglacial stage: Retreat of glacier ice in a warmer phase of the Pleistocene climate.

Interior drainage: A drainage that has no access to the sea.

Interstadial: A shorter and less intense phase of interglacial period.

Intertidal zone: A zone between low and high tides.

Intrusive ice: Ice formed in bedding plane separations, updoming the surface.

Intrusive lava: Lava that penetrated older rocks through cracks and faults and solidified within the earth's crust.

Involution: Intensely deformed sedimentary beds of varying grain size composition in the active layer.

Ion: An atom or molecule carrying positive or negative electrostatic charge.

Ionic potential: Positive charge on a cation divided by its radius in armstrom units.

Isostasy: Theoretical balance in large portions of the earth's crust, such that each portion either rises or subsides until it is buoyantly compensated by the thickness of crust below.

Isotope: One or more forms of a chemical element having the same atomic number but different atomic weight.

Isotropic: Having the same property in all directions.

Jet stream: A band of fast-moving air (in excess of 30 m s^{-1}) in the upper troposphere. The troposphere exists at a height of about 8 to 17 km from the surface of the earth.

Kame: An irregular mound of stratified glaciofluvial sediments deposited on the surface of a stagnant ice sheet.

Kaolinite: A common clay mineral mainly of hydrous aluminum silicate.

Karman–Prandtl equation: A theoretical postulate for logarithmic increase in the wind velocity with height above the surface.

Karren: A group of small-sized features of solution on denuded and mantled karst.

Karst: A term for solution processes and landforms, principally in limestone.

Karst valley: A relict feature of fluvial erosion in horizontally bedded mantled limestone beds.

Katabatic wind: An orographically controlled wind induced by the difference in temperature and pressure between the mountain and lowland below.

Kilobar: A measure of force; 1 kilobar equals 10^6 dynes cm^{-2}.

Kinematic viscosity: The fluid viscosity divided by density. It is a measure of the turbulent mixing in flow.

Laccolith: A concordant lens-shaped magmatic intrusion, which updomes the body of overlying rock.

Lacustrine plain: A flat surface of sediment deposits in former proglacial lakes.

Lag deposits: Coarser particles left behind by selective removal of finer fractions by channel or wind flow.

Lagoon: Shallow stretch of water completely or partly segmented from the sea by spits or bay mouth bars.

Laminar flow: The flow in which parallel layers of fluid move past each other without mixing.

Land capability: A measure of the value of land for agricultural purposes.

Land system: Subdivision of a region into areas of distinct common physical attributes.

Land use planning: A process involving environmental, social, and economic criteria for evolving strategies for rational use of the land resource.

Landscape sensitivity: The extent to which a landscape responds to external stimuli by changing form attributes.

Landslide: A general term for downslope movement of a unit mass of rock or earth materials.

Latent heat: Energy involved in changing the state of matter, such as from water to ice.

Latent heat of fusion: Exchange of heat upon phase change of water among gaseous, liquid, and solid states. Heat of the order of 80 calories per gram of water at $0°C$ is added to the ice to become water.

Lateral accretion: A process of sediment deposition within alluvial channels.

Laterite: A soil of chemical selection, favoring greater mobilization of alkalies and silica than iron and aluminum from the profile depth.

Lava: Magma that emanates from volcanoes or fissures in the earth's surface.

Leachates: Foul-smelling liquid contaminated with bacteria. Leachates form when surface or subsurface water comes in contact with solid waste.

Lichen: A class of plant formed by symbiotic association of fungus and alga.

Lichenometry: A method of relative age determination from the growth rate of lichen on stable features.

Ligand: Any atom or molecule capable of donating a pair of electrons to the central atom.

Limestone: A bedded chemical sedimentary rock comprised mainly of calcium carbonate.

Limnology: A study of freshwater aquatic systems for physical, chemical, and biological aspects.

Limonite: A generic term for hydrous iron oxide.

Linear dune: A long linear dune aligned to the wind in which sand is transported parallel to the dune crest.

Liquidity index: Ratio of the moisture actually present in the sediment to the moisture required for its liquid behavior.

Lithification: A complex process that makes a noncohesive mass of sediments into hard rock.

Lithology: Megascopic physical character of a rock.

Lithosphere: Rigid outer shell of the earth's crust, which is situated above the asthenosphere.

Little climatic optimum: A phase of milder climate in Europe and North America about 750 to 1200 AD.

Little Ice Age: Local advance of mountain glaciers in the Late Holocene period between 300 and 4000 years BP.

Lodgment till: Till lodged directly from beneath the mass of an active glacier onto its bed.

Loess: A homogeneous, nonstratified, and slightly lithified wind-blown dust deposit.

Longshore current: A current moving parallel to the shoreline. It is generated by the wave refraction in shallow coastal waters.

Loo: A local term for strong hot-dry winds at the onset of summer in the Indian subcontinent.

Lopolith: A large-floored intrusive depressed in the form of a basin.

Lunette: A crescent-shaped dune of silt and clay downwind of a pan.

Ma: Million years.

Mafic mineral: A dark-colored mineral composed mainly of iron and magnesium.

Magma: A silica melt released from beneath the earth's crust from which igneous rocks evolve.

Magma chamber: A magma-filled cavity within the lithosphere.

Mangrove: Plants adapted to high salinity of the estuarine environment.

Manning equation: A flow equation relating stream velocity to hydraulic radius, slope, and roughness factors.

Mantle: Main bulk of the earth's interior between the crust and core. It is composed of mafic minerals.

Marble: Metamorphosed limestone rock.

Marker horizon: A key bed of distinctive characteristics in sediments, which can be dated and used for correlation purposes.

Mass movement: Gravity-induced movement of the earth materials. The term is synonymous with mass wasting.

Mass wasting: A general term for slow to rapid gravitative transfer of the earth materials.

Mean annual flood: A statistical probability of channel discharge exceeding the bankful stage once every 2.33 years.

Meltout till: Till released directly from the melting of interstitial ice of a debris-rich stagnant glacier.

Mesa: A flat-topped, steep-sided upland topped by a resistant formation.

Metamorphic rock: A rock of changed mineral composition and texture by the effects of pronounced changes in temperature, pressure, and chemical environment.

Metasomatism: A replacement process that produces new mineral suites by replacing rock minerals with minerals of different chemical composition.

Metastable: Stable with respect to small disturbances of geomorphic systems.

Microbe: A microscopic organism.

Microbial decomposition: See *oxidative carbon dioxide*.

Mid-oceanic ridge: A major linear elevated feature of the sea floor and the source of oceanic crust.

Mineral: A naturally occurring inorganic substance of definite chemical composition and characteristic crystal form.

Misfit stream: A stream of disproportionately small discharge relative to the channel width.

Modulus: A number or quantity that measures a force or function.

Mohrovičić (Moho) discontinuity: Boundary between the earth's crust and mantle. It lies at a depth of 10 to 35 km, and is marked by a rapid increase in the seismic wave velocity to more than 8 km^{-1}.

Molecule: Independently existing smallest particle of any matter.

Moment: The tendency of a force to cause rotation.

Monobasic: A property of acids that donate one hydrogen ion or displace one hydroxyl ion.

Montmorillonite: A group of clay minerals that comprise mainly hydrous aluminosilicates. These minerals swell in contact with water.

Moraine: A general term for ridges and plains formed of glacial till.

Morphogenetic region: A conceptual region of a certain set of climatic conditions that regulate specific geomorphic process activities, evolving typical landscape characteristics.

Morphostructure: A geologic structure produced by cymatogeny.

Moss: A primitive plant that grows on rocks, trees, damp ground, and other man-made facilities.

Moulin: An enlarged drainage opening at the surface of glacier ice.

Mud: Sediment composed of silt and clay fractions.

Mudlion: A yardang cut into soft lakebed sediments.

Multicyclic: Pertaining to more than one cycle of erosion.

Nanobacteria: Submicroscopic organisms that vary between bacteria and viruses in size.

Natural levee: A shallow ridge along the inside bank of meandering channels formed by vertical accretion of sediments.

Neap tide: A tide of lower height produced when the sun and the moon are in quadrature.

Nearshore: A zone seaward of the beach to the limit where waves break first in shallow coastal waters.

Neotectonic: A term for the movements of the earth's crust during the Late Tertiary.

Névé: Upper part of a valley glacier where most of the snow accumulates.

Newtonian substances: Substances in which the strain rate increases linearly with stress. They include fluids and gases, which offer little resistance to deformation.

Nitrogen fixation: A process by which N_2 is reduced to NH_4 and absorbed by plants. A certain group of bacteria help fixation of nitrogen in soils.

Nitrogenase: An enzyme catalyzing the reduction of nitrogen (N_2). It is active under anaerobic conditions and present in nitrogen-fixing organisms.

Nivation: A process of erosion, freeze-thaw, meltwater erosion, and mass wasting around and beneath substantially thick immobile snowpacks.

Nivation hollow: Shallow depression in the rock bed created by nival processes.

Nonbarred (normal) beach: A sand beach of steeper gradient formed by aggradation of the berm and beach face.

Normal fault: A fault at which the hanging wall is depressed relative to the footwall.

Normal stress: A resisting force perpendicular to the shear stress.

Nunatak: An isolated hill or mountain peak standing above the level of continental glaciers and ice caps.

Oceanic crust: A part of the lithosphere formed from the magma that rises from localized hot spots beneath the mid-oceanic ridge system. It is rich in basalt.

Olivine: A common rock-forming plagioclase group of minerals of a general form (Na, Ca) Al (Si, Al) Si_2O_8.

Oölite: Usually spherical to ellipsoidal-shaped calcareous body, up to 2 mm in diameter. Also the limestone composed primarily of cemented oölith.

Open-cavity ice: Ice formed by the sublimation of atmospheric water vapor in an open cavity or a crack in the frozen ground.

Open system: A system regulated by constant supply and removal of energy and mass across its boundary. All geomorphic systems are open systems.

Organic reef: A resistant structure of biogenic carbonate of coral and noncoral organisms.

Organic terrain: See *bog*.

Orogenic belt: See *orogeny*.

Orogeny: A mechanism of crustal thickening, rock deformation, and volcanic activity, forming mountain chains. These mountain belts are called orogens.

Outwash: Stratified glaciofluvial sand and gravel deposits at and beyond retreating ice margins.

Oxidative carbon dioxide: Carbon dioxide released from decomposed organic matter.

Paleoclimate: Average conditions of climate during some past geologic periods.

Paleomagnetism: The earth's ancient magnetic field and positioning of continents from the evidence of remnantal magnetism in ancient rocks.

Paleosol: A soil formed at some period in the past and buried by later deposits.

Paleozoic era: One of the oldest eras of geologic time.

Palsa: Ice-cored ridges and mounds in upheaved frozen bogs of discontinuous permafrost environment.

Palynology: Study of pollen for interpretation of the Quaternary climate and climate change.

Pan: A natural basin or depression holding seasonally stagnant water or mud from which surface and subsurface salts precipitate in the dry season.

Pangea: A term for the entire earth (pan) that had existed in the ancient time (gaia).

Parabolic dune: A U-shaped dune with horns pointing upwind. It evolves in association with vegetation.

Partial pressure: Force per unit area of the weight of a gas of the atmosphere at sea level.

Passive margin: A continental margin affected by only a limited tectonic activity related to divergent plate motion in plate tectonics.

Pathogen: A disease-producing organism or agent.

Patination: See *desert varnish.*

Patterned ground: Nearly symmetrical microrelief forms of circles, polygons, nets, and stripes in the frozen ground environment.

Peat: Partially decomposed remains of mire plants in waterlogged environments.

Pedalfer: A humid region soil characterized by the abundance of iron oxides and clay minerals in the B-horizon.

Pedestal rock: A rock comprising a cap rock of harder stratum above the pedestal of a softer formation in aeolian environments.

Pediment: A slightly concave subaerial and suballuvial bench downslope of the foot of a highland.

Pedocal: A soil rich in the accumulation of calcium carbonate in the A-horizon. It is common to arid environments.

Pelgic: Floating and free-swimming communities of marine organisms.

Peri-desert loess: Loess formed in deserts and deposited at their margins.

Periglacial: Seasonally and perennially cold regions, often synonymous with the permafrost environment.

Permafrost: Frozen ground with or without subsoil ice.

Permafrost table: The upper limit of permafrost.

Permeability: The capacity of a soil or rock for transmitting fluids. It depends on the porosity of materials.

Peroxide compound: A compound that contains a weak oxygen-oxygen bond.

pH: Negative logarithm of the hydrogen ion concentration in aqueous solution, which defines acidity and alkalinity of the environment.

Phase boundary: The line separating any two phase areas or any two liquid surfaces.

Photic zone: Depth of sunlight penetration in water bodies.

Photoautotroph: Organisms utilizing light as an energy source and carbon dioxide as a carbon source for synthesizing complex organic compounds.

Photon: A measure of electromagnetic radiation.

Photosensitizer: A substance that initiates photochemical reactions. It is only a carrier of chemical energy.

Photosynthesis: Conversion of sunlight energy by the chlorophyll to chemical energy.

Phreatic zone: A zone in the aquifer where interstitial pores are saturated with water and the water table appears.

Phytokarst: A feature of biochemical dissolution of karst at the surface of bare limestone rocks.

Phytoplankton: A plant component of plankton, consisting chiefly of microscopic algae. Plankton forms the basis of food for all other forms of aquatic life.

Pingo: An ice-cored conical hill in a permafrost environment.

Planar failure: Failure of a substantially thick mass of cohesive sediments or a rock mass along a planar surface at a shallow subsoil depth.

Plankton: Passively floating microorganisms that live suspended in water.

Plastic deformation: A type of failure producing permanent change in the shape of substances without causing rupture.

Plateau: An elevated tract of comparatively flat tableland rising abruptly from above a lowland area.

Plate tectonics: A theory of global tectonics that studies plate formation, plate movement, plate interaction, and plate consumption.

Playa: Undrained depression in semiarid and arid environments of interior drainage.

Pleistocene epoch: An earlier epoch from 2 Ma to 10,000 years BP of the Quaternary period. Also called the Great Ice Age.

Plucking: Subglacial removal of the bedrock surface from within high-density jointed sections.

Plutonic rock: A rock formed at depth by slow cooling and crystallization of the magma.

Point bar: A deposit of lateral accretion at the outside bend of meandering channels.

Polar glacier: A glacier in which the ice is below pressure melting point throughout the depth of ice.

Polje: A flat-floored karst depression surrounded by steep hills.

Pollen: Microspores of the male element of seed-producing plants (gametophytes).

Polygenetic: Pertaining to the temporal activity of more than process domain.

Polymer: A compound of high molecular weight formed by the joining of a large number of simple molecules of covalent bonds.

Polythermal glacier: High-latitude glacier of a thicker inner ice at pressure melting temperature and an outer thinner ice frozen to the bed. It is intermediate between temperate and polar glaciers in thermal properties of the ice.

Pore pressure: Pressure of water in intergranular pores of unconsolidated sediments or porous rocks. Pore pressure is positive below the water table and negative above it. The negative pore water pressure is called soil moisture tension.

Porosity: Ratio of aggregate volume of pore spaces in soils or rocks to the total volume. It is usually expressed in percentage.

ppm: Abbreviation for parts per million by weight.

Precambrian: A geologic period older than 600 million years before present.

Pressure melting: A fundamental concept stating that the freezing temperature of water decreases with depth in the glacier ice.

Principle of minimum variance: States that open systems react to the imposed disturbance by affecting a minimum change and least work in the adjustment of interrelated variables.

Process (geomorphic): An activity of water, wind, or ice that brings about mechanical and chemical alteration of the earth materials and mobilizes the resulting products as solids and solutes at or near the surface of the earth.

Process domain: A specific process that distinguishes other concurrently functioning processes at the given site.

Proglacial lake: A freshwater lake formed between an ice front and high ground ahead.

Protein: A complex organic compound of numerous amino acids.

Protozoa: Unicellular simple life forms that first evolved during the Precambrian.

Pyrite: A sulfide mineral (FeS_2).

Pyroclastic: Fragmentary products ejected by a volcanic eruption.

Quasi-equilibrium: A state of near dynamic equilibrium in geomorphic systems.

Quaternary: The geologic period from about 2 million years ago to the present. The Quaternary is subdivided into Pleistocene and Holocene (Recent) epochs.

Radioactivity: Emission of charged alpha and beta particles and neutral gamma particles from radioactive elements with attendant generation of heat.

Radiocarbon: See *carbon-14* (^{14}C).

Rate law: A law that predicts negative exponential decay rate for the half-life period of radioactive elements.

Rate process theory: A theory for thermochemical processes predicting that colliding atoms, molecules, or ions yield new reaction products by breaking chemical bonds at a certain activation energy.

Rayleigh number: A quantitative expression for the magnitude of thermal driving of the material to dissipative forces in the fluid motion.

Reaction time: Time lag between the impact of an external stimulus on the functioning of geomorphic systems and the response time of landforms to the stimulus.

Recent: See *Holocene*.

Redox reaction: A reaction involving simultaneous oxidation and reduction processes.

Reduction: A chemical change in which an atom or ion gains one or more electrons.

Reef: Topographic eminence of biogenically produced carbonates at or near the sea level.

Regelation: Freezing of meltwater commonly at the glacier-bedrock interface, forming the regelation ice.

Regolith: A general term for partly decomposed rock material from which soils evolve.

Relaxation time: Time required for geomorphic systems to reach equilibrium with the energy of an external stimulus.

Relict landforms: Landforms of processes that no longer operate in areas where they exist.

Reptation: Sediments moved by the splash impact of saltating particles on a mobile bed of sand.

Residence time: Length of time a solution stays in circulation with limestone. The residence time governs the solute concentration in karst springs and stream flows.

Residual karst: Upstanding residuals of solution in the tropics.

Reynolds number: A dimensionless ratio defining the state of fluid motion as laminar, turbulent, and in transition between the two by the relative magnitude of viscous and inertial forces.

Rheology: A study of flow and deformation behavior of materials.

Ridge and runnel: A microrelief intertidal bar and trough sequence produced by erosion of a sand beach.

Rift: A large strike-slip fault parallel to the regional structure.

Rip current: A narrow seaward flow of water from the breaking of waves in shallow coastal waters. Rip currents can reach a speed of up to 5 m s^{-1}.

Risk map: A map depicting the vulnerability of an area to probable and potential dangers for the safety of existing and proposed structures.

Roches moutonnées: A form of bedrock topography comprising a gentle abraded upglacier slope and a steep shattered downglacier face.

Rock: A naturally occurring coherent mass of mineral matter. This matter distinguishes igneous, metamorphic, and sedimentary rocks.

Rock basin: A feature of subglacial erosion gouged out from joints, other planes of weakness, and preexisting depressions on rock pavements.

Rock burst: A sudden and often violent failure of rock masses in quarry, tunnel, and mine operations. It results from offloading or release of geodynamic stress.

Rock flour: A term for fine-grained abraded sediments at the glacier-bedrock interface.

Rock glacier: A type of valley glacier comprised of clastic sediments with ice cores and segregated ice in frozen ground environments.

Rotational sliding: Failure of cohesive sediments along a spoon-shaped shear plane.

Rudaceous: A sedimentary rock composed of cemented conglomerate and breccia.

Sabkha: A flat depression in deserts, generally close to the water table and covered with a salt crust.

Safety factor: Ratio between the shear strength of a body of earth materials and shear stress acting on it.

Saltation: A near-surface leap-and-bounce movement of grains by the fluid's shear force. The grains lift, propel forward, and descend to the surface describing a parabolic trajectory.

Salt pan: A shallow water body of brackish water.

Sand: Sediment particles between 0.063 and 2.0 mm in diameter.

Sand bank: A submerged ridge of sand in the sea or rivers.

Sand dune: Deposits generally of sand and clay laid down by the wind. They are aerodynamic forms shaped by the deformation of wind flow.

Sand sea: A collection of dunes covering an area of more than 30,000 km^2.

Sand sheet: A vast expanse of sand with few or very low dune forms.

Sanitary landfill: A land site for the burial of municipal solid waste in a manner that causes minimum disturbance of the environment.

Saprolite: Residual clay, silt, or other substance.

Satellite imagery: Digitally scanned imagery of a part of the earth from satellites carrying remote-sensing devices for producing images of the earth's surface.

Scabland: Intensely eroded topography representing rock and rock fragments at the surface.

Sea floor spreading: The process that generates new lithospheric material by the upwelling of magma at mid-oceanic ridges and causes its movement away from the ridge crest as oceanic crust. This process is central to the theory of plate teconics.

Seamount (Guyot): A submarine volcanic cone.

Sea trench: A trench in the ocean crust. It is parallel to the subducting lithospheric plates.

Second law of thermodynamics: A principal law governing energy processes in thermodynamic closed systems. The law states that entropy of the system tends to increase in any chemical or physical process.

Sediment: Solid material deposited at the earth's surface by physical, chemical, or biological agents.

Sedimentary rock: A rock formed by the accumulation and cementation of mineral grains transported by physical agents to the site of deposition or by chemical action or evaporation within a basin of sedimentation.

Segregated ice: Ice formed by adsorption and subsequent freezing of water onto mineral grains.

Seif: A type of linear dune in which crosswinds do not parallel the dune crest.

Seismicity: Distribution or occurrence, or both, of earthquakes in space and time.

Sensible heat: The heat that when added or subtracted from a substance causes a change of temperature and not a change of state.

Settling velocity: The rate at which sediments of a given size settle in fluids.

Shale: A compacted sedimentary rock composed predominantly of clay-sized particles.

Shallow coastal water: Coastal waters with a depth less than one-half of the wavelength of approaching waves.

Shear plane: A plane separating the failed mass from its intact surface.

Shear plane (glacier): A plane of separation caused by the differential flow of ice. Several shear planes develop in the body of active glaciers.

Shear strength: A measure of the ability of materials to resist shear stress. The shear strength of unconsolidated sediments is expressed by the Coulomb equation.

Shear stress: Gravity-controlled driving force that acts along a shear plane.

Sheet flow: Nonchannel flow of water over a surface.

Sheet fracture: Joints formed in massive rock by offloading at the surface and compression deep inside the earth's crust.

Shield: A large rigid mass of Precambrian rock, forming a relatively stable continental block.

Shingle: Loose pebbles often found on mid- and higher-latitudes beaches.

Shoal: A sand bank at the surface of a lake or sea.

Shoreline: Generally a line separating the land from an adjacent body of seawater.

Shore platform: An intertidal platform evolving by abrasion, chemical weathering, and salt-layer weathering processes.

Silt: Clastic sediment between 0.004 and 0.0063 mm in diameter.

Sinkhole: A solution depression with or without a visible opening in the karst. Also called doline.

Sinking creek: A stream in the karst that disappears through a sinkhole or a group of sinkholes.

Slide: Deformation of a unit mass of sediment or rock along shear planes or bedding plane separations.

Slip face: Steeper leeward slope of a sand dune on which sand avalanches to maintain the angle of repose.

Slip-off slope: Gently sloping bank of a stream on the outside bend of a meander.

Slope failure: Instability of slope-forming earth materials under stress.

Slump: Rotational failure of cohesive fine-grained sediments along a spoon-shaped shear plane.

Snow: Solid precipitation composed of single ice crystals or aggregates called flakes.

Snow line: Altitudinal limit for perennial snow on land.

Soil: The earth materials composed of mineral matter and organic remains, which support the growth of rooted plants.

Soil profile: A vertical section of soil, showing the sequence of soil horizons down to the depth of the parent material.

Solubility product: Product of the molar concentration of the ions of calcium or magnesium, or both, and ions of carbonate in solution.

Solution: A homogeneous mixture of two or more substances in which molecules or ions are uniformly dispersed in the solvent.

Solution hardness: A measure of the ionic concentration of calcium or magnesium, or both, in solution.

Solvent: A medium in which the solute is dispersed to provide a homogeneous solution.

Sorbate: Adsorbed or absorbed mineral.

Sorbent: A substance that can absorb or adsorb materials.

Sorption: A process of absorption and adsorption when considered together.

Species: Organisms of a certain group of similar individuals.

Specific heat: Heat required for raising the temperature of a unit weight of fluid by 1°C at constant volume.

Speleology: A study of the processes and development of caves.

Speleothem: Gravity and eccentric forms of calcite precipitation within caves.

Spit: A narrow accumulation of coastal sediment attached at one end to the land. It evolves by wave refraction.

Splay: A fan-shaped accumulation of alluvium outward from an accidental breach in a levee.

Sporadic permafrost: Shallow pockets of frozen ground within the vast expanse of unfrozen ground.

Spring tide: A tide of larger height produced when the sun and the moon are in conjunction.

Star dune: A large pyramidal or dome-shaped sand dune with arms radiating from a high point.

Static (fluid) threshold: A threshold velocity at which grains of a given size are just set in motion on a mobile bed of sand.

Steady state: See *dynamic equilibrium*.

Stoke's law: A law stating that the fall velocity is proportional to the square of particle radius.

Stone pavement: See *desert varnish*.

Straight channel: A type of channel pattern that normally does not exceed five to seven times the local channel width.

Strain: Deformation of a body under stress.

Stratigrapy: A branch of geology that deals with the formation, composition, sequence, and correlation of stratified rocks and sediments.

Stratosphere: A zone in the atmosphere at a height of 10 to 30 km above the earth's surface in which the temperature remains essentially constant.

Stream competence: Described by the largest particle size moved by a stream.

Stream power: The rate of work done, expressed as force times flow velocity.

Stress: Force per unit area expressed in Newton per square centimeter.

Strike: The bearing of an inclined bed. It is perpendicular to the direction of dip.

Strike-slip fault (transcurrent fault): A fault in which the net slip is practically in the direction of a fault strike.

Stromatolites: Laminated sedimentary calcareous reef-like structures of biogenic origin.

Sublimation: Direct conversion of atmospheric water vapor into ice or evaporation from an ice surface.

Sublimation till: The till released directly from the sublimation of interstitial ice.

Supraglacial drift: Nonstratified and angular rock fragments on the surface of valley glaciers and compact fine-grained basal till that had moved through shear planes in the ice to become supraglacial prior to deposition.

Surface tension: A force holding molecules of a liquid to the boundary surface, such that they behave like a stretched membrane.

Surf zone: A shallow water zone in coastal environments where waves approaching a coast oversteepen and break.

Suspended load: Sediments moving in suspension in a fluid as a result of turbulence in the flow.

Sustainable development: A development that satisfies the needs of the present generation without compromising the ability of future generations to meet their own needs.

Swallow hole: A deep vertical opening in the karst through which surface streams disappear underground.

Swash: Wave uprush at an angle to the beach surface.

Swell: Deep water storm waves that disperse outward of the generation area in oceans.

Symbiotic association: An association between two organisms in which both benefit.

Tailings: Bricks and stones placed in the retention walls of dams.

Talik: Unfrozen ground beneath and within the frozen ground environment.

Talus: Slope of loose angular debris against a backwall. It rests at the angle of repose.

Tarn: A small lake occupying the basin of a relict cirque.

Tectonics: A study of broad geologic structures of the earth's lithosphere and their processes of faulting, folding, and warping.

Temperate glacier: A type of glacier in which the ice is at pressure melting temperature throughout its depth.

Terrigeneous: Marine deposits derived from bordering landmass and transported to the sea by bottom currents.

Tertiary: A geologic period from 63 Ma to 2 Ma before present.

Tethys Sea: A sea that had existed in the Paleozoic era between the Mediterranean and Southeast Asia.

Thalassostatic: Quaternary sea level fluctuations.

Thalweg: Channel gradient expressing the fall of bed elevation with distance downchannel.

Thawing index: The sum total of mean positive temperatures in a year. It is expressed in degree-days.

Thermal conductivity: Ratio of heat transmitted per unit area per unit time to temperature gradient, causing heat flow.

Thermal convection: Heat transfer across the medium by unidirectional coherent motion of the material.

Thermokarst: A group of subsidence features related to widespread thaw of the ground ice.

Thermoluminescence (TL): A property of rocks and sediments by which the application of heat or radiation de-traps electrons from crystal lattice. The process emits glow.

Thrust: A low-angled reverse fault.

Tidal bore: A tidal wave of considerable height, which rushes up narrow estuaries of large tidal range.

Tidal current: The movement of tidal water through an estuary or a tidal inlet, generating flood at high and ebb at low tide in the channel.

Tidal flat: Unvegetated part of the intertidal zone.

Tidal inlet: A break in a barrier island through which tidal currents pass.

Tidal marsh: Vegetated upper part of the intertidal zone that typically rims estuaries.

Tidal prism: The amount of water moving in and out of estuary or tidal inlet with the movement of tide.

Tidal range: Difference in the height of sea surface level at high and low tides.

Tide: A periodic rise and fall of the sea surface level due to gravitational attraction of the sun and the moon on the earth.

Till: A general term for nonstratified and nonsorted glacial deposits.

Topset beds: Horizontal sedimentary layers deposited in deltaic and aeolian environments.

Trace element: The nutrient element required by plants in minute quantities for growth and metabolic activity.

Transform fault: A type of fault associated with the mid-oceanic ridge system.

Transverse dune: An asymmetric dune ridge with crests perpendicular to the dominant wind.

Tropical cyclone: A cyclonic vortex system across tropical seas, which develops from July to October in the northern hemisphere and from January to March in the southern hemisphere. The vortex system moves northwest in the northern hemisphere and southeast in the southern hemisphere, generating strong winds and widespread heavy rain along the track.

Troposphere: The lower atmosphere to a height of 10 km from the earth's surface, in which the average temperature decreases with height at the rate of about 6°C/km, clouds form, and precipitation occurs.

Tsunami: Disruptive sea waves generated by submarine earthquakes and volcanic activity in the sea.

Tufa: A chemical sedimentary rock comprising calcium carbonate or silica. It is deposited from the solution of karst springs, lakes, or a percolating groundwater system.

Tundra: A vast treeless marshy region, usually associated with the permafrost environment.

Turbulent flow: A type of flow in which fluid layers thoroughly mix in the vertical by eddy motion superimposed on the main forward flow.

Ultraviolet radiation: Electromagnetic radiation in 4 and 400 nm (nm = nanometer = 10^{-6} m) wavelength. It is shorter than the visible light but longer than x-rays.

Upwelling: The rise of deep and cold water by diverging ocean currents or by offshore movement of coastal currents.

Uvala: A fairly deep and unevenly floored elongated depression created by coalescent sinkholes.

Vadose zone: A subsurface zone of partly air- and partly water-filled intergranular pores in soils.

Valley fill: Alluvium deposited within the confines of a valley.

Valley glacier: A glacier confined to a valley.

Varve: A sedimentary sequence comprised of a thicker layer of silt beneath an upper thinner layer of clay laid down by the meltwater flow in proglacial lakes.

Vector unit: Magnitude of the drift potential.

Vein ice: The ice formed by the freezing of water in frost cracks of the active layer.

Ventifact: A loose stone or a boulder or a lump of cohesive sediments with typical wind-abraded faces.

Vertical accretion: A process in which sediments are carried overbank and deposited at and beyond channel margins. It is one of the processes of flood plain development.

Virus: A small intracellular parasite, often composed of nucleic acids and proteins.

Viscosity: A property of fluids resisting flow. Viscous materials deform at all levels of applied stress.

Volcano: A vent in the earth's crust through which molten lava erupts. Also a landform built of the lava ejected through a vent in the earth's crust.

Warm-based glacier: See *temperate glaciers*.

Water table: A subsurface depth of water-saturated pore spaces in sediments and permeable rocks.

Wave base: A depth at which waves begin to "feel the bottom" and seabed sediments begin to be moved by the waves.

Wave climate: A term for overall wave condition given by the length, period, and height of waves in shallow coastal waters.

Wave reflection: Reflection of swell from islands and man-made facilities.

Wave refraction: A phenomenon of the bending of waves in shallow coastal waters.

Wave steepness: Ratio between wave height and wavelength.

Waves of translation: Waves formed by the breaking of waves in shallow coastal waters.

Weathering: *In situ* irreversible change in the earth materials by mechanical, chemical, and biochemical stress of the environment.

Wisconsin glaciation: The last of four Pleistocene glacial events in North America.

X-ray: High-energy short-wavelength radiation in the electromagnetic spectrum.

Yardang: An aerodynamically shaped wind-eroded feature in softer cohesive sediments and resistant rocks.

Zooplankton: Microscopic animals, which float or feebly swim.

Index